T0258759

ANALYSIS AND OPTIMUM DESIGN OF METAL STRUCTURES

1. Analysis

Analysis and Optimum Design of Metal Structures

JÓZSEF FARKAS & KÁROLY JÁRMAI
University of Miskolc, Hungary

A.A.BALKEMA/ROTTERDAM/BROOKFIELD/1997

Authorization to photocopy items for internal or personal use, or the internal or personal use of specific clients, is granted by A.A. Balkema, Rotterdam, provided that the base fee of US$1.50 per copy, plus US$0.10 per page is paid directly to Copyright Clearance Center, 222 Rosewood Drive, Danvers, MA 01923, USA. For those organizations that have been granted a photocopy license by CCC, a separate system of payment has been arranged. The fee code for users of the Transactional Reporting Service is: 90 5410 669 7/97 US$1.50 + US$0.10.

Published by
A.A. Balkema, P.O. Box 1675, 3000 BR Rotterdam, Netherlands
Fax: +31.10.4135947; E-mail: balkema@balkema.nl; Internet site: http://www.balkema.nl

A.A. Balkema Publishers, Old Post Road, Brookfield, VT 05036-9704, USA
Fax: 802.276.3837; E-mail: info@ashgate.com

ISBN: 978-90-5410-669-2

Table of contents

Part 2: Optimum design

About the authors

Dr József Farkas is a professor of Metal Structures. He graduated in 1950 at the Faculty of Civil Engineering of the Technical University of Budapest. He teaches at the University of Miskolc since 1950, as a university professor since 1975. His academic scientific degrees: Candidate of technical science 1966, doctor of technical science 1978. His research field is the optimum design of metal structures, residual welding stresses and distortions, tubular structures, stiffened plates, vibration damping of sandwich structures. He has worked out expert opinions for many industrial problems, especially in the case of storage tanks, silos, cranes, welded press frames and other machine structures. He is the author of a university textbook about metal structures, an English book 'Optimum design of metal structures' (Ellis Horwood, Chichester 1984) and about 140 scientific articles in journals and conference proceedings. He is a Hungarian delegate of the International Institute of Welding (IIW), member of the International Society for Structural and Multidisciplinary Optimization (ISSMO) and a presidential member of the Welding Division of the Hungarian Scientific Society of Mechanical Engineers.

Dr Károly Jármai is a professor in the Faculty of Mechanical Engineering at the University of Miskolc, Hungary. He graduated as a certified mechanical engineer and received his doctorate (dr. univ.) in 1979 at the University of Miskolc. He teaches design of steel structures, welded structures, composite structures and optimization in Hungarian and in English language for foreign students. His research interest includes structural optimization, mathematical programming techniques and expert systems. He made his C.Sc. (Ph.D.) dissertation at the Hungarian Academy of Science in 1988. He became European Engineer (Eur. Ing., FEANI, Paris) in 1990. He has got the habilitation (dr habil.) at the University of Miskolc, in 1995. He has defended his D.Sc. thesis at the Hungarian Academy of Science with the economic design of different steel structures, optimization and expert systems in 1995. He has published over 120 technical papers, student aids, textbook chapters and conference papers. He is member of ISSMO and the Hungarian delegate in IIW. He is the Deputy Secretary General of the Scientific Society of Mechanical Engineers (GTE) in Hungary and president of this society at the University of Miskolc.

List of symbols

a, a_x, a_y	Spacing of ribs (mm)
a_{fc}, a_{gc}	Time parameters
a_{tc}, a_{sp}	Time parameters
a_w	Weld dimension (mm)
A	Cross-sectional area (mm^2)
A_f	Area of a flange (mm^2)
A_p	Sectional property (Chapter 2)
A_s	Surface area (m^2)
A_T	Thermal impulse due to welding
A_T'	Thermal impulse due to welding in the case of initial strains
A_w	Area of the web (mm^2)
A_y, A_{y1}, A_{yt}	Area of the plastic zone around the weld, due to single and multi-pass welding (mm^2)
b, b_0	Plate width (mm)
b_e	Effective plate width (mm)
b_f	Flange width (mm)
b_{fc}	Time parameter
b_x, b_y	Side lengths of a rectangular plate (mm)
b_y	Average width of the plastic zone around a weld (mm)
B	Bending stiffness (Nmm^2)
B_f, B_s	Bending stiffnesses of a sandwich beam (Nmm^2)
B_x, B_y	Bending stiffnesses of an orthotropic plate (Nmm^2)
B_{xy}, B_{yx}	Torsional stiffnesses (Nmm^2)
B_{f2}	Shear stiffness of a sandwich beam (Nmm^2)
B_ω, B_ϖ	Bimoments (Nmm^2)
c^*	Allowable ratio of deflection
c_0	Specific heat
c_m	Spring constant
c_{fx}, c_{fy}	Constants
$c_{x\sigma}, c_{y\sigma}$	Constants
C	Curvature (1/mm)
C_w	Deflection constant
D	Diameter (mm)
e, e_1, e_2	Distances, eccentricities (mm)
E	Modulus of elasticity (GPa)

$E_1 = E/(1 - \nu^2)$ Modulus of elasticity for plates (GPa)

E_A, E_S Modulus of elasticity for Al-alloys and steels, respectively (GPa)

f Frequency (Hz)

f_0 Eigenfrequency (Hz)

f_y Yield stress (MPa)

F Force (N)

F' Tendon force (N)

g_0 Equation (4.71)

g_j Inequality constraint functions

G Mass (kg)

G Permanent action

G Shear modulus (GPa)

G_{Al}, G_S Mass of Al-alloy and steel, respectively (kg)

G_s, G_d Static and dynamic shear modulus, respectively (GPa)

h Height (mm)

h_k Equality constraint functions

H Torsional stiffness of an orthotropic plate (Nmm^2)

H Approximate Hessian matrix (Eq. 6.55)

i Index

$i_u = \sqrt{-1}$ Imaginary unit

I Unit matrix

I Arc current (Eq. 1.24) (A)

I_x, I_y Moments of inertia (mm^4)

I_p Polar moment of inertia for warping torsion (mm^4)

I_t Torsional constant (mm^4)

I_ω, I_{ϖ} Warping constants (mm^6)

j Index

k_m, k_f Cost coefficients for material and fabrication, respectively ($/kg, $/hr)

k_σ, k_τ Buckling factors

K Cost ($)

K_m, K_f Costs of material and fabrication, respectively ($)

l Index

L Span length, length (m)

L_w Weld length (mm)

m Mass (kg)

m_{red} Reduced mass (kg)

m_t Correction parameter for multi-pass welding (Eq. 1.29)

m_x, m_y Bending moments per unit length (Nm/m)

m_{xy} Twisting moment per unit length (Nm/m)

M Bending moment (Nm)

M_p Plastic limit bending moment (Nm)

M_t Twisting moment (Nm)

n Constant for different welding technologies

n_φ, n_z Membrane forces in a shell (Chapter 14)

N Axial force (N)

p Uniform normal load intensity (N/m)

Q	Variable action
Q_T, Q_{T1}, Q_{T2}	Specific heat input caused by welding and by the first and second weld
r	Radius of gyration (mm)
r_k	Response factors (Eq. 6.51)
R	Radius of a circular silo (Chapter 14)
S	Static moment
S_i	Step length (Eq. 6.34)
t	Thickness (mm)
t_0	Prescribed minimal plate thickness (mm)
t_f, t_w	Flange and web thickness, respectively (mm)
t_r	Rib thickness (mm)
t_S	Stiffener thickness (mm)
$\bar{t}$	Time at forced vibration (s)
T_1, T_2, T_3	Time, fabrication times in Chapter 5 (s)
T_{e1}, T_{e2}	Temperatures in Chapter 1 (°C)
T_p	Preheating temperature at welding (°C)
T_R	Transmissibility (Eq. 4.33)
U	Arc voltage (Eq. 1.24) (V)
v	Shear flow
v_w	Welding speed of travel (Eq. 1.24) (mm/s)
V	Volume (mm^3)
V	Shear force
w	Deflection (mm)
$w*$	Allowable deflection (mm)
W, W_x	Elastic section moduli (mm^3)
W_p	Plastic section modulus (mm^3)
W_0	Required section modulus (mm^3)
x	Co-ordinate
X	Shear parameter in Chapter 4
x_i, x_1, x_2	Variables
y	Co-ordinate
y_A	Distance between point A and center of gravity
y_T	Distance between weld and center of gravity
z	Co-ordinate
z_0	Width of HAZ for welded Al-alloy structures
Z	Seismic zone coefficient (Chapter 14)
$\alpha = \sqrt{GI_t / EI_\omega}$	Chapter 2
α	Angle of inclination
$\alpha = a/h$	Parameter (Eq. 12.14)
α_0	Coefficient of thermal expansion
β	Limiting web slenderness
β	Convergence parameter (Eq. 6.9)
γ	Angular distortion
γ_S	Safety factor
δ	Limiting flange plate slenderness

$\delta_c = D/t$	Diameter/thickness ratio for a circular hollow section
δ	Logarithmic decrement (Eq. 4.7)
ε	Specific strain
$\varepsilon = \sqrt{235/f_y}$, $\varepsilon = \sqrt{250/p_0}$	Modifying factors for steels and Al-alloys, respectively (Chapter 3)
ε_A	Specific yield strain at point A
ε_I	Initial strain
ε_y	Specific yield strain
ε_G	Specific strain at gravity centre (Eq. 1.6)
ζ	Equation (8.7)
$\eta = b_y/t$	Figure 3.13
η_d	Loss factor (Eq. 4.14)
η_0	Coefficient of efficiency (Eq. 1.23)
κ	Number of structural elements assembled for welding
$\vartheta = 100D/L$	Parameter for circular hollow sections
$\vartheta = a/t_f$	Parameter (Chapter 12)
$\vartheta_S = b/t$	Plate slenderness
Θ_d	Fabrication difficulty factor
λ	Column slenderness ratio
$\bar{\lambda}$	Reduced slenderness ratio
λ^*	Optimum step length (Eq. 6.35)
λ_j	Lagrange multipliers (Eq. 6.43)
λ_p	Plate slenderness ratio
μ	Parameter (Chapter 12)
υ	Poisson's ratio
ρ	Material density (kg/m^3)
ρ_0	Radius of curvature (1/m)
$\rho_\perp$	Stress in a fillet weld (MPa)
θ	Angle of rotation (Chapter 3)
σ	Normal stress (MPa)
σ_a	Residual compression stress (MPa)
σ_b	Residual tension stress (MPa)
σ_e	Tension stress from external force (MPa)
σ_{adm}	Admissible normal stress (MPa)
$\sigma_\omega, \sigma_\varpi$	Normal stresses due to warping torsion in open and closed sections, respectively (MPa)
$\Delta\sigma_N$	Fatigue stress range (MPa)
τ	Shear stress (MPa)
φ	Angle of rotation (Chapter 2)
φ	Parameter (Chapter 12)
$\xi = x/L$	Variable (Chapter 11)
χ	Overall buckling factor
ψ_d	Dynamic factor
ψ_e	Effective width factor
ω_T, ϖ_T	Sectorial coordinates for open and closed sections, respectively (Chapter 2) (m^2)

Mathematics symbols

∂ / ∂	Partial derivative
Δ	Derivative or range
∇	Second derivative
∞	Infinity
$\in$	Element of a region
Σ	Sum
Π	Product

Abbreviations

CHS	Circular hollow section
EC3	Eurocode 3
ES	Expert system
SHS	Square hollow section
IIW	International Institute of Welding
HAZ	Heat affected zone

Introduction

Analysis is the procedure of understanding the behaviour of structures.

Verification is the check of a given structure for fulfilling the design constraints expressing the safety of the structure.

Optimum design is the search for better solutions which minimize the cost function and fulfil the design constraints. The search for optimum is a natural demand of thinking people.

Optimum design means that we understand the behaviour of structures based on analytical results and we are able to synthesize all important aspects of design, fabrication and economy, and, by using mathematical methods, to develop safe and economic structural versions.

Optimum design is a *structural synthesis* which adds the aspects of economy to design constraints. The objective function is defined as the cost or mass of a structure, which should be minimized considering the design constraints on the safety against failure, instability, fatigue, vibration, etc.

The three main phases of the structural synthesis are as follows:

1. *Analytical phase*: Preparation and carrying out necessary analyses by the designer:

– Selection of structural versions, i.e. materials, profiles, type of structure, production technology;

– Formulation of the design constraints and the cost function,

2. *Mathematical phase*: Constrained function minimization: Minimization of the cost function with fulfilment of the design constraints,

3. *Phase of evaluation by the engineer*: Sensitivity analysis, investigation of the effect of main parameters by comparisons, elaboration of design aids, implementation into expert systems.

The synthesis is inverse of analysis: Analysis is like a turned out coat, the synthesis turns it back to make it usable for designers.

Our aim is to bridge the gap between the optimization theory and fabrication practice. It is very important to include fabrication aspects into design constraints and to consider also fabrication costs in the cost function. In the International Institute of Welding (IIW) a new subcommission investigates *the interaction of design and fabrication*. With such a synthesis will be it possible to produce better structural versions.

Optimum design should play an important role in the *education* as well, since it can give a wide view necessary for engineers. The widest the application range the

1

deepest the understanding of the optimum design, therefore we try to show as much applications as possible.

The main part of the book is devoted to the optimum design. In the first part the most important *analytical problems* are treated. In some cases we do not separate a structural synthesis to individual chapters of analysis and optimum design, since the reader can perhaps better follow the synthesis, when the analytical part needed to define the design constraints is treated together with optimization.

For structural optimization it was a retrograde period when some authors have shown that the 'naiv' optimization method which treats the stability constraints as simultaneously active and does not consider the initial imperfections, leads to imperfection-sensitive solutions. It will be shown (Chapter 9) that the use of stability constraints which consider the initial imperfections, results in solutions which are less imperfection-sensitive, safe and economic.

The optimum design of metal structures is like a slimming cure: The best way to decrease the weight of structures is to *decrease the plate thicknesses*. This is limited by several phenomena which should be avoided. Thus, in order to know the limits of the thickness minimization, the following problems should be investigated:

– Too thin plates exhibit too large fabrication imperfections due to shrinkage of welds after welding, thus, residual welding stresses and distortions should be analyzed (Chapter 1),

– Thin-walled structures, mainly open sections are very sensitive to twist. Restrained warping causes additional normal stresses which can cause failures. Thus, the behaviour of thin-walled rods due to bending, shear and torsion should be treated (Chapter 2),

– Too thin plates can buckle due to in-plan compression, bending or shear. Thin-walled columns and beams can fail due to overall buckling or lateral-torsional buckling. Thus, the main problems of structural stability should be investigated (Chapter 3),

–Too thin plates can vibrate due to pulsating loads, since their eigenfrequency is very low. Effective damping can be achieved by using sandwich structures with layers of materials with high damping capacity (Chapter 4),

– Decreasing plate thicknesses means to create structural parts with higher stress concentrations which, in the case of pulsating loads, can cause fatigue failure of welded joints (Chapter 10),

– To avoid residual welding distortions, buckling and vibration, stiffening can be used. Stiffened and cellular plates should be treated (Chapter 12),

– Stiffening by welded ribs can be too costly, therefore the fabrication costs should be analyzed (Chapter 5).

We have investigated several mathematical methods of optimization and adapted some software in this field as well as worked out aspects of expert systems (Chapters 6 and 7).

Our aim is to collect our recent research results in the field of optimum design of metal structures published in various conference proceedings and journal articles. We have participated in international symposia on tubular structures (Chapters 9, 11 and 13) and worked out some documents for the IIW, e.g. about the welded silos (Chapter 14).

Our recent research was concentrated to the application of the new design rules of the Eurocode 3 (EC3) for steel structures, although these rules do not contain many application fields (e.g. cranes, silos, aluminium structures). The new British Standard BS 8118 for aluminium structures made it possible to extend the optimum design also for simple aluminium structural components.

We have worked out some educational aids in the frame of a TEMPUS JEP in years 1992-1993. The financial support of the Hungarian Fund for Scientific Research (grants OTKA 4479 and 4407) in years 1992-1995 enabled us to work out some conference papers.

Most parts of the English book of the first author (Farkas 1984) are actual also now and do not need any actualization. The chapter about the optimum design of welded I- and box beams loaded in bending and shear is modified considering the new rules of EC3 (Chapter 8).

The second author's research including the dissertation for the academic degree of doctor of technical science made it possible to incorporate in the book the multiobjective optimization and the expert systems (Chapters 6 and 7).

If the repetition is mother of studies, then the comparison can be called as father of research, since it helps designers to select the most suitable optimum versions. Optimum design enables the comparison, since only optimized versions can be realistically compared to each other.

For design constraints we should use the standard rules, but it is not our aim to describe these rules in detail, we prefer the derivations of some important formulae used in standards. Mainly the EC3 formulae are used but other standards (DIN, BS, API etc.) are also applied.

In our further research work we plan to apply the optimum design methods for other types of structures such as semi-rigid frames, stiffened shells and fiber-reinforced plastic structures.

The figures have been drawn by Mrs Flórián Kiss, in the administrative work helped engineer István Bedross, we acknowledge their valuable work.

Optimum designers are always optimistic, since they hope to find suitable optima. We hope also that engineers can use this book to improve their creative work with success.

CHAPTER 1

Residual welding stresses and distortions

When steel structures are constructed by welding, deformations and welding residual stresses could occur as a result of the high heat input and subsequent cooling (Boley & Weiner 1960). The welding process can create significant locked-in stresses and deformations in fabricated steel structures (Rykalin 1951).

The residual stresses and initial imperfections can have an important influence upon the behaviour of the structure under the variable loading (Gurney 1979). It is well known that these initial imperfections due to welding reduce the ultimate strength of the structure.

Even though various efforts have been made in the past to express the deflection of panels from experimental aspects and measurements of actual structures, it may be said that there are few investigations from the theoretical point of view. In order to find out a practical estimation method for the welding distortion of a panel, the following analyses have been carried out by Okerblom (1955).

To determine the thermal effect on the structure it is advisable to investigate some simple structures (Chang Doo Jang & Seung Il Seo 1995). Finite element calculations can help to determine these residual stresses (Josefson 1993, Wikander et al. 1994).

1.1 SIMPLE EXAMPLES OF THERMOELASTICITY

We assume the following:
– The coefficient of thermal expansion and the Young modulus are independent from the temperature,
– The deflections are in the elastic range, the Hooke-law is valid,
– The cross sections of the beam will be planar after deflection,
– The cross section is uniform,
– The beam is made of one material grade,
– The thermal distribution is uniform along the length of the beam and steady state.

The change in length (ΔL) and deflection (w_{max}) of a simply supported beam, as shown in Figure 1.1 due to linear heat distribution is as follows:
– The strain in the centre of gravity is $\varepsilon_G = \alpha_0 T_{es} = \alpha_0 (T_{e1} - \Delta T_e \, e_1 / h)$. Here $\Delta T_e = T_{e1} - T_{e2}$ is the temperature difference and α_0 is the coefficient of thermal expansion. The change in length (ΔL) caused by heat expansion is $\Delta L = \varepsilon_G L$. The heat

7

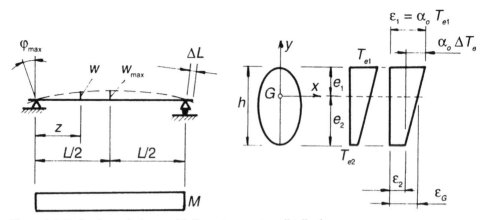

Figure 1.1. Deflections of a beam with linear temperature distribution.

distribution is non-uniform at the cross section, so a curvature occurs at the beam. The radius of curvature is ρ_0.

The curvature is $C = 1/\rho_0 = \alpha_0 \Delta T_e / h$. There is a relation between radius of curvature, bending moment and bending stiffness as $1/\rho_0 = M/EI_x$. So the curvature can be considered as the effect of a uniform bending moment across the length as $M = \alpha_0 \Delta T_e EI_x / h$.

The rotation of the angle is $\varphi(z) = M/EI_x(L/2 - z)$ so the maximum value is $\varphi_{max} = CL/2$.

The deflection is $w(z) = (M/2EI_x)(Lz - z^2)$, so the maximum value is $w_{max} = CL^2/8$.

Another problem is, when the thermal distribution is nonlinear as shown in Figure 1.2. The thermal strain would be different at different points of cross section if they were independent form each other: $\varepsilon = \alpha_0 T_e(y)$. Because they are connected to each other, we assume that the cross section remains planar, only a linear strain can occur in the cross section. This linear strain is characterized by the strain of the gravity centre and the curvature of the beam: $\varepsilon = \varepsilon_G + Cy$. The differences between the theoretical thermal strain and the linear strain cause the stresses:

$$\sigma = E\varepsilon = E\left[\varepsilon_G + Cy - \alpha_0 T_e(y)\right] \tag{1.1}$$

There is no external loading on the beam, so the internal stresses caused by thermal difference are in equilibrium,

$$\int_A \sigma dA = 0 \quad \text{and} \quad \int_A \sigma y dA = 0 \tag{1.2}$$

By inserting Equation (1.1) to Equation (1.2), we get

$$\varepsilon_G = \frac{1}{A}\int_{e_1}^{e_2} \alpha_0 T_e(y)t(y)dy \quad \text{and} \quad C = \frac{1}{I_x}\int_{e_1}^{e_2} \alpha_0 T_e(y)yt(y)dy \tag{1.3}$$

If the thickness is constant, i.e. $t(y) = t$ we can define the thermal shrinkage impulse A_T as

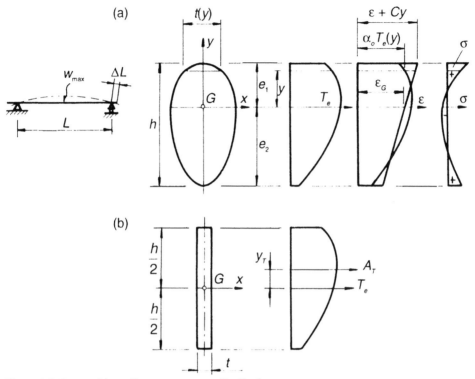

Figure 1.2. Beam with nonlinear temperature distribution.

$$A_T = \int_{e_1}^{e_2} \alpha_0 T_e(y) dy \tag{1.4}$$

The thermal impulsive moment is defined as follows

$$A_T y_T = \int_{e_1}^{e_2} \alpha_0 T_e(y) y dy \tag{1.5}$$

Using these definitions the strain at the center of gravity and the curvature are as follows

$$\varepsilon_G = \frac{A_T t}{A} \tag{1.6}$$

$$C = \frac{A_T t y_T}{I_x} \tag{1.7}$$

1.2 THE OKERBLOM'S ANALYSIS

When a structural section is welded, it undergoes distortion as a result of thermal

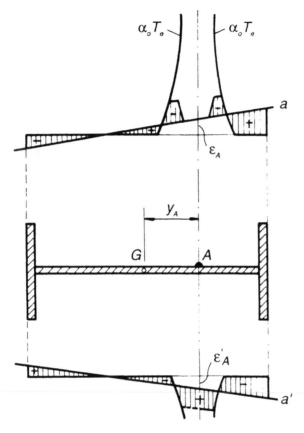

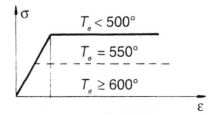

Figure 1.3. Strain distribution during and after welding.

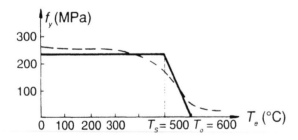

Figure 1.4. The yield stress in the function of the temperature and strain.

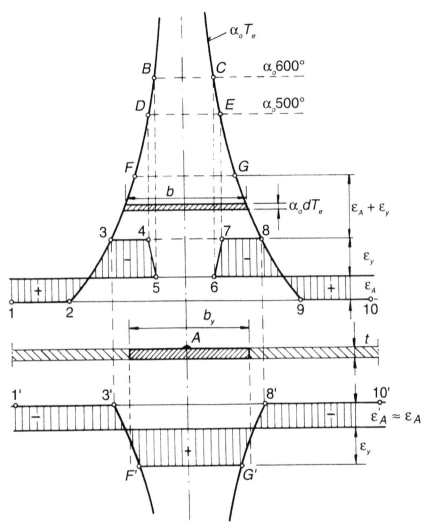

Figure 1.5. Distribution of thermal strains during and after welding.

shrinkage along the axis of the weld. For example, an edge-welded bar section shortens (ΔL) and deflects (w_{max}). Experiments indicate that the Okerblom's analysis provides excellent prediction for longitudinal deflections caused by thermal shrinkage along the weld (Okerblom 1955, Okerblom et al. 1963, Farkas 1969).

The analytical heat-transfer theory of welding was developed by Rykalin (1951). Essentially, Okerblom utilises the analytical heat-transfer theory of moving heat sources to establish the thermal strain and stress distributions around the weld. The primary objective of the Okerblom's analysis is to predict the beam shrinkage (ΔL) and deflection (w) as shown in Figure 1.1. I investigate the effect of a longitudinal weld welded on the fiber A of an elastic I-beam (Fig. 1.3).

In the Okerblom's analysis the material is linearly elastic and ideally plastic. The yield stress is constant till 500°C, and between 500 and 600°C it decreases to zero. If the temperature is larger than 600°C, there is no measurable stress in material (Fig. 1.4).

The approximation of the T_e temperature suggested by Okerblom is as follows

$$T_e = \frac{0.4840 Q_T}{c_0 \rho t (2y)} \tag{1.8}$$

where c_0 is the specific heat, ρ is the material density, t is the thickness of the plate.

The thermal impulse can be calculated according to Figure 1.5. The diagram determined by points 1-10 shows the stress distribution during welding. It can be obtained by projection of points B and C to the line ε_A which occurs due to elastic deformation of the structure during welding. Points B and C are determined by the line 600°C α_0, so between points 5 and 6 no stresses occur. Points 3-4 and 7-8 are obtained by the line parallel to line ε_A in a distance ε_y, with projection to this line the points D and E determined by the line 500°C α_0. It can be seen that plastic strains occur during welding between points 3 and 8. These restrained strains cause residual stresses after cooling.

The residual stress diagram after cooling can be obtained projecting points 3 and 8 onto the basic line 1'-10'. Considering the elastic deformation $\varepsilon'_A \approx \varepsilon_A$ during cooling and the line of ε_y, one obtains the residual stress diagram 1'-3'-F'-G'-8'-10'. The area 3'-F'-G'-8' characterizes the thermal shrinkage impulse A_T which causes the residual stresses and deformations in the structure.

Since the line parts of 3'-F' and 8'-G' are the same as parts 3-F and 8-G, the A_T can be calculated by investigating the area 3-F-G-8 in the diagram drawn for the state during welding:

$$A_T = \int_{\varepsilon_A + \varepsilon_y}^{2(\varepsilon_A + \varepsilon_y)} b \alpha_0 dT_e = \frac{0.4840 \alpha_0 Q_T}{c_0 \rho t} \int_{T_{e1} = (\varepsilon_A + \varepsilon_y)/\alpha_0}^{T_{e2} = 2 T_{e1}} \frac{dT_e}{T_e} \tag{1.9}$$

$$A_T = \frac{0.4840 \alpha_0 Q_T}{c_0 \rho t} \ln 2 = \frac{0.3355 \alpha_0 Q_T}{c_0 \rho t} \tag{1.10}$$

where $Q_T = \eta_0 (UI / v_w) = q_0 A_w$, U is arc voltage, I is arc current, v_w is speed of welding, c_0 is specific heat, η_0 is coefficient of efficiency, q_0 is the specific heat for the unit welded joint area (1 mm^2), A_w is the cross-sectional area of a weld.

For a mild or low alloy steels, where $\alpha_0 = 12*10^{-6}$ (1/°C), $c_0 \rho = 4.77*10^{-3}$ (J/mm^3/°C), the thermal impulse is

$$A_T t \ (\text{mm}^2) = 0.844 * 10^{-3} Q_T \ (\text{J / mm})$$

Inserting this into Equations (1.6) and (1.7), we get the basic Okerblom formulae

$$\varepsilon_G = \frac{A_T t}{A} = -0.844 * 10^{-3} \frac{Q_T}{A} \tag{1.11}$$

The minus sign means shrinkage.

$$C = \frac{A_T t \, y_T}{I_x} = -0.844 * 10^{-3} \frac{Q_T \, y_T}{I_x} \tag{1.12}$$

Note that the distorted form can be determined by view. y_T and C have opposite signs (Fig. 1.15).

The elastic strain in the weld can be calculated using the previous two expressions

$$\varepsilon_A = \varepsilon_G + C y_T \tag{1.13}$$

The average width of the plastic tension zone around the weld is

$$b_y = \frac{A_T}{\varepsilon_A + \varepsilon_y} \tag{1.14}$$

At the region of weld the residual tensile stress after welding reaches the yield stress (Fig. 1.5). The area of the plastic zone is

$$A_y = b_y t = \frac{A_T t}{\varepsilon_A + \varepsilon_y} \tag{1.15}$$

By using Equations (1.6), (1.7) and (1.13) one obtains

$$\frac{1}{A_y} = \frac{1}{A} + \frac{y_T^2}{I_x} + \frac{\varepsilon_y}{A_T t} \tag{1.16}$$

If no crookedness is developed in beam during welding, as for example in the case of a symmetrical weld arrangement, Equation (1.16) takes the form

$$\frac{1}{A_y} = \frac{1}{A} + \frac{\varepsilon_y}{A_T t} \tag{1.17}$$

For mild steels

$$\frac{1}{A_y} = \frac{1}{A} + \frac{y_T^2}{I_x} + \frac{1.43}{Q_T} \text{ (J / mm)} \tag{1.18}$$

If the structure can be regarded as a very stiff one, when $\varepsilon_A = 0$, area of plastic zone is

$$\frac{1}{A_y} = \frac{\varepsilon_y}{A_T t} \tag{1.19}$$

For mild steels

$$A_y = \frac{Q_T}{1.43} \tag{1.20}$$

The equilibrium equation for a section with tension and compression stresses is according to Figure 1.6

$$(b - b_y)\sigma_c = b_y f_y \tag{1.21}$$

Using Equation (1.17) one can compute the residual compressive stress,

$$\sigma_c = \frac{A_T t f_y}{A \varepsilon_y} = \frac{A_T t}{A} E = \frac{0.3355 \alpha_0 \eta_0 UIE}{c_0 \rho v_w b t} \tag{1.22}$$

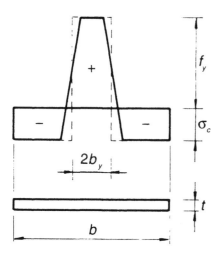

Figure 1.6. Approximate stress distribution for a plate with a single weld at the middle.

With data $\alpha_0 = 11*10^{-6}$, $c_0\rho = 3.53*10^{-3}$ (J/mm³ °C), $E = 2.05*10^5$ (MPa), used by White, the Okerblom formula is

$$\sigma_c = \frac{0.214\eta_0 UI}{v_w bt} \tag{1.23}$$

White (1977a, b) proposed an approximate formula based on own experiments

$$\sigma_c = \frac{0.2\eta_0 UI}{v_w bt} \tag{1.24}$$

It can be seen that Okerblom's formula is in agreement with White's experimental results.

The formulae above are valid for symmetrically arranged welds $y_A = 0$ when

$$\frac{Q_T}{A} \le 2.50 \, (\text{J} / \text{mm}^3) \tag{1.25}$$

for eccentric welds ($y_A \ne 0$) when

$$\frac{Q_T}{A} \le 0.63 \, (\text{J} / \text{mm}^3) \tag{1.26}$$

For approximate calculations one can use the simple formula

$$\varepsilon_y = \frac{f_y}{E} \tag{1.27}$$

where f_y is the yield stress of the parent material. The weld metal may have a different yield stress. This discrepancy arises due to the electrode material. In this case it is important to measure the yield stress of the weld metal. Using high strength steels, the yield stress of the weld metal can be smaller that of the parent material. Therefore the residual stresses are relatively smaller, than in the case of mild steel.

For some simple cross sections the residual stress distribution can be seen on Figure 1.7.

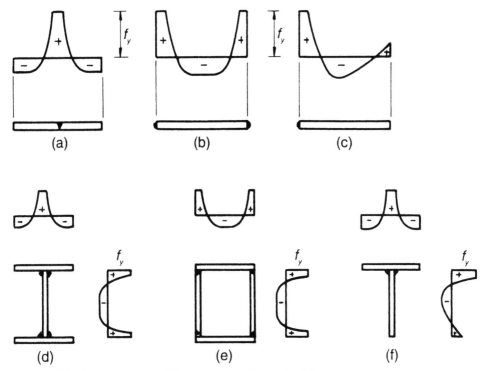

Figure 1.7. Residual stresses in different cross sections and weld positions.

1.3 MULTI-PASS WELDING

The basic Okerblom formulae are valid for single pass-welding. For multi-pass welding it is necessary to modify Equation (1.10), because the new weld pass resolves the plastic zone, made by the previous weld pass. For the value of residual stresses that weld pass is governing, which causes the largest plastic zone. Introducing a parameter for the correction of thermal shrinkage impulse

$$A_T = m_t \frac{0.3355\alpha_0 Q_T}{c_0 \rho t} \tag{1.28}$$

where

$$m_t = \frac{A_{yt}}{A_{y1}} \tag{1.29}$$

A_{y1}, A_{yt} the areas of plastic zone due to single- and multi-pass welding, respectively.

For example at a two-pass (equal passes) butt joint $m_t = 1$. For a double fillet weld for thin plates, where the welds are welded one after the other $m_t = 1.2 - 1.3$. For intermittent fillet welds $m_t = L_w / L_u$, where L_w and L_u are the distances of the welded and unwelded part at intermittent fillet weld.

White (1977c) suggested to calculate the tendon force from the parameters of that pass, which has the greatest section area.

The effect of preheating can be taken account with a correction parameter

$$F' = \left(1.1 - \frac{T_p}{1000}\right) F; \quad F = \sigma_c bt \tag{1.30}$$

where the temperature of preheat is $T_p > 100°C$.

1.4 EFFECT OF INITIAL STRAINS

In the previous calculations it was assumed, that there are no initial strains and stresses in the structure. In practice there are usually some strains and stresses before welding, or previous welds cause initial strains and stresses for the next weld(s). Preheating, flame cutting and pre-stressing have the same effect.

The strain diagram is similar to Figure 1.5 except of the initial tensile strain ε_I.

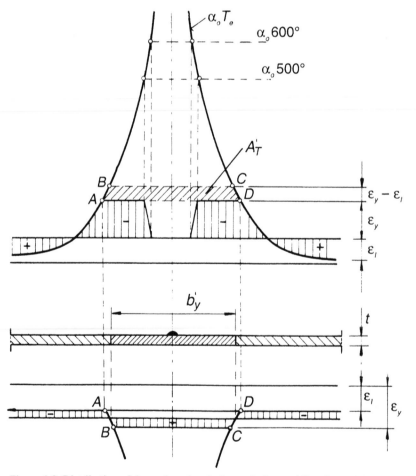

Figure 1.8. Distribution of thermal strains during and after welding due to initial strains.

Figure 1.8 shows the strain distribution during welding and after welding. The final deformations after welding are caused by the difference of $\varepsilon_y - \varepsilon_I$. The effective zone is between ABCD points. The thermal impulse can be computed as follows:

$$A'_T = \int_{\varepsilon_I + \varepsilon_y}^{2\varepsilon_y} b\alpha_0 dT_e = \frac{0.4840\alpha_0 Q_T}{c_0\rho t} \int_{T_{e1}=(\varepsilon_I+\varepsilon_y)/\alpha_0}^{T_{e2}=2\varepsilon_y/\alpha_0} \frac{dT_e}{T_e} \tag{1.31}$$

$$A'_T = \frac{0.4840\alpha_0 Q_T}{c_0\rho t} \ln\frac{2\varepsilon_y}{\varepsilon_y + \varepsilon_I} \tag{1.32}$$

To consider the effect of initial deformation we introduce a modifying parameter v_m, which is the ratio of the thermal impulse with and without initial strain.

$$v_m = \frac{A'_T}{A_T} = 1 - \frac{\ln[1+(\varepsilon_I/\varepsilon_y)]}{\ln 2} \approx 1 - \frac{\varepsilon_I}{\varepsilon_y} \tag{1.33}$$

The approximate formula is valid when $(\varepsilon_I/\varepsilon_y) \geq 0$.

Figure 1.9 shows v_m in the function of $\varepsilon_I/\varepsilon_y$. Without initial deformation no modification is necessary, so if $\varepsilon_I = 0$, then $v_m = 1$. If there is a tension in the elastic zone, $0 < \varepsilon_I < \varepsilon_y$, then $1 > v_m > 0$. If the initial strain is equal to the yield strain, $\varepsilon_I = \varepsilon_y$, $v_m = 0$, there is no residual stresses and deformations after welding. If the initial strain is negative (compression), $\varepsilon_I < 0$, $v_m > 1$, this strain increases the deformation, but the approximate formula cannot be used.

The thermal shrinkage impulse is according to Figure 1.8

$$A'_T = \frac{b'_y}{\varepsilon_y - \varepsilon_I} \tag{1.34}$$

the area of plastic zone is according to Figure 1.10

$$A'_y = b'_y t = \frac{A'_T t}{\varepsilon_y - \varepsilon_I} = \frac{v_m A_T t}{\varepsilon_y - \varepsilon_I} \tag{1.35}$$

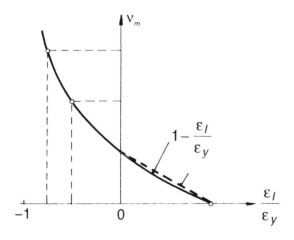

Figure 1.9. Modifying parameter in the function of initial strain.

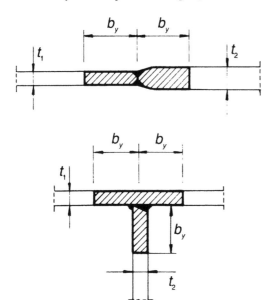

Figure 1.10 The area of plastic zone.

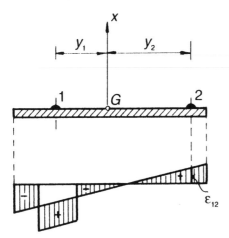

Figure 1.11. Initial strain in the place of the second weld due to the first weld.

if $\varepsilon_i > 0$ then $A'_y = (A_T t / \varepsilon_y)$ and for a normal grade steel $A'_y = (Q_T / 1.43)$ (J/mm).

By using the modifying parameter it is possible to determine the right welding sequence. If there are more welds on a structure, the welding sequence can be very important, its effect on the final strain is significant. Calculating v_m, one can compute the effect of the welds on each other, how large is the initial deformation at the place of the other weld, what is this effect on the total deformation, what is the effect of changing the welding sequence (Fig. 1.11).

The strain at the gravity centre line and the curvature are as follows

$$\varepsilon_{G1} = \frac{A_{T1} t}{A}, \quad C_1 = \frac{A_{T1} t y_T}{I_x} \tag{1.36}$$

$$\varepsilon_{I12} = \varepsilon_{G1} + C_1 y_2 = A_{T1} t \left(\frac{1}{A} + \frac{y_1 y_2}{I_x} \right) \tag{1.37}$$

the modifying factor

$$v_{m12} = 1 - \frac{\ln\left[1 + \left(\varepsilon_{I12}/\varepsilon_y\right)\right]}{\ln 2} \approx 1 - \frac{\varepsilon_{I12}}{\varepsilon_y} \tag{1.38}$$

expresses the effect of the first weld at the place of the second weld.

The final strain and curvature caused by two welds is

$$\varepsilon_{G(1+2)} = \varepsilon_{G1} + v_{m12} \varepsilon_{G2} = \varepsilon_{G1} \left(1 + v_{m12} \frac{Q_{T2}}{Q_{T1}} \right) \tag{1.39}$$

$$C_{1+2} = C_1 + v_{m12} C_2 = C_1 \left(1 + v_{m12} \frac{Q_{T2} y_2}{Q_{T1} y_1} \right) \tag{1.40}$$

Changing the welding sequence $\varepsilon_{G(2+1)}$ and C_{2+1} can be calculated using Equations (1.39) and (1.40), changing the subscripts:

$$\varepsilon_{G(2+1)} = \varepsilon_{G2} + v_{m21} \varepsilon_{G1} = \varepsilon_{G2} \left(1 + v_{m21} \frac{Q_{T1}}{Q_{T2}} \right) \tag{1.41}$$

$$C_{2+1} = C_2 + v_{m21} C_1 = C_2 \left(1 + v_{m21} \frac{Q_{T1} y_1}{Q_{T2} y_2} \right) \tag{1.42}$$

Comparing the two strains and curvatures, the smallest absolute value gives the better welding sequence.

If there are two longitudinal welds in an asymmetric I-beam and the 1st weld is closer to the gravity center, $y_1 < |y_2|$, it means $C_1 < |C_2|$, so the 2nd weld has greater effect that the 1st one. The modifying parameter is always less then 1 in this case, $0 < v_m < 1$. The conclusion is, that the closer weld should be made first, because the final deformation will be less, $C_{1+2} < |C_{2+1}|$.

The maximum deflection caused by welding is

$$w_{max} = C \frac{L^2}{8}, \quad w_{1+2} = C_{1+2} \frac{L^2}{8}, \quad w_{2+1} = C_{2+1} \frac{L^2}{8} \tag{1.43}$$

1.5 THE EFFECT OF EXTERNAL LOADING ON THE WELDING RESIDUAL STRESSES

To investigate the effect of static tension forces on residual stresses, we simplify the distribution of residual stresses according to Figure 1.12.

The stresses in the tension field are σ_b, in the compression field σ_a. If the sum of tension stresses due to the external force and the residual stress is less than the yield stress, $\sigma_e < \sigma' = f_y - \sigma_b$, the resulting stress on the width part (b) of the plate is $\sigma'_b = \sigma_b + \sigma_e$, on the width part (a) of the plate $\sigma'_a = \sigma_a + \sigma_e$. If the tension stresses due to the external force are larger $\sigma' \le \sigma_e \le f_y$, in this case on the width part (b) of

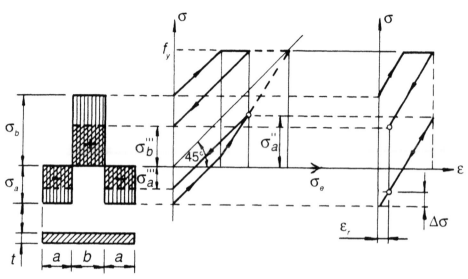

Figure 1.12. Simplified residual stress distribution and the decrease of initial stresses due to static loading.

the plate $\sigma_b'' = f_y$, on the width part (a) of the plate

$$\sigma_a'' = \sigma_a + \sigma' + (\sigma_e - \sigma')\frac{2a+b}{2a} \tag{1.44}$$

During unloading, all fibres deform elastically

$$\sigma_a''' = \sigma_a'' - \sigma_e = \sigma_a + (\sigma_e - f_y + \sigma_b)\frac{b}{2a} \tag{1.45}$$

$$\sigma_b''' = \sigma_b'' - \sigma_e = f_y - \sigma_e \tag{1.46}$$

It can be seen, that the residual stresses decrease. If the external force were $F_y = f_y t\,(2a + b)$, then no residual stresses remain (Fig. 1.12) and if the material is ideally elastic-plastic, then the residual stresses will decrease to zero but residual deformations will occur.

1.6 REDUCTION OF RESIDUAL STRESSES

There are several ways to reduce the deformation and the residual stresses in the welded structures. They are as follows:

Reduction techniques in the design stage
- – Cross-section symmetrical to the gravity center,
- – Symmetrical welded joints,
- – Suitable choice of welding sequence,
- – Suitable choice of welding parameters,
- – Welding in clamping device,

– Welding in prebent state in clamping device.

The deformation is much larger, if the cross-section is asymmetrical, or the welded joint is only on one side of the cross section. An opposite weld can reduce the deformation. The welding sequence can also be important for the final deformation of the structure. The best welding sequence is to start with the welded joint closest to the gravity center of the cross-section and continue with an opposite joint to reduce the final deformation. There is a choice of welding parameters, such as voltage, current and welding speed among the technological limits. The use of different heat input for different welded joints can decrease the final deformation.

Welding in a clamping device
The production sequence: Tacking, clamping, welding, loosening (Fig. 1.13).

During welding the deformation w would occur, but it is restrained by clamping moments M. The bending moment necessary to keep the beam straight against the welding deformations is

$$M = I_\xi EC \tag{1.47}$$

where I_ξ is the moment of inertia for the elastic part of the cross-section area, calculated without the plastic zone A_y, C is the curvature of the beam caused be welding in free state.

It is assumed, that the beam material is ideally elasto-plastic, that means that the tensile stress in the plastic zone cannot be larger than the yield stress, so this zone cannot be loaded beyond this limit.

The loosening of the clamped state acts as the bending moments M with opposite sign. These moments cause compressive stresses in the plastic zone which behaves elastically during this unloading, thus one can calculate with the moment of inertia of the whole cross-section I_x. Thus, after the loosening of the clamped state the following curvature occurs

$$C' = \frac{M}{EI_x} = C \frac{I_\xi}{I_x} \tag{1.48}$$

where I_x is the moment of inertia for the total elastic section area.

It can be concluded that, using a clamped state, the residual welding deformations cannot be totally eliminated, they can be decreased only in a measure of I_ξ / I_x,

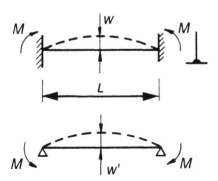

Figure 1.13. Welding in a clamping device.

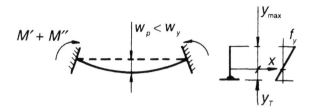

Figure 1.14. Welding in prebent state in clamping device.

$C' < C$; $C' \neq 0$. The ratio between the two curvatures depends on the area of the plastic zone.

Welding in elastically prebent state in clamping device
The production sequence: Tacking, prebending, clamping, welding, loosening (Fig. 1.14)

To prevent very large deformations and cracks, it is advisable to use prebending moments not larger than

$$M_y = \frac{f_y I_x}{y_{max}} \tag{1.49}$$

The curvature and deformation caused by M_y are

$$C_y = \frac{M_y}{EI_x}, \quad w_y = \varepsilon_y \frac{L^2}{8 y_{max}} \tag{1.50}$$

The prebending $w_p < w_y$ causes a tensile prestrain in the place of the longitudinal weld

$$\varepsilon_P = C_p y_T = w_p \frac{8 y_T}{L^2} \tag{1.51}$$

the corresponding modifying factor is

$$v_m = 1 - \frac{\varepsilon_P}{\varepsilon_y} \tag{1.52}$$

The bending moment necessary to keep straight the beam after welding consists of two parts as follows: The moment which is necessary for prebending

$$M' = I_\xi EC_p = 8w_p \frac{EI_\xi}{L^2} \tag{1.53}$$

and the moment which is necessary to eliminate the residual welding deformations

$$M'' = v_m I_\xi EC = 8v_m w \frac{EI_\xi}{L^2} \tag{1.54}$$

These moments act opposite after the loosening and decrease the prebending deformations,

$$M = M' + M'' = I_\xi EC_p + v_m I_\xi EC \tag{1.55}$$

so that the remaining final deformations can be expressed as

$$w_f = w - w_p = \frac{M' + M''}{8EI_x} L^2 - w_p \qquad (1.56)$$

$$w_f = (w_p + v_m w)\frac{I_\xi}{I_x} - w_p \qquad (1.57)$$

where $v_m = 1 - (8w_p y_T / L^2 \varepsilon_y)$, I_x is the moment of inertia for the elastic section area, I_ξ is the moment of inertia for the elastic section area, reduced by the plastic zone, $\overset{\circ}{C}$ is the curvature of the beam caused be welding in free state, v_m is the modifying factor according to Equation (1.52).

The prebending w_p necessary to totally eliminate the residual welding deformations can be calculated from the condition $w_f = 0$.

$$w_p = \frac{w}{\left(I_x / I_\xi\right) + \left(8 y_T w / L^2 \varepsilon_y\right) - 1} \qquad (1.58)$$

Reduction techniques after production
- Straightening welded plates,
- Pressing different welded shapes,
- Vibration (Wozney & Crawmer 1968),
- Heat treatment,
- Weld geometry modification methods (see Section 1.8.1),
- Residual stress methods (see Section 1.8.2).

1.7 NUMERICAL EXAMPLES

1.7.1 *Suitable welding sequence in the case of a welded asymmetric I-beam*

Find the best welding sequence for two welded joints in an I-beam (Fig. 1.15)

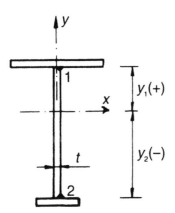

Figure 1.15. Welding sequemce for an I-beam.
Widths: upper flange 0.8*h*, lower flange 0.4*h*.

Section dimensions

$$t = 10 \text{ mm}, \quad h = y_1 + |y_2| = 600 \text{ mm}, \quad L = 6 \text{ m}, \quad \text{steel grade Fe 360}$$

Welding parameters

$$Q_{T1} = Q_{T2} = 60.7 \, A_{w1} \text{ (J/mm)}$$

$$A_{w1} = A_{w2} = 100 \text{ mm}^2$$

Determination of the center of gravity

$$\int_A y \, dA = 0$$

$$0.8ht\left(\frac{h+t}{2} - y_0\right) - hty_0 - 0.4ht\left(\frac{h+t}{2} + y_0\right) = 0$$

$$y_0 = 55.5 \text{ mm}$$

Determination of the moment of inertia

$$I_x = \int_{(a)} y^2 dA = \frac{h^3 t}{12} + hty_0 + \frac{0.8ht^3}{12} + 0.8ht\left(\frac{h+t}{2} - y_0\right)^2 + \frac{0.4ht^3}{12}$$

$$+ 0.4ht\left(\frac{h+t}{2} + y_0\right)^2$$

$$I_x = 8.881 * 10^{-4} \text{ mm}^4$$

$$\varepsilon_{G1} = -0.844 * 10^{-3} \frac{Q_{T1}}{A} = -3.881 * 10^{-4}$$

$$A = 0.8*600*10 + 600*10 + 0.4*600*10 = 1.32*10^4 \text{ mm}^2$$

$$C_1 = -0.844 * 10^{-3} \frac{Q_{T1} y_{T1}}{I_x} = -0.844 * 10^{-3} \frac{60.7 * 100 * 244.5}{8.09247 * 10^8} =$$
$$-1.548 * 10^{-6} \text{ (1 / mm)}$$

$$\varepsilon_{f12} = \varepsilon_{G1} + C_1 y_2 = -3.881 * 10^{-4} - 1.548 * 10^{-6} * (-355.5) = 1.6216 * 10^{-4}$$

the modifying factor is

$$v_{m12} = 1 - \frac{\varepsilon_{f12}}{\varepsilon_y} = 1 - \frac{1.6216 * 10^{-4}}{1.119 * 10^{-3}} = 0.855$$

$$\varepsilon_{G2} = |\varepsilon_{G1}| = 3.881 * 10^{-4}$$

$$C_2 = -0.844 * 10^{-3} \frac{Q_{T2} y_{T2}}{I_x} = -0.844 * 10^{-3} \frac{60.7 * 100 * (-355.5)}{8.09247 * 10^8} =$$
$$2.251 * 10^{-6} \text{ (1 / mm)}$$

The final strain and curvature after the two welds in sequence 1st then 2nd joint are

$$\varepsilon_{G(1+2)} = \varepsilon_{G1} + v_{m12}\varepsilon_{G2} = -3.881*10^{-4} + 0.855*3.881*10^{-4} = -5.627*10^{-4}$$

$$C_{1+2} = C_1 + v_{m12}C_2 = -1.548*10^{-6} + 0.855*2.251*10^{-6} = 3.77*10^{-7} \ (1/mm)$$

The welding sequence the 2nd then the 1st joint

$$\varepsilon_{I21} = \varepsilon_{G2} + C_2 y_1 = 3.881*10^{-4} + 2.251*10^{-6}*244.5 = 9.3847*10^{-4}$$

the modifying factor is

$$v_{m21} = 1 - \frac{\varepsilon_{I21}}{\varepsilon_y} = 1 - \frac{9.3847*10^{-4}}{1.119*10^{-3}} = 0.1613$$

$$\varepsilon_{G(2+1)} = \varepsilon_{G2} + v_{m21}\varepsilon_{G1} = 3.881*10^{-4} + 0.1613*(-3.881*10^{-4}) = 3.2549*10^{-4}$$

$$C_{2+1} = C_2 + n_{m21}C_1 = 2.21*10^{-6} + 0.1613*(-1.548*10^{-6}) = \\ 2.001*10^{-6} \ (1/mm)$$

$$y_1 < y_2, \quad C_1 < C_2, \quad C_{1+2} < C_{2+1}$$

Changing the welding parameters to reduce the final deformation to zero. From

$$C_{1+2} = C_1 + v_{m12}C_2 = C_1 + v_{m12}C_1\frac{Q_{T2}y}{Q_{T1}y_1} = C_1\left(1 + v_{m12}\frac{Q_{T2}y}{Q_{T1}y_1}\right) = 0$$

$$\frac{Q_{T1}}{Q_{T2}} = -v_{m12}\frac{y_2}{y_1} = -0.855*\frac{-355.5}{244.5} = 1.243$$

so if $Q_{T1} = 6.07*10^3$ and $Q_{T2} = 4.88*10^3$ (J/mm) then the final deformation $C_{1+2} = 0$.

Choosing a good welding parameter ratio one can reduce the deformation of the welded structure.

1.7.2 *Welding in a clamping device*

Figure 1.16 shows an asymmetric I-section welded in free state, in clamping device and in prebent state.

Data: $Q_T = 950$ (J/mm), $L = 7$ m, $E = 2.1*10^5$ MPa, $f_y = 240$ (MPa).

Determination of the gravity center

$$\int_{(A)} y \, dA = 0, \quad y_T = 26 \text{ mm}, \quad y_0 = 24 \text{ (mm)}$$

Determination of the moment of inertia

$$I_x = \int_{(A)} y^2 dA, \quad I_x = 1.1381*10^8 \ (mm^4)$$

Welding in free state

$$C = -0.844*10^{-3}\frac{Q_T y_T}{I_x} = 1.827*10^{-6} \ (1/mm)$$

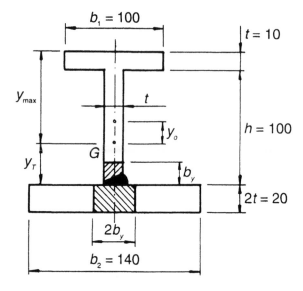

Figure 1.16. Welding of an I-section made in a free state, in clamping device and in prebent state.

$$w = \frac{CL^2}{8} = 11.19 \text{ (mm)}$$

Welding in clamping device

$$w' = w\frac{I_\xi}{I_x}$$

Cross section area and width of the yield zone

$$A_y = \frac{Q_T}{1.43} = 5b_yt = 664 \text{ (mm}^2)$$

$$b_y = 13.3 \text{ (mm)}$$

Determination of the gravity center

$$\int\limits_{(A)} y\,dA = 0$$

The moment of inertia decreasing the section with the yield zone

$$I_\xi = \int\limits_{(A)} y^2\,dA = 1.033*10^8 \text{ (mm}^4)$$

$$C = C\frac{I_\xi}{I_x} = 1.658*10^{-6} \text{ (1/ mm)}$$

$$w' = C'\frac{L^2}{8}$$

$$w' = w\frac{I_\xi}{I_x} = 10.15 \text{ (mm)}$$

It can be seen that $w' < w$ always, but $w \neq 0$.

1.7.3 *Welding in prebent state in a clamping device*

Consider the same I-section in Figure 1.16.

$$w_f = 0, \quad w_P = \frac{w}{\left(I_x / I_\xi\right) + \left(8 y_T w / L^2 \varepsilon_y\right) - 1}$$

With the data $y_T = 26$ mm, $w = 11.19$ mm, $\varepsilon_y = 1.119 * 10^{-3}$

$$w_P = 77.605 \text{ (mm)}$$

The prebending should be in the elastic zone.
   The limit prebending deflection is as follows

$$w_y = C_y \frac{L^2}{8} = \varepsilon_y \frac{L^2}{8 y_{max}} = 83.12 \text{ (mm)}$$

where $y_{max} = 84$ (mm).
   Since the prebending deflection is less than the yield deflection, so the result is suitable.

1.8 WELD IMPROVEMENT METHODS

There are many commonly used improvement techniques to increase the fatigue strength of welded steel joints (Haagensen 1985, 1996). There is no strong correlation between fatigue strength and yield or tensile strength for welded joints because the fatigue life is dominated by crack propagation and the fatigue crack growth rate is practically independent of the steel strength (Barsom 1971, Maddox 1991).
   The crack growth rate is slower in improved welds. The decrease of crack growth rate can be achieved either by the removal of large defects, by a reduced stress concentration at the weld toe, or the crack growth is retarded by compressive residual stresses.
   The reasons of fatigue cracks are illustrated in Figure 1.17. The stress concentra-

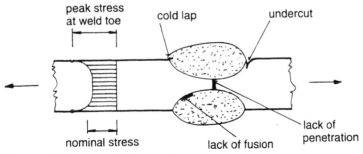

Figure 1.17. Reasons of fatigue cracks.

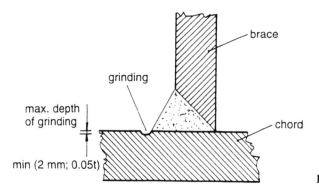

Figure 1.18. Stress concentration.

tion or notch effect is one reason why cracks initiate at the weld toe. The detailed stress analysis shows that the stress concentration factor of the welded joint in Figure 1.18 is lower than that for the plate with a hole. Other weld defects, such as cold laps as shown in Figure 1.17, reduce the fatigue strength. A third contribution to the reduction of the length of the crack initiation stage are the high tensile residual stresses. The waviness of the weld toe in the length direction has considerable influence on the fatigue life (Chapetti & Otegui 1995).

Methods to extend crack initiation life, are as follows:
– Reducing the stress concentration factor of the weld by improving its shape,
– Removing the crack-like defects at the weld toe, and
– Removing the harmful tensile welding residual stresses or introducing beneficial compressive stresses in the toe region.

Modification of the weld shape also reduce the severity of the weld toe defects, weld improvement methods can be placed in two broad groups:
1. Weld geometry modification methods,
2. Residual stress methods.

We describe only some methods, which are mainly used in industrial applications.

1.8.1 Weld geometry modification methods

Weld profile control
Weld profile control, i.e. performing the welding in such a manner that the overall weld shape gives a low stress concentration and the weld metal blends smoothly with the plate.

Improved welding techniques, such as weld profiling and the use of special electrodes can reduce the weld toe defects.

Improved weld profiles means that a low stress concentration factor is aimed for by controlling the overall shape of the weld to obtain a concave profile and by requiring a gradual transition at the weld toe.

Weld toe grinding
Grinding can be carried out with a rotary burr grinder or disc grinder, the former requiring much more time and therefore incurring higher costs. The lower stress con-

centration factor and the removal of crack-like defects at the weld toe generally give large increases in fatigue life for transverse welds, around 50-100% at long lives (N > 1 million cycles).

Tungsten inert gas (TIG) or plasma arc remelting
Remelting of the weld toe using either TIG or plasma welding equipment generally results in large gains in fatigue strength. The smoother weld toe transition reduces the stress concentration factor; the slag inclusions and undercuts are almost completely removed, and the increased surface hardness in the remelted zone contribute to the higher fatigue strength (Kado et al. 1988, Ohta 1988). In addition the original residual stress field is altered.

Standard TIG dressing equipment is used, usually without any filler material. TIG remelting also introduces a residual stress field, which like the welding process used for depositing the weld metal, usually gives tensile stresses at the surface (Lieurade et al.1992, Lopez Martinez & Blom 1995). The magnitude of the improvement depends primarily on the joint severity and base material strength. The improvements for medium strength steels are about 50 -70% over the as-welded strength.

Plasma dressing generally gives better results than TIG dressing, due to the higher heat input and the larger pool of melted metal.

1.8.2 *Residual stress methods*

Improvement in fatigue behaviour can be obtained by removing welding residual stresses by postweld heat treatment, especially if the applied load cycle is wholly or partly in compression. The largest benefits are obtained if compressive residual stresses are introduced.

Hammer peening
Hammer peening is usually performed with a pneumatic hammer fitted solid tool with a rounded tip of 6-14 mm radius. A similar technique consists of using a wire bundle instead of a solid tool. The solid tool gives a far more severe deformation and gives better improvements than either wire bundle or shot peening (Gurney 1979). Optimum results for hammer peening are obtained after four passes.

Ultrasonic peening
This is a new method, and as for other improvement methods, the magnitude of improvement varies with material, type of joint and type of loading, but the improvement is about 50-200% (Trufiakov 1995).

Shot peening
In the shot peening process the surface is blasted with small steel or cast iron shots in a high velocity air stream, producing residual surface stresses of about 70-80% of the yield stress. Results from fatigue tests on shot peened welded joints show substantial improvements for all types of joints, the magnitude of the improvements varying with type of joint and static strength of the steel. The improvement is about 30-100% increase in fatigue lives in the long life region (Grimme et al. 1984, Haagensen 1992).

Even larger improvements may be obtained when techniques from the two main groups are combined. The effectiveness of most improvement methods is highest in the long life, low stress part of the S-N curve. In the short life region, where the local stress at the weld toe exceeds the yield limit the effect of most improvement methods is small or non-existing, this is particularly true for the residual stress methods. Since most structures are designed to the long life part of the S-N curve these improvement techniques can have a great advantages. Fatigue strength improvement is best measured by the fatigue strength at 2 million cycles.

Comparison of costs (Godfrey & Hicks 1987)
If we choose hammer peening as unity, then shot peening is approximately 1.5, disc grinding is 5 and TIG dressing is 3 times more expensive. If we compare the efficiency of the techniques we find that hammer peening and grinding combined with hammer peening gives the largest improvements, in excess of 100% increase. A 60% increase over the as-welded design curve is therefore proposed.

CHAPTER 2

Thin-walled rods

2.1 INTRODUCTION

As mentioned earlier the best way to achieve lighter structures is the decrease of thicknesses of plate elements. In thin-walled structures the thicknesses are much smaller than the other dimensions. In thin-walled structures special stress, deformation and instability phenomena occur which should be taken into account in the optimum design. In this chapter the main problems of the theory of thin-walled structures are summarized.

In order to simplify the formulae only sections of single symmetry are treated. The more general theory is treated in related literature (e.g. Murray 1984, Chen & Atsuta 1977). The books of Yu (1991), Hancock (1994), Umanskiy (1961), Vlasov (1963), US Manual (1986), the German Design Rules (DASt 1986) and the journal 'Thin-walled Structures' are also worth mentioning.

Our aim is to emphasize the significant difference between the torsional behaviour of the open and closed sections and to show the disadvantages of open sections for structures in which torsional phenomena can occur.

Two loading cases are treated here as follows:

1. In the case of bending and shear the determination of shear center is described,

2. In the case of torsion the stresses and deformations are calculated for open and closed sections.

The compression is treated in Chapter 3. In the present chapter the following assumptions are valid: The material is elastic and homogeneous, the rods are long and prismatic.

2.2 BENDING AND SHEAR, SHEAR CENTER

The T point is defined (a) as shear center which is the action point of the resultant of the shear flow $v = \tau t$

$$\tau = \frac{V_y S_x}{I_x t} \tag{2.1}$$

where S_x is the statical moment and I_x is the moment of inertia.

When the shear force V is acting at point T, torsion does not occur. (b) Another

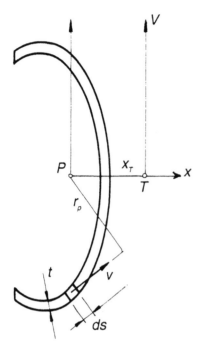

Figure 2.1. Shear center T of an open section.

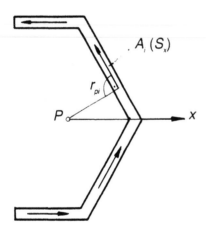

Figure 2.2. Section consisting of planar plate elements.

definition is that, in the case of torsion, the section rotates about the longitudinal axis z corresponding to the point T.

Consider first a singly symmetric open section (Fig. 2.1). The distance of the shear center from an optional point P in the axis x can be calculated on the basis of the moment equilibrium equation

$$x_T V_y = \int\limits_{(A)} r_p v \, ds = \frac{V_y}{I_x} \int r_P S_x \, ds \qquad (2.2)$$

or for sections consisting of planar plate elements (Fig. 2.2)

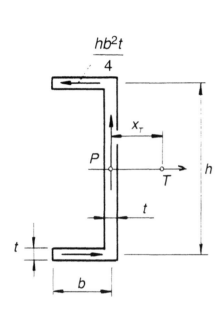

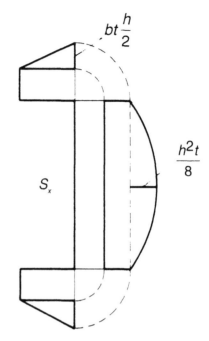

Figure 2.3. Shear center of a channel section.

$$x_T = \frac{1}{I_x}\sum_i r_{Pi} A_i(S_x)$$ (2.3)

where $A_i(S_x)$ is the area of the ith part of the statical moments diagram.

In the case of a channel section the S_x-diagram is shown in Figure 2.3.

$$A_1 = A_3 = b^2th/4$$ (2.4)

$$x_T = A_1h/I_x = b^2th^2/(4I_x)$$ (2.5)

Another example is a cross-section of a crane runway girder (Fig. 2.4). The areas A_i are as follows:

$$A_1 = A_3 = b^2t_1h_1/4 \quad \text{and} \quad A_4 = h_2^3t_2/12$$ (2.6)

$$x_T = (A_1h_1 - A_4H)/I_x$$ (2.7)

For some sections the T point can easily be found as shown in Figure 2.5. In these cases the resultant of shear flows acts at the intersection of planes of plate elements.

The determination of T point for *closed sections* is more complicated. The displacements in the direction of coordinate axes z and s are w and u, respectively. The specific angular shear distortion of a shell element is expressed by (Fig. 2.6)

$$\gamma = \frac{\partial w}{\partial s} + \frac{\partial u}{\partial z}$$ (2.8)

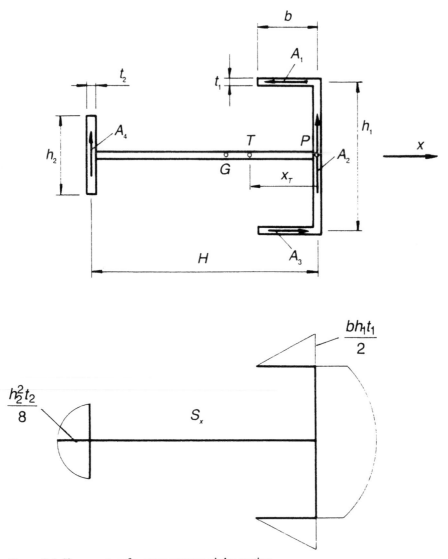

Figure 2.4. Shear center of a crane runway girder section.

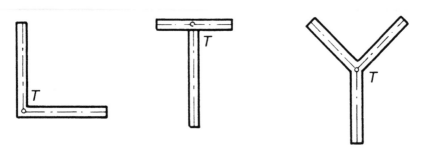

Figure 2.5. Shear center of profiles consisting of plate elements with a common intersection point.

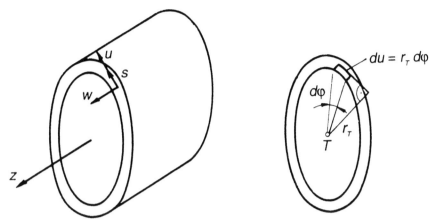

Figure 2.6. Displacements of a point in a closed section.

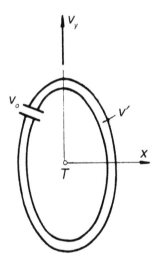

Figure 2.7. Two shear flow components in a closed section.

As shown in Figure 2.6

$$du = r_T d\varphi \tag{2.9}$$

$$\frac{\partial u}{\partial z} = r_T \frac{d\varphi}{dz} \tag{2.10}$$

When the shear force acts on T point, torsion does not occur, thus

$$\frac{\partial u}{\partial z} = 0 \tag{2.11}$$

and

$$\gamma = \frac{v}{tG} = \frac{\partial w}{\partial s} \tag{2.12}$$

furthermore

$$\oint \frac{v}{tG} ds = \oint \frac{\partial w}{\partial s} ds = 0 \tag{2.13}$$

The shear flow v can be divided in two parts as follows

$$v = v_0 + v_1 \tag{2.14}$$

where v_0 is the shear flow in the point where the section is cut off (Fig. 2.7), and v_1 is the shear flow of the open section

$$v_1 = V_y S_x / I_x \tag{2.15}$$

v_0 can be calculated using Equations (2.13) and (2.14)

$$v_0 = -\frac{V_y}{I_x} \frac{\oint (S_x / t)}{\oint (ds / t)} \tag{2.16}$$

and, with Equations (2.14-2.16) one obtains

$$v = \frac{V_y}{I_x} \left(S_x - \frac{\oint (S_x / t) ds}{\oint (ds / t)} \right) \tag{2.17}$$

The moment equilibrium equation is now

$$x_T V_y = \oint r_p v ds \tag{2.18}$$

and, using Equation (2.17)

$$x_T = \frac{1}{I_x} \left(\oint r_p S_x ds - \frac{2A_p}{\oint (ds / t)} \oint \frac{S_x}{t} ds \right) \tag{2.19}$$

where

$$A_p = \oint r_p ds \tag{2.20}$$

is the area enclosed by the profile periphery. For profiles consisting of planar plate elements Equation (2.19) is modified to

$$x_T = \frac{1}{I_x} \left(\sum r_{Pi} A_i (S_x) - \frac{2A_p}{\sum b_i / t_i} \sum A_i (S_x) \right) \tag{2.21}$$

where b_i and t_i are the width and thickness of the ith plate element, respectively.

As an example consider a singly symmetric box section as shown in Figure 2.8.

$$A_1 = A_5 = h^3 t_1 / 48 \tag{2.22}$$

$$A_2 = A_4 = h^2 t_1 b / 8 + h b^2 t_2 / 4 \tag{2.23}$$

$$A_p = hb \tag{2.24}$$

$$A_3 = h^3 t_1 / 8 + h^2 b t_2 / 2 + h^3 t_2 / 12 \tag{2.25}$$

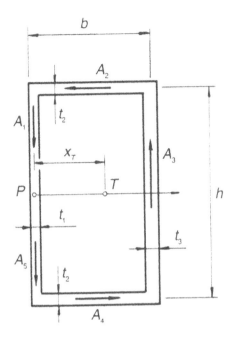

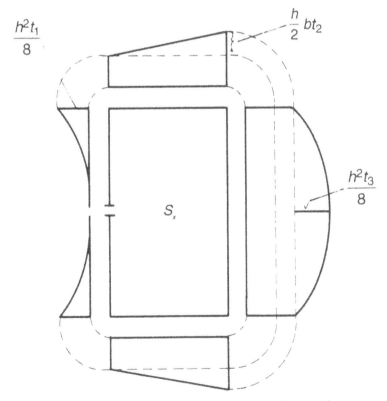

Figure 2.8. Calculation of the shear center location of a closed section.

$$\sum r_{P_i} A_i \left(S_x \right) = h A_2 + b A_3 \tag{2.26}$$

$$\sum b_i / t_i = h / t_1 + 2b / t_2 + h / t_3 \tag{2.27}$$

$$\sum A_i \left(S_x \right) / t_i = 2 A_1 / t_1 + 2 A_2 / t_2 + A_3 / t_3 \tag{2.28}$$

2.3 TORSION

As illustrated in Figure 2.9, large deformations can occur in thin-walled rods due to torsion.

The cross-sections do not remain planar since nonuniform longitudinal strains occur. The form of cross-section deforms as well, but this deformations are neglected in the following. Two cases of torsional behaviour are treated as follows: *Saint Venant torsion* without restraints (called also uniform or free torsion), and *warping torsion* when the longitudinal distortions of cross-section are restrained (called also nonuniform torsion).

The torsional behaviour of rods is important not only in the case of pure torsion but also for the analysis of several instability phenomena such as flexural torsional buckling of compressed struts or lateral torsional buckling of beams loaded in bending (Chapter 3).

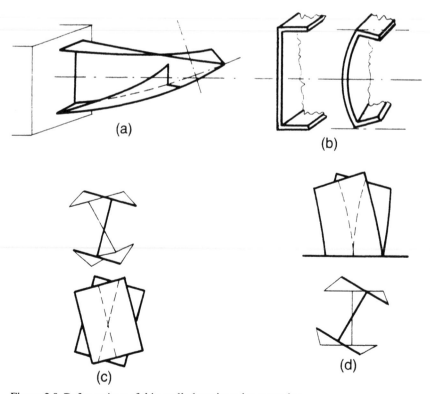

Figure 2.9. Deformations of thin-walled sections due to torsion.

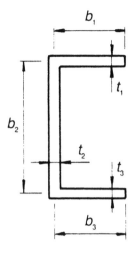

Figure 2.10. Example of an open section.

2.3.1 *Saint Venant torsion*

Open sections
The shear stress in the *i*th plate element of a profile shown in Figure 2.10 is given by

$$\tau_i = M_t\, t_i/I_t \tag{2.29}$$

where M_t = the twisting moment and

$$I_t = \alpha_0 \sum b_i t_i^3 / 3 \tag{2.30}$$

is the torsional constant. The specific rotation is expressed by

$$\frac{d\varphi}{dz} = \frac{M_t}{GI_t} \tag{2.31}$$

G is the shear modulus, α_0 is a correcting factor, for rolled I-section it is 1.20, for welded I-section 1.50. Note that the shear flow can be created very difficult across the small thicknesses, thus, the torsional resistance of open sections (I_t) is very small and therefore the shear stresses and deformations are large.

Closed sections
Based on the equilibrium of a shell element it can be stated that the shear flow is constant in the whole section. Thus, from the moment equality

$$M_t = \oint r_T v\, ds = v \oint r_T\, ds = 2A_p v \tag{2.32}$$

we obtain

$$v = \frac{M_t}{2A_p} \quad \text{or} \quad \tau_i = \frac{M_t}{2A_p t_i} \tag{2.33}$$

In this case, in Equation (2.8) the derivative of u is not zero, thus

$$\frac{v}{tG} = \frac{\partial w}{\partial s} + r_T \frac{d\varphi}{dz} \tag{2.34}$$

Expressing the derivative of w and integrating, the result should be zero:

$$\oint \frac{\partial w}{\partial s} ds = 0 = \frac{v}{G} \oint \frac{ds}{t} - \frac{d\varphi}{dz} \oint r_T ds \tag{2.35}$$

Equation (2.35) yields

$$\frac{d\varphi}{dz} = \frac{M_t}{GI_{tc}} \tag{2.36}$$

I_{tc} = the torsional constant for closed sections

$$I_{tc} = \frac{4A_p^2}{\sum b_i / t_i} \tag{2.37}$$

Since A_p is large, I_{tc} is also large, thus, the torsional resistance of closed sections is much more larger than that of open sections. A comparison of the torsional behaviour of open and closed sections is treated in Chapter 8.

2.3.2 Warping torsion

Open sections
Based on the Vlasov's hypothesis (Vlasov 1963) that the angular distortion in the middle surface is zero ($\gamma = 0$), from Equations (2.8) and (2.10) we get

$$\frac{\partial w}{\partial s} = -r_T \frac{d\varphi}{dz} \tag{2.38}$$

The displacement in z-direction in the section point in distance s from the initial point K can be obtained integrating Equation (2.38)

$$w - w_0 = -\frac{d\varphi}{dz} \int_0^s r_T ds = -\frac{d\varphi}{dz} \omega_T \tag{2.39}$$

where

$$\omega_T = \int_0^s r_T ds \tag{2.40}$$

is the sectorial coordinate, which can be calculated as twice of the sectorial area corresponding to the arc distance s (Fig. 2.11).

When the warping is restrained, warping normal stresses occur as follows

$$\sigma_\omega = E\varepsilon_\omega = E \frac{\partial w}{\partial z} = -E\omega_T \frac{d^2\varphi}{dz^2} \tag{2.41}$$

If there is not an external force, then

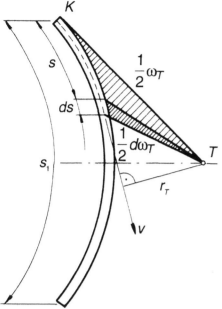

Figure 2.11. Sectorial coordinate for the calculation of warping.

$$\int_A \sigma_\omega dA = -E\frac{d^2\varphi}{dz^2}\int_A \omega_T dA = 0 \qquad (2.42)$$

Based on the equilibrium of a shell element, it can be written that

$$t\frac{\partial\sigma_\omega}{\partial z} + \frac{\partial v_\omega}{\partial s} = 0 \qquad (2.43)$$

v_ω is the shear flow due to warping torsion. From Equation (2.43) one obtains

$$\frac{\partial v_\omega}{\partial s} = -t\frac{\partial\sigma_\omega}{\partial z} = E\omega_T t\frac{d^3\varphi}{dz^3} \qquad (2.44)$$

and

$$v_\omega = E\frac{d^3\varphi}{dz^3}\int_0^s \omega_T dA = ES_\omega\frac{d^3\varphi}{dz^3} \qquad (2.45)$$

S_ω is the warping statical moment. The moment of v_ω is expressed by

$$M_\omega = \int_A r_T v_\omega ds = E\frac{d^3\varphi}{dz^3}\int r_T\left(\omega_T dA\right)ds \qquad (2.46)$$

Using the integration by parts one obtains

$$\int_A r_T\left(\int \omega_T dA\right)ds = \omega_T\int \omega_T dA - \int \omega_T^2 dA \qquad (2.47)$$

The first member is zero, the second is the warping constant

$$I_\omega = \int_A \omega_T^2 \, dA \tag{2.48}$$

and

$$M_\omega = EI_\omega \frac{d^3\varphi}{dz^3} \tag{2.49}$$

The torsional moment due to Saint Venant torsion is (see Eq. 2.36)

$$M_v = GI_t \frac{d\varphi}{dz} \tag{2.50}$$

Thus, in the case of warping torsion, the external twisting moment is in equilibrium with M_ω and M_v

$$GI_t \frac{d\varphi}{dz} - EI_\omega \frac{d^3\varphi}{dz^3} = M_t \tag{2.51}$$

This is the basic differential equation of warping torsion from which the function of angular twist $\varphi(z)$ can be determined.

When the external twisting moment varies, another equation holds

$$GI_t \frac{d^2\varphi}{dz^2} - EI_\omega \frac{d^4\varphi}{dz^4} = \frac{dM_t}{dz} \tag{2.52}$$

Defining the bimoment

$$B_\omega = \int_A \sigma_\omega \omega_T \, dA = -EI_\omega \frac{d^2\varphi}{dz^2} \tag{2.53}$$

one obtains

$$\sigma_\omega = \frac{B_\omega}{I_\omega} \omega_T \tag{2.54}$$

which is analogous with the formula for bending stress

$$\sigma = \frac{M}{I_x} y \tag{2.55}$$

Equation (2.51) can be written in the following form

$$\frac{d^3\varphi}{dz^3} - \alpha^2 \frac{d\varphi}{dz} = -\frac{M_t}{EI_\omega} \qquad \alpha^2 = \frac{GI_t}{EI_\omega} \tag{2.56}$$

The solution of Equation (2.56) is

$$\varphi(z) = C_1 + C_2 z + C_3 \sinh \alpha z + C_4 \cosh \alpha z \qquad C_2 = \frac{M_t}{GI_t} \tag{2.57}$$

The constants C_1, C_3 and C_4 can be calculated using the boundary conditions.

The ω_T diagram for an I-section is shown in Figure 2.12. Thus, for an I-section

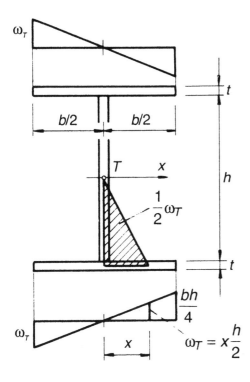

Figure 2.12. Sectorial coordinates of an I-section.

$$\omega_T = x\frac{h}{2}$$

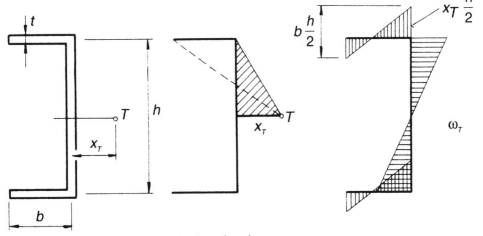

Figure 2.13. Sectorial coordinates of a channel section.

$$I_\omega = 4 \int_0^{b/2} \omega_T^2 t_f \, dx = \frac{h^2 b^3 t_f}{24} \tag{2.58}$$

The ω_T diagram for a channel section is given in Figure 2.13.

The integration results in

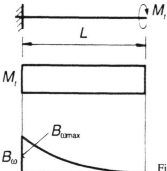

Figure 2.14. Torsion of a cantilever.

$$I_\omega = \frac{h^2 b^3 t}{12} \frac{2+3b/h}{1+6b/h} \tag{2.59}$$

Example of a cantilever (Fig. 2.14)
The boundary conditions are as follows:

$$\varphi(z=0)=0 \tag{2.60}$$

$$w = 0 \; (z=0) \quad \text{or} \quad \frac{d\varphi}{dz} = 0 \; (z=0) \tag{2.61}$$

$$z = L \quad \sigma_\omega = 0 \quad \text{or} \quad \frac{d^2\varphi}{dz^2} = 0 \tag{2.62}$$

For these boundary conditions one obtains

$$\varphi = \frac{M_t}{GI_t}\left[z - \frac{\sinh \alpha z}{\alpha} + \frac{\tanh \alpha L}{\alpha}\left(\cosh \alpha z - 1 \right) \right] \tag{2.63}$$

the maximal rotation at $z = L$ is

$$\varphi_{max} = \frac{M_t}{GI_t}\left(1 - \frac{\tanh \alpha L}{\alpha L} \right) \tag{2.64}$$

The maximal bimoment at $z = 0$ is

$$B_{\omega max} = M_t \frac{\tanh \alpha L}{\alpha} \tag{2.65}$$

When $\alpha L < 0.5$, the following approximation may be used: $\tanh \alpha L \approx \alpha L$ and

$$B_{\omega max} \approx M_t L \tag{2.66}$$

The maximal normal stress due to warping torsion in an I-section, using Equations (2.54) and (2.58) and $\omega_{T max} = hb/4$, is given by

$$\sigma_{\omega max} \approx \frac{6 M_t L}{b^2 h t_f} \tag{2.67}$$

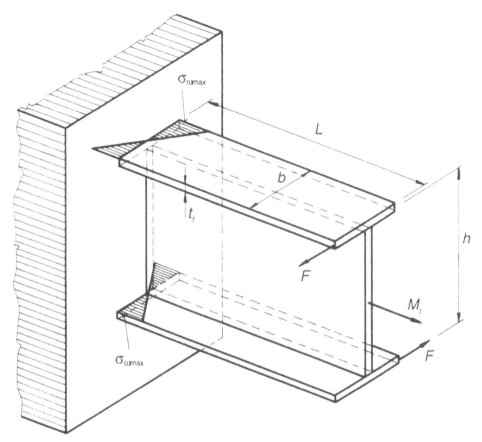

Figure 2.15. Normal stresses in an I-section cantilever due to warping torsion.

This equation can be obtained also by treating the warping torsion as a bending of flanges due to the couple of forces $F = M_t/h$ acting in the plane of flanges of section modulus $W = b^2 h/6$ (Fig. 2.15)

$$\sigma_{\omega\max} = \frac{FL}{W} = \frac{6M_t L}{b^2 ht_f} \tag{2.68}$$

This is valid only for symmetric I-sections, but this view is suitable for determining the sign of normal stresses due to warping torsion.

The magnitude of the normal stresses caused by warping torsion can be characterized using section characteristics of a welded I-beam optimized for bending (see Chapter 8).

Closed sections
The longitudinal displacement (warping) can be obtained by integration of Equation (2.34)

$$w - w_0 = \frac{v}{G} \int_0^s \frac{ds}{t} - \frac{d\varphi}{dz} \int_0^s r_T ds \tag{2.69}$$

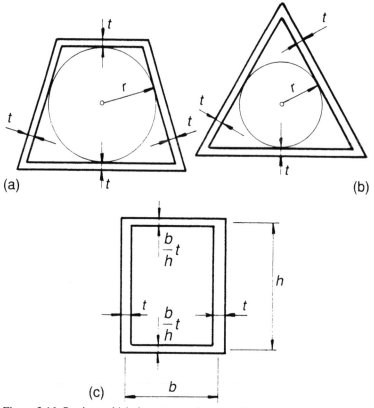

Figure 2.16. Sections which do not warp due to torsion.

From Equation (2.35) one obtains

$$\frac{v}{G} = \frac{d\varphi}{dz} \frac{2A_p}{\oint (ds/t)} \tag{2.70}$$

Thus

$$w - w_0 = -\frac{d\varphi}{dz}\left(\omega_T - \frac{2A_p}{\oint (ds/t)} \int \frac{ds}{t}\right) = -\frac{d\varphi}{dz}\varpi_T \tag{2.71}$$

ϖ_T is the sectorial coordinate for a closed section. This sectorial coordinate is for some special sections zero. Figure 2.16 shows such sections. For these sections it is

$$\omega_T = br \qquad 2A_p = r\sum b_i \qquad \int \frac{ds}{t} = \frac{b}{t} \qquad \oint \frac{ds}{t} = \frac{\sum b_i}{t}$$

therefore ϖ_T is zero.

The warping can be expressed as

$$w = \psi(z)\varpi_T(s) \tag{2.72}$$

$\psi(z)$ is the warping torsion function for closed sections. According to Equation (2.41)

$$\sigma_{\varpi} = E\frac{\partial w}{\partial z} = E\frac{d\psi}{dz}\varpi_T \tag{2.73}$$

The bimoment is defined by

$$B_{\varpi} = \oint_A \sigma_{\varpi}\varpi_T dA = EI_{\varpi}\frac{d\psi}{dz} \tag{2.74}$$

where

$$I_{\varpi} = \oint_A \varpi_T^2 dA \tag{2.75}$$

and the normal stress due to warping torsion is given by

$$\sigma_{\varpi} = \frac{B_{\varpi}}{I_{\varpi}}\varpi_T \tag{2.76}$$

The basic system of differential equations for constant twisting moment is as follows (see e.g. Murray 1984)

$$\frac{d^3\varphi}{dz^3} - \alpha_1^2\frac{d\varphi}{dz} = -\frac{\mu M_t}{2EI_{\varpi}} \qquad \alpha_1^2 = \frac{\mu GI_{tc}}{EI_{\varpi}} \qquad \psi = \frac{M_t}{2\mu GI_p} - \frac{1}{\mu}\frac{d\varphi}{dz}$$

$$\mu = 1 - \frac{I_{tc}}{I_p} \qquad I_p = \oint_A r_T^2 dA \tag{2.77}$$

The solution for a simple beam with fork supports (Fig. 2.17) is as follows:
   The boundary conditions:

$$z = 0: \qquad \varphi = \frac{d\psi}{dz} = \frac{d^2\varphi}{dz^2} = 0$$
$$z = L: \qquad \psi = 0$$

$$\varphi = \frac{M_t}{GI_{tc}}\left(z - \frac{\mu\sinh\alpha_1 z}{\alpha_1\cosh\alpha_1 L}\right) \tag{2.78}$$

$$B_{\varpi} = \frac{\mu M_t}{\alpha_1\cosh\alpha_1 L}\sinh\alpha_1 z \tag{2.79}$$

$$B_{\varpi\,max} = \frac{\mu M_t}{\alpha_1}\tanh\alpha_1 L \tag{2.80}$$

For closed sections it generally holds that $\alpha_1 L > 1$, then $\tanh\alpha_1 L \approx 1$ and

$$B_{\varpi\,max} \cong \mu M_t / \alpha_1 \tag{2.81}$$

For a doubly symmetric welded box section (Fig. 2.18) the formulae are as follows:

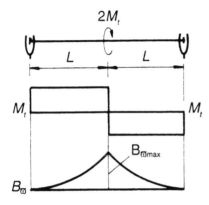

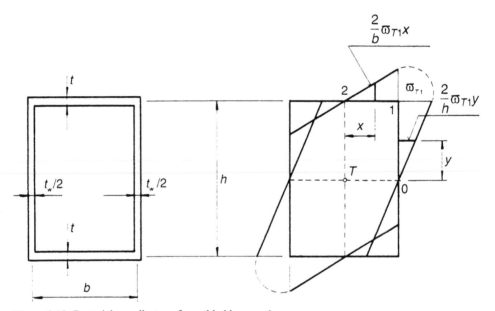

Figure 2.17. Example of a simply supported beam with fork supports.

Figure 2.18. Sectorial coordinates of a welded box section.

$$I_p = ht_w b^2 / 4 + h^2 bt / 2 \tag{2.82}$$

$$\omega_{T1} = \frac{hb}{4} - \frac{2hb}{(2b/t) + (4h/t_w)} \frac{h}{t_w} \tag{2.83}$$

$$I_\omega = 4 \int_0^{b/2} \left(\frac{2\omega_{T1}}{b} \right)^2 x^2 t dx + 4 \int_0^{h/2} \left(\frac{2\omega_{T1}}{h} \right)^2 y^2 \frac{t_w}{2} dy = \omega_{T1}^2 \left(\frac{2}{3} bt + \frac{1}{3} ht_w \right) \tag{2.84}$$

The magnitude of normal stress is investigated for optimal designed sections in Chapter 8. It is shown that normal stresses caused by warping torsion in closed sections are much smaller than those in open sections.

CHAPTER 3

Stability

3.1 INTRODUCTION

Stability is one of the most important problems in the design of metal structures, since the instability causes in many cases failure or collapse of structures. As a result of several conference series a world view has been worked out by an international group of scientists (Beedle 1991). In this book the results obtained in Australia, China, Eastern and Western Europe, Japan and North America are summarized in 12 chapters as follows: Compression members, built-up members, beams, plate- and box girders, beam-columns, frames, arches, triangulated structures, tubular structures, shells, cold-formed members and composite (steel + concrete) members.

The world view shows the complexity of problems and the differences among the solutions and design rules. From many books on structural stability it is worth mentioning the following: Kollár & Dulácska (1984) about the stability of shells, Petersen (1980) with many numerical examples, a detailed Japanese handbook on stability Handbook (1970), Chen & Lui (1991) about the stability of frames, Rondal et al. (1992) about stability of structures with hollow sections, Waszczyszyn et al. (1994) treat the stability problems by finite element methods.

3.2 CLASSES OF CROSS-SECTIONS

Consider a simply supported beam loaded in bending and shear (Fig. 3.1a). In the plastic design it is supposed that a plastic hinge forms if the rotation of the cross-section at the maximum bending moment is sufficient. Figure 3.1b shows the relationship $M - \theta$ and the stress distributions in a welded I-beam corresponding to the limit states defined by EC3 for four classes of cross-sections as follows:

– Class 1: A plastic hinge can be formed with sufficient rotational capacity without local buckling,

– Class 2: The plastic moment resistance M_p can be developed but the rotational capacity is limited because of local buckling,

– Class 3: The yield stress f_y can be reached in the extreme compression fibre without local buckling,

– Class 4: The yield stress cannot be reached without local buckling, so the effective width concept should be used.

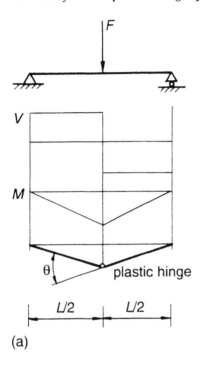

(a)

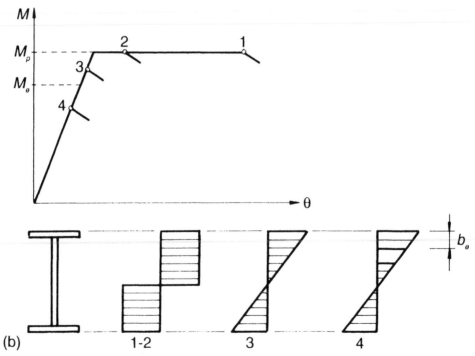

(b)

Figure 3.1. Classification of cross-sections according to EC3. a) Plastic hinge in a simply supported beam loaded in bending; b) Bending moment versus rotation angle, limit states of the cross-section at the plastic hinge affected by local buckling characterizing the classes of cross-sections.

3.3 COMPRESSION MEMBERS

3.3.1 *Flexural buckling*

The development in the calculation of overall buckling of compressed struts shows clearly how the model was refined considering the fabrication aspects.

In the first phase Euler (1778) has solved the differential equation for a straight strut to obtain the critical force

$$F_E = \pi^2 EI_x / (KL)^2$$

or stress

$$\sigma_E = \pi^2 E / \lambda^2, \quad \lambda = KL / r \tag{3.1}$$

where $r = \sqrt{I_x / A}$ is the radius of gyration, K is the effective length factor, A is the cross-sectional area, E is the modulus of elasticity, L is the strut length, I_x is the moment of inertia. Figure 3.2 shows that the Euler hyperbola is valid only in the elastic range, when $\sigma \leq \sigma_0$ where σ_0 is the elastic limit stress. Later the plastic buckling has been described by several authors.

In the second phase Ayrton & Perry (1886) have taken into account the initial crookedness as regards the initial imperfections. It is worth describing their model since this model is the basis of the EC3 overall buckling formula.

The differential equation for a centrally compressed strut with pinned ends and a sinusoidal initial crookedness (Fig. 3.3)

$$a = a_0 \sin (\pi z / L) \tag{3.2}$$

is

$$\frac{d^2 y}{dz^2} = -\frac{M}{EI_x} = -\frac{N(a + y)}{EI_x} \tag{3.3}$$

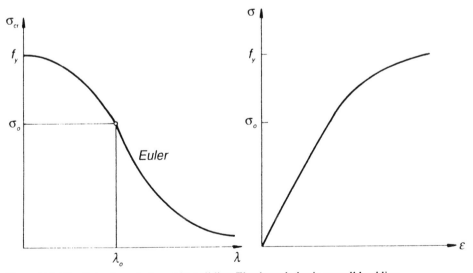

Figure 3.2. The Euler-hyperbola and its validity: Elastic and plastic overall buckling

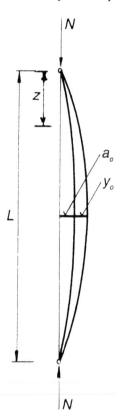

Figure 3.3. The Ayrton-Perry model of an initially crooked compression strut and its elastic second-order deformation.

N is the compressive force. Searching the solution in the form of

$$y = y_0 \sin (\pi z / L) \tag{3.4}$$

we get

$$y_0 = \frac{a_0}{F_E / N - 1} \tag{3.5}$$

The overall buckling formula can be derived on the basis of the check for eccentric compression

$$\frac{N}{A} + \frac{N(a_0 + y_0)}{W_x} \le f_y \tag{3.6}$$

In the third phase the effect of residual welding stresses has been considered. The European overall buckling curves have been determined for various welded sections based on statistical evaluation of many test results (Beer & Schulz 1970). The tests have shown that the residual welding stresses significantly influence the buckling strength mainly in the case of welded I-sections since compressive residual stresses occur at the edges of flanges (see Chapter 1, Fig. 1.5) which decrease the buckling strength.

EC3 applies the formula proposed by Maquoi & Rondal (1978). This formula can be derived from Equation (3.6) considering one parameter which expresses the effect of initial imperfection and residual welding stresses as follows.

With notations of

$$\sigma = N / A, \quad \sigma_E = F_E / A, \quad \eta_b = a_0 A / W_x$$

Equation (3.6) can be written in the form

$$\left(f_y - \sigma\right)\left(\sigma_E - \sigma\right) = \eta_b \sigma \sigma_E \tag{3.7}$$

This equation can be transformed using the following relationships

$$\sigma / f_y = \chi, \quad \sigma_E / f_y = \pi^2 E / \left(f_y \lambda^2\right) = 1 / \overline{\lambda}^2 \tag{3.8}$$

$$\overline{\lambda} = \lambda / \lambda_E, \quad \lambda_E = \pi \sqrt{E / f_y}$$

to obtain

$$(1 - \chi)\left(\frac{1}{\overline{\lambda}^2} - \chi\right) = \frac{\chi \eta_b}{\overline{\lambda}^2} \tag{3.9}$$

This leads to the following quadratic equation

$$\chi^2 - \left(1 + \frac{\eta_b}{\overline{\lambda}^2} + \frac{1}{\overline{\lambda}^2}\right)\chi + \frac{1}{\overline{\lambda}^2} = 0 \tag{3.10}$$

The solution of Equation (3.10) is

$$\chi = \frac{\phi - \sqrt{\phi^2 - \overline{\lambda}^2}}{\overline{\lambda}^2} = \frac{1}{\phi + \sqrt{\phi^2 - \overline{\lambda}^2}} \tag{3.11}$$

where

$$\phi = 0.5\left(1 + \eta_b + \overline{\lambda}^2\right) \quad \text{and} \quad \eta_b = \alpha(\overline{\lambda} - 0.2)$$

For $\overline{\lambda} \le 0.2$ it is $\chi = 1$, α is the imperfection factor given for various buckling curves in Table 3.1.

The strut should be checked for

$$N \le \chi A f_y / \gamma_{M1} \tag{3.12}$$

where $\gamma_{M1} = 1.1$ is the safety factor for buckling.

Table 3.1. Imperfection factors.

Buckling curve	a	b	c	d
Imperfection factor	0.21	0.34	0.49	0.76

Note: selection of buckling curve for cross-sections according to EC3: a = for hot rolled hollow sections; b = for cold formed hollow sections, welded box sections, welded I-sections buckling about the strong axis with flange thickness smaller than 40 mm; c = for welded I-sections buckling about the weak axis with flange thickness smaller than 40 mm, for welded I-sections buckling about the strong axis with flange thickness larger than 40 mm, for U-, L-, T- and solid sections; d = for welded I-sections buckling about the weak axis flange thickness larger than 40 mm.

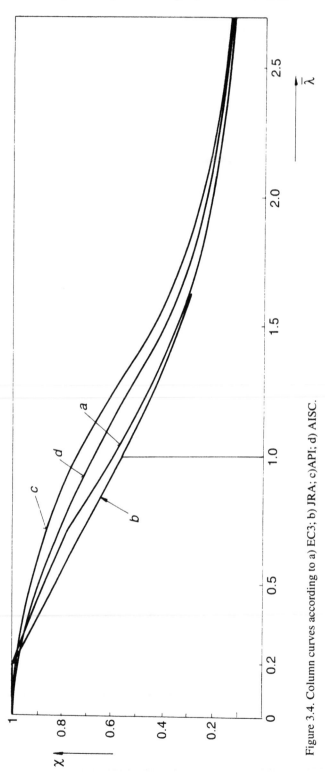

Figure 3.4. Column curves according to a) EC3; b) JRA; c)API; d) AISC.

It should be mentioned that the EC3 formula is too complicate for design (non-computerized optimization) purposes. There exist another column curves used in other countries which can be applied instead of EC3 curves. Figure 3.4 shows some other buckling curves in addition to the EC3 curve 'b'. It can be seen that the curve of the Japan Railroad Association (JRA) gives values near the EC3 curve 'b'. The JRA curve is described by the following formulae

$$\chi = 1 \qquad \text{for} \quad \overline{\lambda} \le 0.2$$
$$\chi = 1.109 - 0.545\overline{\lambda} \quad \text{for} \quad 0.2 \le \overline{\lambda} \le 1 \qquad\qquad (3.13)$$
$$\chi = 1/\left(0.773 + \overline{\lambda}^2\right) \quad \text{for} \quad \overline{\lambda} \ge 1$$

The curve of American Petroleum Institute (API) is defined by

$$\chi = 1 - 0.25\overline{\lambda}^2 \quad \text{for} \quad 0 \le \overline{\lambda} \le 1.41$$
$$\chi = 1/\overline{\lambda}^2 \qquad \text{for} \quad \overline{\lambda} \ge 1.41 \qquad\qquad (3.14)$$

The curve of American Institute of Steel Construction (AISC) mainly for round tubes is given by

$$\chi = 1 - 0.091\overline{\lambda} - 0.22\overline{\lambda}^2 \quad \text{for} \quad \overline{\lambda} \le 1.41 \qquad\qquad (3.15a)$$

$$\chi = 0.015 + 0.834/\overline{\lambda}^2 \quad \text{for} \quad \overline{\lambda} \ge 1.41 \qquad\qquad (3.15b)$$

In the fourth phase the experiments carried out on thin-walled rectangular hollow section compression struts in the University of Liège have shown the interaction of local buckling of plate elements and the overall buckling. When the plate element loaded in maximum compressive stress buckles, the overall buckling strength decreases. Braham et al. (1980) have proposed a reduction factor for this case which is included in EC3 as follows: Equation (3.12) is modified to

$$N \le \beta_A \chi A f_y / \gamma_{M1} \quad \text{and} \quad \overline{\lambda} = \lambda \sqrt{\beta_A} / \lambda_E \qquad\qquad (3.16)$$

where $\beta_A = 1$ for cross-sections of class 1, 2 and 3, $\beta_A = A_{eff}/A$ for class 4 cross-sections. The area of the effective cross-section A_{eff} can be calculated using the effective width formulae for compression elements (see Section 3.6).

For aluminium alloy compression members the rules of BS 8118 (1991) can be used. This standard contains overall buckling formulae which are the same as the EC3 formulae. The initial imperfection factor should be taken as follows: For symmetric sections unwelded $\alpha = 0.2$, welded 0.45.

Summarizing it can be concluded that the development of the strength calculation leads from the Euler's differential equation to the EC3 formulae which take into account the initial imperfections, residual welding stresses and the interaction of local and overall buckling.

Note that the interaction of two instability phenomena plays an important role in the optimum design (see Chapter 9).

The effective length factor K expresses the effect of end restraints. The classical values are given in Figure 3.5. Different values are used in frame and truss analysis.

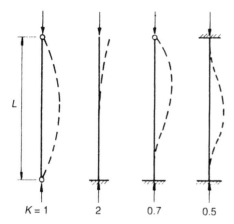

$K = 1$ 2   0.5   0.5   Figure 3.5. Values of the effective length factor K.

Note that in the case of loads fluctuating in tension-compression the strut should be designed also against fatigue failure of its connections (e.g. welded connections to gusset plates or other structural members) (see Chapter 10).

3.3.2 *Flexural-torsional buckling*

The system of differential equations for a rod loaded in bending, compression and torsion (Fig. 3.6) is derived by Vol'mir (1967). For a doubly symmetric section (for notations see Chapter 2)

$$EI_x v'''' + Nv'' - M_y \varphi'' = p_y$$

$$EI_y u'''' + Nu'' - M_x \varphi'' = p_x \tag{3.17}$$

$$EI_\omega \varphi'''' + \left(\frac{NI_p}{A} - GI_t \right) \varphi'' - M_y v'' - M_x u'' = M_t'$$

The prime ($'$) superscripts denote derivatives with respect to z.
    For the twist of a centrally compressed rod one obtains

$$\varphi'''' + \frac{1}{EI_\omega} \left(\frac{NI_p}{A} - GI_t \right) \varphi'' = 0 \tag{3.18}$$

Since $B_\omega = EI_\omega \varphi''$, Equation (3.18) can be rewritten as

$$B_\omega'' + \alpha^2 B_\omega = 0 \qquad \alpha^2 = \frac{1}{EI_\omega} \left(\frac{NI_p}{A} - GI_t \right) \tag{3.19}$$

For boundary conditions assuming fork supports at the ends of the rod $z = 0$, $z = L$. $B_\omega = 0$ the solution is $B_\omega = C \sin \alpha z$. Since $C \neq 0$, from $\sin \alpha z = 0$ we get $\alpha L = m\pi$. The minimum value corresponds to $m = 1$, thus

$$\sigma_{\omega cr} = \frac{\pi^2 EI_\omega}{L^2 I_p} + \frac{GI_t}{I_p} \tag{3.20}$$

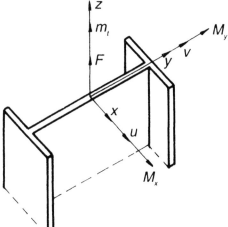

Figure 3.6. To the calculation of flexural-torsional buckling.

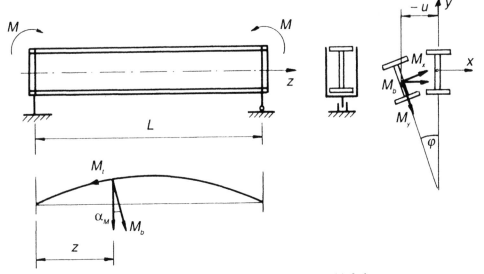

Figure 3.7. Lateral-torsional buckling of a simply supported I-beam with fork supports.

3.4 LATERAL-TORSIONAL BUCKLING OF BEAMS LOADED IN BENDING

Lateral-torsional buckling occurs when open-section beams of small torsional rigidity are loaded in bending without lateral restraint against twist. The compressed top flange may buckle in lateral direction due to additional twist from the second-order twisting moment components as shown in Figure 3.7. in the case of an I-beam with fork support loaded by bending moments at the ends.

The cross-section in distance z is loaded by a bending moment $M_b = M \cos \alpha_M$ and by a twisting moment $M_t = M \sin \alpha_M$. Since the deformations are small, the moment components may be calculated as

$$M_x = M_b \cos \varphi \approx M \tag{3.21a}$$

$$M_y = M_b \sin \varphi \approx M\varphi \tag{3.21b}$$

The differential equation of the bending about the y axis is

$$u'' = \frac{M}{EI_y} \varphi \tag{3.22}$$

The differential equation of twist is (see Chapter 2)

$$GI_t \varphi'' - EI_x \varphi'''' = M_t' = -Mu'' \tag{3.23}$$

Substituting Equation (3.22) into Equation (3.23) we get

$$\varphi'''' - 2\alpha\varphi'' - \beta\varphi = 0 \qquad \alpha = \frac{GI_t}{2EI_\omega}, \quad \beta = \frac{M^2}{E^2 I_y I_\omega} \tag{3.24}$$

The boundary conditions are as follows: At $z = 0$ and $z = L$, $\varphi = \varphi'' = 0$. The solution satisfying the boundary conditions is

$$\varphi = C \sin mz \tag{3.25}$$

Substituting Equation (3.25) into Equation (3.24) one obtains

$$m = \sqrt{-\alpha + \sqrt{\alpha^2 + \beta}} \tag{3.26}$$

Since $C \neq 0$ from the condition $z = L$, $\varphi = 0$ we get $\sin mL = 0$. The minimum value of m is π, thus $m = \pi / L$. Substituting here the expression (Eq. 3.26) we obtain the critical lateral-torsional bending moment

$$M_{cr} = \frac{\pi}{L} \sqrt{EI_y GI_t \left(1 + \frac{\pi^2 EI_\omega}{L^2 GI_t}\right)} \tag{3.27}$$

or in other form given by EC3 for doubly symmetric cross sections with fork support, zero end moment loading and for transverse loads applied at the shear center

$$M_{cr} = C_1 \frac{\pi^2 EI_y}{L^2} \sqrt{\frac{I_\omega}{I_y} + \frac{L^2 GI_t}{\pi^2 EI_y}} \tag{3.28}$$

where the values of C_1 are as follows:
    – For a concentrated force applied at midspan $C_1 = 1.365$,
    – For a uniformly distributed load $C_1 = 1.132$.
    The check of the lateral-torsional buckling according to EC3 should be carried out using the formulae analogous with those for the flexural buckling:

$$M \leq M_b = \chi_{LT} \beta_w W_{pl.y} f_y / \gamma_{M1} \tag{3.29}$$

where $\beta_w = 1$ for class 1 or class 2 cross-sections, $\beta_w = W_{eff.y}/W_{pl.y}$ for class 3 cross-sections, $\beta_w = W_{eff.y}/W_{pl.y}$ for class 4 cross-sections,

$$\chi_{LT} = \frac{1}{\phi_{LT} + \sqrt{\phi^2 - \overline{\lambda}_{LT}^2}} \quad \text{but} \quad \chi_{LT} \leq 1$$

$$\phi_{LT} = 0.5\left[1 + \alpha_{LT}\left(\overline{\lambda}_{LT} - 0.2\right) + \overline{\lambda}_{LT}^2\right] \quad \text{for} \quad \overline{\lambda}_{LT} \leq 0.2, \quad \chi_{LT} = 1$$

$\alpha_{LT} = 0.21$ for rolled sections, $\alpha_{LT} = 0.49$ for welded sections:

$$\overline{\lambda}_{LT} = \sqrt{\beta_w W_{pl.y} f_y / M_{cr}} = \lambda_{LT}\sqrt{\beta_w} / \lambda_E, \quad \lambda_{LT} = \sqrt{\pi^2 E W_{pl.y} / M_{cr}}$$

$W_{el.y}$ and $W_{pl.y}$ are the elastic and plastic section moduli for bending about y axis, respectively, $W_{eff.y}$ is the effective section modulus considering only the effective plate widths.

3.5 BEAM-COLUMNS

In the design of rods subject to bending and axial compression the second-order elastic deformation should be considered. This depends on the bending moment distribution along the rod and on the class of cross-section. Instead of complicate exact solution (see e.g. Chen & Atsuta 1977, Trahair 1993) several approximate design formulae are given in design standards. We give here the EC3 formulae only for class 3 cross-sections and the Duan-Chen formulae (Duan & Chen 1989, Duan 1990) proposed for CHS beam-columns. These formulae consider the second-order effect by means of amplification factors for bending moments.

EC3 formulae for members which do not fail due to lateral-torsional buckling (hollow sections) are as follows:

$$\frac{N}{\chi_{\min} A f_{y1}} + \frac{k_x M_x}{W_{el.x} f_{y1}} + \frac{k_y M_y}{W_{el.y} f_{y1}} \leq 1 \tag{3.30}$$

where

$$f_{y1} = f_y / \gamma_{M1}, \quad \gamma_{M1} = 1.1$$

the amplification factors for bending moments about x and y axis, respectively are expressed by

$$k_x = 1 - \frac{\mu_x N}{\chi_x A f_y} \quad \text{but} \quad k_x \leq 1.5$$

$$k_y = 1 - \frac{\mu_y N}{\chi_y A f_y} \quad \text{but} \quad k_y \leq 1.5$$

$$\mu_x = \overline{\lambda}_x (2\beta_{Mx} - 4) \quad \text{but} \quad \mu_x \leq 0.90$$

$$\mu_y = \overline{\lambda}_y (2\beta_{My} - 4) \quad \text{but} \quad \mu_y \leq 0.90$$

$\beta_{Mx,y}$ are the equivalent uniform moment factors expressing the effect of moment distribution along the rod. For end moments only (linear distribution) considering the

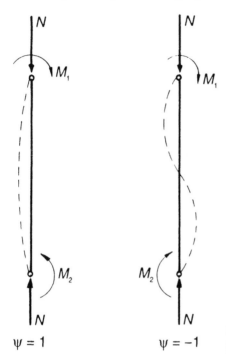

$\psi = 1$ $\psi = -1$

Figure 3.8. Limit cases of end moments for a beam-column.

maximum end moment M_1 and the other end moment ψM_1, where $-1 \leq \psi \leq 1$

$$\beta_M = 1.8 + 0.7\psi \tag{3.31}$$

Figure 3.8 shows two limit cases of end moments.

For moments due to a lateral concentrated load acting at midspan of a simply supported beam $\beta_M = 1.4$, in the case of a uniformly distributed load acting on a simply supported beam $\beta_M = 1.3$. EC3 gives these factors also for another moment distributions.

For open section rods which can fail due to lateral-torsional buckling, the EC3 formula is as follows

$$\frac{N}{\chi_y A f_{y1}} + \frac{k_{LT} M_x}{\chi_{LT} W_{el.x} f_{y1}} + \frac{k_y M_y}{W_{el.y} f_{y1}} \leq 1 \tag{3.32}$$

where

$$k_{LT} = 1 - \frac{\mu_{LT} N}{\chi_y A f_y} \qquad \text{but} \quad k_{LT} \leq 1$$

$$\mu_{LT} = 0.15 \overline{\lambda}_y \beta_{My} - 0.15 \quad \text{but} \quad \mu_{LT} \leq 0.90$$

For CHS beam-columns Sohal et al. (1989) have proposed the following interaction equation

$$\left(\frac{N}{\chi A f_y}\right)^{\alpha} + B_1 \frac{M_{max}}{M_p} \leq 1 \tag{3.33}$$

where, using notation $\delta_C = D/t, \vartheta = 100\ D/L$ (D is the mean diameter, t is the thickness)

$$\alpha = 1.75 + 0.01\lambda_m \geq 1.3, \quad \lambda_m = -KL\psi/r = -100K\sqrt{8}\psi/\vartheta$$

and the plastic bending moment is

$$M_p = f_y D^2 t = f_y D^3 / \delta_C \tag{3.34}$$

The amplification factor is given by

$$B_1 = \frac{1 + 0.25(N/F_E) - 0.6(N/F_E)^{1/3}(1-\psi)}{1 - N/F_E} \geq 1 \tag{3.35}$$

where F_E is given by Equation (3.1). Note that the amplification factor should not be smaller than 1. Equation (3.33) is extended also for members subject to external hydrostatic pressure for offshore applications.

3.6 PLATE BUCKLING

3.6.1 *Classic results for plate buckling*

As shown in Section 3.3, the effect of initial imperfections and residual welding stresses should be considered in all stability problems. Thus, the classic results for plate buckling (Timoshenko & Gere 1961) should also be modified.

Consider an isotropic elastic rectangular plate without initial imperfections and residual stresses loaded in its plane by specific forces N_x, N_y and N_{xy} (Fig. 3.9). The partial differential equation for the displacements w in direction z is expressed as

$$\frac{\partial^4 w}{\partial x^4} + 2\frac{\partial^4 w}{\partial x^2 \partial y^2} + \frac{\partial^4 w}{\partial y^4} + \frac{1}{B}\left(N_x \frac{\partial^2 w}{\partial x^2} + 2N_{xy}\frac{\partial^2 w}{\partial x \partial y} + N_y \frac{\partial^2 w}{\partial y^2}\right) = 0 \tag{3.36}$$

where the bending stiffness of the plate is defined by

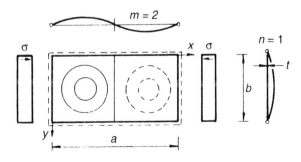

Figure 3.9. Simply supported plate loaded in unidirectional uniform compression.

$$B = \frac{Et^3}{12(1-\upsilon^2)} \qquad (3.37)$$

t is the plate thickness.

In the special case when $N_{xy} = N_y = 0$, $N_x = -\sigma t$ and the edges are simply supported (Fig. 3.9), the solution of Equation (3.36) is sought in the form of

$$w = \sum_m \sum_n w_{mn} \sin\frac{m\pi x}{a} \sin\frac{n\pi y}{b}, \quad m = 1, 2, 3..., \quad n = 1, 2, 3... \qquad (3.38)$$

Substituting Equation (3.38) into Equation (3.36) we get the basic plate buckling formula

$$\sigma_{cr} = k_\sigma \frac{\pi^2 E}{12(1-\upsilon^2)}\left(\frac{t}{b}\right)^2 \qquad k_\sigma = \left(\frac{m}{\alpha} + n^2\frac{\alpha}{m}\right)^2 \qquad (3.39)$$

k_σ is the plate buckling coefficient, which, in general, depends on the following parameters:

 – m and n numbers of half waves of buckling pattern,
 – $\alpha = a/b$ ratio of plate dimensions,
 – The in-plane loading: Compression, bending, shear,
 – Boundary conditions: Simply supported, fixed, free or elastically supported,
 – The form of plate: Rectangular, circular, triangular, etc.

Figure 3.10 gives the values of the plate buckling coefficient in the case of a plate shown in Figure 3.9 and for $n = 1$. Simplifying this diagram for design purposes it can be stated that

$$k_\sigma = 4 \qquad \text{for} \quad \alpha \geq 1 \qquad (3.40a)$$

$$k_\sigma = \left(\frac{1}{\alpha} + \alpha\right)^2 \text{ for } \quad \alpha \leq 1 \qquad (3.40b)$$

For bending $k_\sigma = 23.9$.

For shear of a plate with simply supported edges

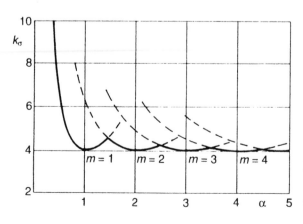

Figure 3.10. Values of plate buckling coefficient in function of $\alpha = a/b$ for the case shown in Figure 3.9.

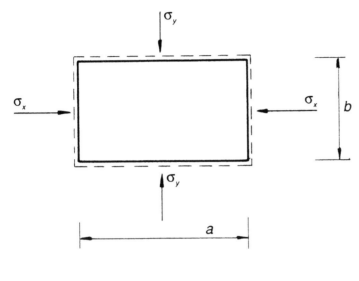

Figure 3.11. Simply supported plate compressed in two directions.

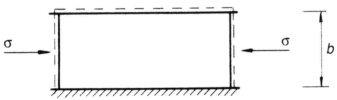

Figure 3.12. A longitudinally compressed plate strip with three edges simply supported and one edge free.

$$\tau_{cr} = k_\tau \frac{\pi^2 E}{12(1-\upsilon^2)}\left(\frac{t}{b}\right)^2$$

where

$$k_\tau = 5.34 + 4/\alpha^2 \quad \text{for} \quad \alpha \geq 1 \tag{3.41a}$$

$$k_\tau = 4 + 5.34/\alpha^2 \quad \text{for} \quad \alpha \leq 1 \tag{3.41b}$$

For a rectangular plate with simply supported edges loaded in compression in two directions (Fig. 3.11) (Vol'mir 1967)

$$k_\sigma = \frac{\left[(m/\alpha)^2 + n^2\right]^2}{(m/\alpha)^2 + \varphi n^2} \qquad \varphi = \sigma_y / \sigma_x \tag{3.42}$$

For outstands (three edges simply supported, fourth free Fig. 3.12) (Vol'mir 1967) $k_\sigma = 0.43$.

3.6.2 *Post-buckling behaviour of compressed plates*

Consider the simply supported plate in Figure 3.12. When $\sigma_{max} \geq \sigma_{cr}$ (Eq. 3.39), a part of the plate buckles but the other parts show a further load bearing capacity, the stress distribution will be non-uniform. The post-critical behaviour of the plate may be described by means of the effective width b_e:

$$b_e \sigma_{max} = b \sigma_{av} \tag{3.43}$$

Taking $k_\sigma = 4, \upsilon = 0.3$ in Equation (3.39), assuming that Equation (3.39) is also valid for $\sigma_{max} - b_e$ and introducing the notation $\vartheta_S = b / t, \psi_e = b_e / b$ we obtain

$$\sigma_{max} = 3.6152 E (t / b_e)^2 = 3.6152 E / (\vartheta_S \psi_e)^2 \tag{3.44}$$

from which

$$\psi_e = 1.9014 / \lambda_p, \quad \lambda_p = \vartheta_S \sqrt{\sigma_{max} / E} \tag{3.45}$$

Equation (3.45) is called as Karman's formula and is valid for elastic behaviour, i.e. if

$$3.6152 E / \vartheta_S^2 \le \sigma_0 \tag{3.46}$$

$\sigma_0 = r_0 f_y$ is the structural proportional limit in compression, for the base material $r_0 = 0.75$-0.80, for welded structural parts $r_0 = 0.5$-0.6. Rearranging Equation (3.46) we get

$$\lambda_p \ge \lambda_{p0} = 1.9014 \sqrt{\sigma_{max} / \sigma_0} \tag{3.47}$$

For plates with initial imperfections and residual welding stresses an empirical formula proposed by Faulkner et al. (1973) can be used instead of Equation (3.45):

$$\psi_e = \frac{2}{\lambda_p} - \frac{1}{\lambda_p^2} - \frac{\sigma_C(\vartheta_S)}{f_y} \tag{3.48}$$

where σ_C is the residual compressive stress. According to Chapter 1 (Fig. 1.5), in the case of the residual stress distribution shown in Figure 3.13, we have

$$\frac{\sigma_C}{f_y} = \frac{2\eta}{\vartheta_S - 2\eta} \tag{3.49}$$

η can be calculated according to Chapter 1 or can simply be taken as 3 and 4.5 for lightly or heavily welded parts, respectively. With $\eta = 3, \sigma_{max} = f_y = 235, E = 2.1 * 10^5$ MPa Equation (3.48) can be written in the form

$$\psi_e = \frac{2}{\lambda_p} - \frac{1}{\lambda_p^2} - \frac{6}{30\lambda_p - 6} \tag{3.50}$$

Faulkner's formulae for plastic zone, based on tangent modulus, result in intricate expressions. Farkas (1977) has proposed a second-degree parabola

$$\psi_e = 1 - (1 - \psi_{e0})(\lambda_p / \lambda_{p0})^2 \quad \psi_{e0} = \frac{2}{\lambda_{p0}} - \frac{1}{\lambda_{p0}^2} - \frac{\sigma_{C0}(\vartheta_{S0})}{f_y} \tag{3.51}$$

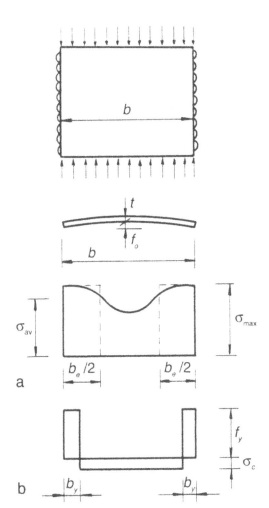

Figure 3.13. A simply supported plate. a) Stress distribution in the post-buckling region; b) Simplified residual welding stress distribution due to longitudinal edge welds.

where

$$\vartheta_{S0} = 1.9041\sqrt{E/\sigma_0}$$

Usami & Fukumoto (1982) have proposed a very simple formula of

$$\psi_e = 1.426/\lambda_p \qquad \psi_e \leq 1 \tag{3.52}$$

The EC3 formula is as follows

$$\psi_e = \frac{1.9}{\lambda_p} - \frac{0.7955}{\lambda_p^2} \tag{3.53}$$

Note that EC3 defines another plate slenderness $\overline{\lambda}_p = \lambda_p/1.9$.

The above mentioned curves are shown in Figure 3.14.

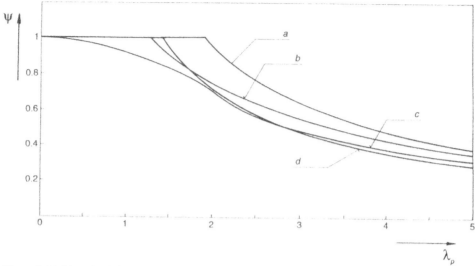

Figure 3.14. Plate buckling curves according to a) Karman; b) EC3; c) Faulkner; d) Usami-Fukumoto.

3.6.3 *Limiting plate slendernesses*

It is useful for design to define limiting plate slendernesses, since in this case it is not necessary to calculate with effective widths and the cross-section can be categorized in class 3. In the optimum design the local buckling constraints can be expressed by means of the limiting plate slendernesses. For the definition the basic formula of Equation (3.39) can be used:

$$\sigma_{cr} = k_\sigma \frac{\pi^2 E}{12(1-\upsilon^2)} \left(\frac{t}{b}\right)^2 \geq \sigma_{max} \tag{3.54}$$

where σ_{max} is the design stress, usually the yield stress, but in the case when the deflection or the fatigue constraint is active, the actual maximum static stress can be used. From Equation (3.54) one obtains for limiting slenderness

$$\left(\frac{b}{t}\right)_L = \sqrt{\frac{k_\sigma \pi^2 E}{12(1-\upsilon^2)\sigma_{max}}} \tag{3.55}$$

In EC3 the stress of 235 MPa is selected for basis and the ratio of $\varepsilon = \sqrt{235/f_y}$ is introduced. With values $E = 2.1*10^5$ MPa and $\upsilon = 0.3$ Equation (3.55) takes the form of

$$\left(\frac{b}{t}\right)_L = 28.42\varepsilon\sqrt{k_\sigma}\sqrt{\frac{f_y}{\sigma_{max}}} \tag{3.56}$$

For a simply supported uniformly compressed plate (e.g. a flange of a box girder) $k_\sigma = 4.0$ and

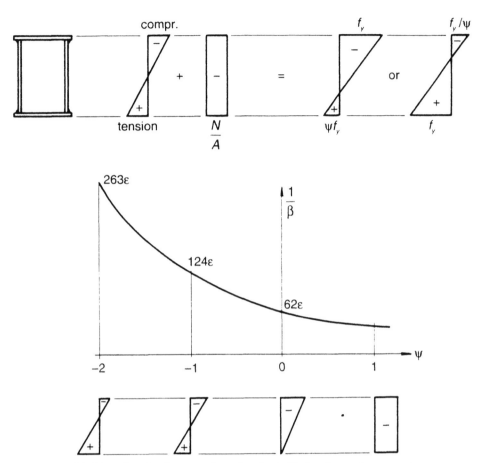

Figure 3.15. Limiting plate slendernesses for a plate loaded in bending and compression.

$$(b/t)_L = 56.84\varepsilon\sqrt{f_y/\sigma_{max}} \tag{3.57}$$

Since this value does not contain the effect of initial imperfections and residual welding stresses, EC3 gives instead of 56.84 the value of 42.

For a uniformly compressed plate with three edges simply supported and the fourth free (e.g. half width of flanges of a welded I-section) with $k_\sigma = 0.456$

$$(b/2t)_L = 19.19\varepsilon \quad \text{in} \quad \text{EC3 } 14\varepsilon \tag{3.58}$$

For a simply supported plate loaded in bending (web of a doubly symmetric welded I-beam) $k_\sigma = 23.9$

$$\left(h/t_w\right)_L = 138.94\varepsilon \quad \text{in} \quad \text{EC3 } 124\varepsilon \tag{3.59}$$

Note that in EC3 values are given for class 1 and 2 cross-sections and for other stress distributions.

For a plate element of a beam loaded in bending and compression (Fig. 3.15) for $-1 \le \psi_b \le 1$ $\quad \left(\dfrac{b}{t}\right)_L = \dfrac{42\varepsilon}{0.67+0.33\psi_b}$ $\hspace{2cm}$ (3.60a)

for $\psi_b \le -1$ $\quad (b/t)_L = 62\varepsilon\left(1-\psi_b\right)\sqrt{-\psi_b}$ $\hspace{2cm}$ (3.60b)

For CHS: Class 1 cross-section $D/t \le 50\varepsilon^2$

$\hspace{3.4cm}$ Class 2 $\hspace{2.3cm}$ $D/t \le 70\varepsilon^2$ $\hspace{1.5cm}$ (3.61)

$\hspace{3.4cm}$ Class 3 $\hspace{2.3cm}$ $D/t \le 90\varepsilon^2$

CHAPTER 4

Vibration and damping, sandwich structures

The aim of optimum design is to reduce the mass of structures. Using thin plates leads to vibration problems which need damping. In structural design one should take into account such kind of design objectives or constraints as low noise, light weight, long life, and increased reliability against structural resonance. One method to reduce high vibration and noise levels due to structural resonance is to build high-energy-dissipating elements into the structure during fabrication.

Structural damping means the capacity of a structure or structural component to dissipate a part of vibration energy. This removed energy may be converted directly into heat and transferred to connected structures and to the air. Damping has effect in controlling the amplitude of resonant vibrations, but also has other effects, such as reducing structural fatigue, and increasing structural life.

A combination of viscoelastic damping material and metal will provide larger strength and rigidity with a lower response to vibration.

Classification of damping, characterization of viscoelastic materials, effects of environmental factors (temperature, frequency, cyclic strain amplitude, and static preload) on the viscoelastic material properties, fundamentals of damping material properties, and vibration control techniques such as damping treatments will be discussed.

Some books and many papers exist on this subject. Books of Nashif et al. (1985), Sun & Lu (1995) contain the fundamentals of vibration and damping, and many practical applications.

4.1 MEASURES OF DAMPING

There are many parameters used to measure the effectiveness of damping. These parameters are generally defined by the damping parameters of a simple spring-mass system (one degree of freedom) and certain observable properties of the motion of a system in free or forced vibrations. Common damping quantities are as follows:

– *Hysteresis loop*, a plot of the amplitude of instantaneous force versus instantaneous displacement (or stress versus strain) during a steady-state forced vibration. The area of loop is proportional to the energy dissipated in the vibratory system.

Two types of structural damping exist: Material and frictional damping. The material damping depends on the amplitude of stress as it is defined by Lazan (1968)

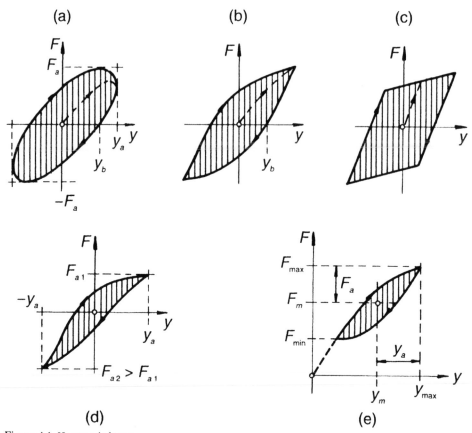

Figure 4.1. Hysteresis loops.

$$D = J\sigma_a^n \tag{4.1}$$

where J and n are material constants.

Plotting the external force of the vibrating system in the function of displacement we get nonlinear curves. For a total vibrating period it forms a hysteresis loop. Figure 4.1 shows some hysteresis loops. The hysteresis loops that may be used to quantify damping properties of materials are very useful in understanding damping. For conventional structural materials the hysteresis loop is very small, unless the material is stressed in plastic range.

If the damping is *linear*, the hysteresis loop is an ellipse and $n = 2$. This is the approximate damping of metals in the small and medium stress range, if $2\sigma_a < 0.8\Delta\sigma_N$. $\Delta\sigma_N$ is the fatigue stress range (Fig. 4.1a).

If the damping is *nonlinear*, the hysteresis loop has a sharp peak or other shape and $n > 2$. This is the case of metals stressed beyond the elastic limit (Fig. 4.1b), or frictional damping (Fig. 4.1c).

Within the elastic range the loop can be observed only in the case of high damping alloys and viscoelastic materials. For nonlinear damping, the hysteresis loop is a closed but not elliptical curve.

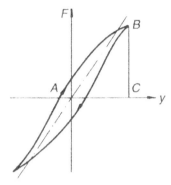

Figure 4.2. Hysteresis loop. Calculation of the specific damping capacity.

The measure of damping is the area of hysteresis loop (Fig. 4.2):

$$D = \oint F dy \tag{4.2}$$

where F is the force, y is the displacement.

– *Logarithmic decrement of free vibrations of the system*, δ, is defined as the natural logarithm of the ratio of the amplitudes of any two cycles of the decaying vibration.

For a single degree of freedom spring-mass system in the case of linear damping the equation of motion is as follows:

$$m\ddot{y} + r\dot{y} + \frac{y}{c_m} = 0 \tag{4.3}$$

where m is the mass, c_m is the spring constant, r is the viscosity of the dashpot.

Equation (4.3) can be written as

$$\ddot{y} + 2\beta r\dot{y} + \alpha^2 y = 0 \tag{4.4}$$

where $\alpha^2 = 1 / mc_m$, $\beta = r / 2m$.

For weak damping, if $\beta < \alpha$ the solution of Equation (4.4) is the following

$$y(\bar{t}) = Ae^{-\beta \bar{t}} \sin(\gamma \bar{t} + \varepsilon_0) \tag{4.5}$$

where $\gamma = \sqrt{\alpha^2 - \beta^2}$, $\bar{t}$ is the time.

β is relatively small compared to α, so $\gamma \approx \alpha$, the damping does not change the eigenfrequencies.

The rate of reduction of the vibration level during vibration is exponential. The rate of amplitudes of any two following cycles of the decaying vibration is (Fig. 4.3)

$$\frac{A_1}{A_2} = \frac{e^{-\beta \bar{t}}}{e^{-\beta(\bar{t}+T)}} = e^{\beta T} \tag{4.6}$$

The natural logarithm of this ratio is the logarithmic decrement

$$\delta = \ln\frac{A_1}{A_2} = \beta T = \frac{r}{2m}T = r\,\pi\sqrt{\frac{c_m}{m}} \tag{4.7}$$

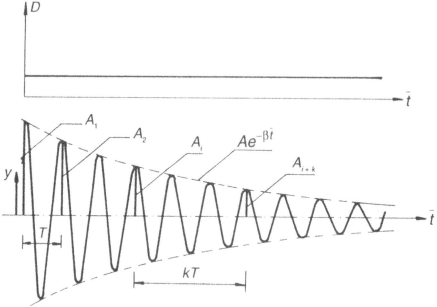

Figure 4.3. Logarithmic decrement due to constant damping.

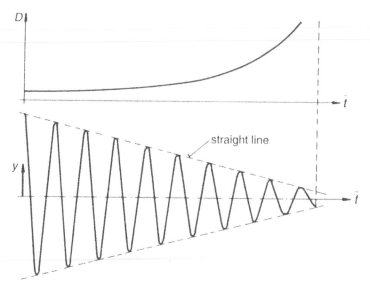

Figure 4.4. Vibration amplitudes due to in-creasing damping.

since the oscillation time is

$$T = \frac{2\pi}{\gamma} \approx \frac{2\pi}{\alpha} = 2\pi\sqrt{mc_m} \tag{4.8}$$

The logarithmic decrement can be computed from the rate of amplitudes of any two successive cycles of the decaying vibration

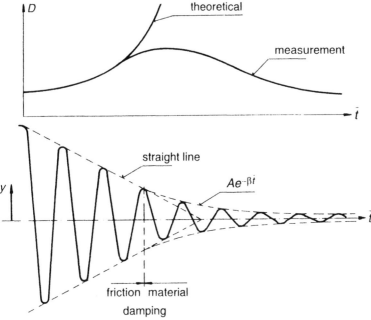

Figure 4.5. Vibration amplitudes due to friction damping at the beginning and later material damping.

$$\delta = \frac{1}{k} \ln \frac{A_i}{A_{i+k}} \tag{4.9}$$

– *Specific damping capacity of a vibrating system,* ψ is the ratio of dissipated and total energies (Fig. 4.2)

$$\psi = \frac{D}{U} \tag{4.10}$$

where U is the area between ABC points.

$$\psi = \int_{U_i}^{U_{i+T}} \frac{dU}{U} = \ln \frac{U_i}{U_{i+T}} \tag{4.11}$$

The potential energy is proportional to the square of amplitude

$$U = \frac{y^2}{2c_m} \tag{4.12}$$

so the specific damping capacity is as follows

$$\psi = \ln \frac{U_i}{U_{i+T}} = \ln \frac{A_i^2}{A_{i+T}^2} = 2\ln \frac{A_i}{A_{i+T}} = 2\delta \tag{4.13}$$

– *Loss factor* η, is defined as a fraction of the system's vibrational energy that is dissipated per radian of the vibratory motion at resonances and proportional to δ.

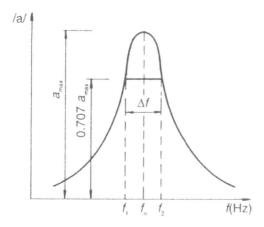

Figure 4.6. Measurement of loss factor by the half-power-bandwidth according to Oberst.

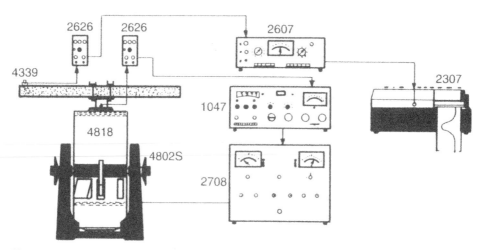

Figure 4.7. Brüel-Kjaer measurement equipment.

$$\eta = \frac{\delta}{\pi} \qquad (4.14)$$

To measure the loss factor on resonant frequency, the half-power-band-width technique, suggested by Oberst (1952) is applicable. In this case we measure the sharpness of resonance. It is applicable only if the vibrating system is linear and damping is relatively small. This is defined as the ratio of the difference of two frequencies $(f_2 - f_1)$ on both sides (3 dB down from peak amplitudes) of resonance and the resonance frequency f_n (Fig. 4.6)

$$\eta = \frac{f_2 - f_1}{f_n} = \frac{\Delta f}{f_n} \qquad (4.15)$$

Using the Brüel & Kjaer dynamic measuring devices, it is possible to measure the loss factors by the half-power-band-width technique. The equipment can be seen in Figure 4.7. The resonant curves are quite different if the damping is different. If the

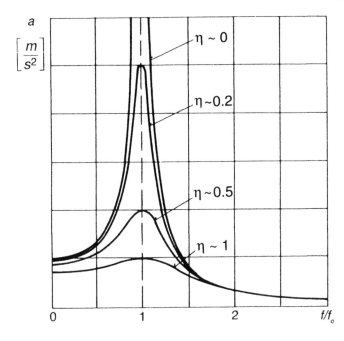

Figure 4.8. Some resonant
curves due to different
damping.

damping is zero, the vibration amplitude is infinite. Increasing the damping, the amplitudes are smaller and smaller (Fig. 4.8).

4.2 RELATIONS BETWEEN THE MEASURES OF DAMPING

All these parameters are related to one another and are described as follows:

$$\delta = \pi \eta = \frac{\psi}{2} \tag{4.16}$$

$$\psi = 2\pi \eta = 2(\delta - \delta^2) \quad \text{or} \quad 2\delta \tag{4.17}$$

$$\eta = \frac{\delta}{\pi} = \frac{\psi}{2\pi} \tag{4.18}$$

The loss factor and the damping capacity are defined directly in terms of the cyclic energy dissipation. Also, none of the measures of damping discussed above depend on how the energy is dissipated, and these measures make no reference to any damping mechanism within a cycle. Most of the commonly used measures of damping are defined on the basis of viscous damping. For linear viscously damped systems, all of the measures of damping are independent of amplitude, provided that the damping is small.

4.3 CLASSIFICATION OF DAMPING

In general, damping can be classified into two basic categories: Material damping and nonmaterial damping.

– *Material damping.* Every material has an internal damping, but it can be very different. Comparing different materials under exactly same boundary conditions, same geometrical dimensions, the same magnitude of periodic forcing with the same frequency of excitation we find, that one material may oscillate shorter with smaller amplitude than the other. This is due to the difference in material properties. The damping force due to internal molecular friction in material is more than at the other material. This kind of damping is called material damping.

– *Nonmaterial damping.* Another types of damping are friction damping, viscous and acoustic radiation damping. For example, a vibratory structural system will oscillate much longer in the air than in water. This kind of damping is called viscous damping. The viscous damping force depends on the property of the surrounding medium and the velocity of motion. The friction damping is caused by friction in the joints.

4.4 MATERIAL DAMPING

All materials dissipate energy during cyclic deformation. Some materials, such as rubber, plastics, foams and elastomers, dissipate much more energy per cycle of deformation than steels or aluminium. For conventional structural materials, the energy dissipation per unit-volume per cycle is very small compared to certain high damping alloys, polymer matrix composites, and rubberlike materials. These are viscoelastic materials.

4.4.1 *Characterization of viscoelastic materials*

The name viscoelastic material shows, that it has both elastic and viscous features. The basic models are:
  – The Maxwell model,
  – The Kelwin-Voigt model, and
  – The standard linear model.
    The elastic element can be modelled by a linear spring and the viscous element by a dashpot.

The Maxwell model
If the two elements are combined in series, it is known as a Maxwell model (Fig. 4.9), c_m is the spring constant and r the viscosity. When a force F is applied to this model, the elongation is equal to the sum of the extensions in the elastic and viscous elements, that is,

$$u = u_e + u_v \qquad\qquad (4.19)$$

where u represents the total elongation of the Maxwell model, u_e the elongation of

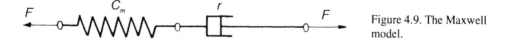

Figure 4.9. The Maxwell model.

the spring element, and u_v, the elongation of the viscous element.

The force F is the same in the elastic and viscous elements

$$F_e = F_v \tag{4.20}$$

The Kelwin-Voigt model

If the two elements are combined in parallel manner, it is known as a Kelwin-Voigt model (Fig. 4.10). When a force F is applied to this model, the sum of the forces in the spring and in the dashpot is equal to the applied force F, that is,

$$F = F_e + F_v \tag{4.21}$$

where F_e is the force in the elastic spring and F_v is the force in the viscous dashpot. The displacements of the spring and dashpot are equal

$$u_e = u_v \tag{4.22}$$

Standard linear model

It is a combination of a Maxwell model and a linear spring connected in parallel (Fig. 4.11). From equilibrium equation when a force F is applied to this model, the sum of the forces in the spring and in the Maxwell model is equal to the applied force F, that is,

$$F = F_e + F_m \tag{4.23}$$

where F_e is the force in the elastic spring and F_m is the force in the Maxwell model. Since the displacements of the spring and Maxwell model are equal

$$u_e = u_m \tag{4.24}$$

where u_m represents the total elongation of the Maxwell model, u_e the elongation of the spring element.

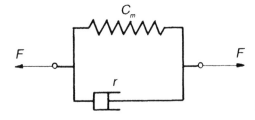

Figure 4.10. The Kelwin-Voigt model.

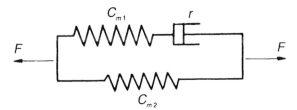

Figure 4.11. Standard linear model.

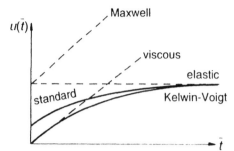

Figure 4.12. Displacement at the different models.

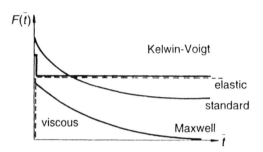

Figure 4.13. Force at the different models.

The displacements for the three models in the function of time, which is the creep function of the models are shown in Figure 4.12.

The force for the three models in the function of time, which is the relaxation function of the models are shown in Figure 4.13.

The real models of different viscoelastic materials are the combinations of the above mentioned models.

4.5 MATERIAL PROPERTY MEASUREMENTS

Considering that Poisson's ratio is approximately constant for most conventional materials, the property of metallic materials can be represented by only one variable, either Young's modulus or the shear modulus. However, for many rubberlike or linearly viscoelastic materials, the dynamic material properties, primarily Young's modulus, shear modulus and the loss factor, are functions of frequency, temperature, strain amplitude, and prestress or prestrain. Temperature and frequency are generally the most dominating variables. In the resonant beam test, for instance, both the drive signal and the response signal at a given point of the specimen are measured. With this information, the resonance frequencies and the modal damping values are calculated. These values are then used to calculate the Young's modulus, the shear modulus and the loss factor of the damping material. In all test methods, intermediate response information is measured and then the desired ultimate material properties are calculated.

To control effectively vibration problems in structures, it is essential to know accurate damping material properties for specific applications. There are various techniques for determining the dynamic properties of damping materials. The most notable methods are the resonant beam test, the dynamic mechanical analyzer and the resonant test.

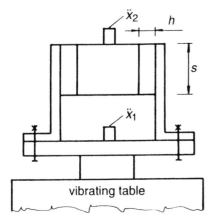

Figure 4.14. Jones measurements.

– The resonant beam technique for determining damping properties of materials means, that after measuring and drawing the resonant curve it is possible to find the resonant frequencies and to compute the loss factors according to Oberst (1952) as mention above. The disadvantage of this technique, that one can compute the loss factor only at the resonant frequencies (see Fig. 4.6).

– The measurement of damping due to shear is proposed by Jones & Parin (1972) (Fig. 4.14). This method is useful not only to get the loss factor, but the dynamic shear modulus as well, which is important in constrained layer damping.

We define the dynamic shear modulus as a complex value:

$$G_d^* = G_d(1 + i_u \eta_d) \tag{4.25}$$

The sizes of the measured material according to Figure 4.14 are the length s, the width b_0 and the thickness h. If we glue these two pieces to steel plates and fix a mass m into it, then the measured material is loaded by shear. We should measure either the displacements, or the accelerations on the mass and on the vibrating table $\ddot{x}_1, \ddot{x}_2$.

The movement of the vibrating table is described as

$$x_1 = x_{10} e^{i_u \omega t} \tag{4.26}$$

The movement of the mass m is

$$x_2 = x_{20}^* e^{i_u \omega t} \tag{4.27}$$

where x_{20}^* is a complex value due to the phase shift.

The equilibrium equation on the m mass is expressed by

$$m\ddot{x}_2 = kG_d^*(x_1 - x_2) \tag{4.28}$$

where

$$k = \frac{A}{s} = 2b_0 \tag{4.29}$$

$$A = 2sb_0 \tag{4.30}$$

Substituting Equations (4.26) and (4.27) into Equation (4.28) we get

$$\frac{x_{20}^*}{x_{10}} = \frac{kG_d^*}{kG_d^* - \omega^2 m} = \frac{1 + i_u \eta_d}{1 - (\omega^2 m / kG_d) + i_u \eta_d} \tag{4.31}$$

Introducing ω_0 and G_d which are valid only at resonant frequencies

$$G_{d0} = \frac{m\omega_0^2}{k} = \frac{m\omega_0^2}{2b_0} \tag{4.32}$$

and the transmissibility defined by Snowdon (1968) is

$$T_R = \left| \frac{x_{20}^*}{x_{10}} \right| = \left| \frac{\ddot{x}_2}{\ddot{x}_1} \right| = \frac{(1 + \eta_d^2)^{1/2}}{\left\{ \left[1 - (\omega / \omega_0)^2 (G_{d0} / G_d) \right]^2 + \eta_d^2 \right\}^{1/2}} \tag{4.33}$$

At resonant frequency

$$T_{R0}^2 = \frac{(1 + \eta_d^2)}{\eta_d^2} \tag{4.34}$$

$$\eta_d = \frac{1}{\sqrt{T_{R0}^2 - 1}} \tag{4.35}$$

Measuring the acceleration at two places one can calculate both the dynamic shear modulus and the loss factor.

The material damping depends on the following: Type or grade of the material, temperature, frequency, and stress level in the damping material.

– *Type or grade of the material*. A viscoelastic material sometimes is called material with memory. This implies that a viscoelastic material's behaviour depends not only on the current loading conditions, but also on the loading history.

– *Temperature* is an important factor affecting the properties of viscoelastic materials. Temperature can be divided into four different regions (Sun & Lu 1995) according to the behaviour of E and η: Glassy, transition, rubberlike and flow region. The glassy region usually occurs at room temperature. In this region, the storage modulus E decreases slightly as the temperature increases and the loss factor increases sharply as the temperature increases. This trend will continue into the transition region, until the transition temperature T_t is reached. At transition temperature T_t, the loss factor η reaches its maximum value while the slope of the storage modulus E is almost a constant. In the rubberlike region, both E and η nearly remain constants. This trend continues into the flow region in which the values of E decrease and η increases. In engineering applications structures are used below the transition temperature T_t. The range of the transition region may vary from about 20 to 200°C. The loss factor η is low in glassy region and reaches high value at the transition temperature.

– *Frequency* has also great effect on damping. The effect of frequency on the storage modulus E and the loss factor η again can be divided into four different regions: Glassy, transition, rubberlike and flow region. In general, E increases as the frequency increases. However, the rate of increase is small in the glassy and rubberlike regions, and the greatest rate of increase is in the transition region. The loss factor η decreases as frequency increases in the glassy region, reaches its maximum in the transition region at the transition temperature and increases in the rubberlike region. In the flow region, E decreases and η increases as the temperature increases.

– *Stress level in the damping material* is the most important factor. According to Warnaka & Miller (1968), the effects of cyclic strain amplitude on the damping behaviour are very complex. First, high dynamic strain amplitude will cause more energy dissipation in the material, and high energy dissipation will increase the temperature of the material. Therefore, the two effects on damping behaviour are coupled. In the glassy region, the variation of E and η with strain amplitude is small; E decreases sharply and η reaches its peak value in the transition region. It should be mentioned that the temperature effect on the modulus and loss factor become small in the rubberlike region. Thus the dynamic strain amplitude effects on E and η in the rubberlike region are more severe than in the other regions. The effects of tensile static preload will increase the modulus and decrease the loss factor.

A good approximation of the dependence of the material damping from the stress amplitude is given by Lazan (1968) (Eq. 4.1 and Fig. 4.15).

The practical stress range according to Lazan is the following

$$0.05 < \frac{2\sigma_a}{\Delta\sigma_N} < 0.8 \tag{4.36}$$

If the stress is near to the fatigue stress range $2\sigma_a > 0.8\Delta\sigma_N$ then the fatigue danger is larger, n increases from 2-3 to 8 due to microcracks in the material.

The loss factors η of different materials according to Lazan are as follows:
– Steel with 0.2% C $(0.3\text{-}1.8)10^{-3}$
– Spherical graphic cast iron $(1.4\text{-}6.3)10^{-4}$
– Lamellar graphic cast iron $(1.9\text{-}16)10^{-3}$
– Aluminium alloy (6063T6 USA) $(0.5\text{-}5)10^{-2}$
– Mg-Cu alloy, containing (12-64)% Cu $10^{-2}\text{-}1$
– Ni-Ti-Nol $0.3\text{-}0.4$
– PVC foams $10^{-2}\text{-}1.8$

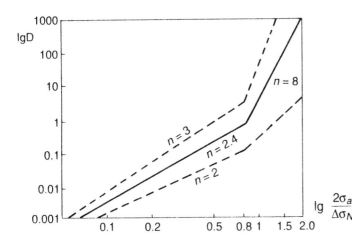

Figure 4.15. Material damping in the function of stress.

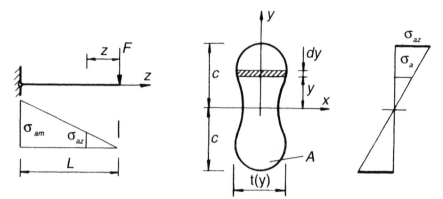

Figure 4.16. Stress distribution.

4.6 MATERIAL DAMPING IN STRUCTURES

If the stress distribution is not homogeneous, the geometry of the structure affects the damping.

The material damping in structures is defined as

$$D_S = \int_V D dV \tag{4.37}$$

for the homogeneous stress distribution the integral of D on the volume of the structure.

For a bent cantilever beam the stresses can be expressed by the maximum stress σ_{am} (Fig. 4.16)

$$\sigma_a = \sigma_{az}\frac{y}{c} = \sigma_{am}\frac{z}{L}\frac{y}{c} \tag{4.38}$$

substituting Equations (4.1) and (4.37) and multiplying and dividing by the volume $V = LA$, we get

$$D_S = D_{am}V\alpha_c\alpha_L \tag{4.39}$$

here

$$D_{am} = J\sigma_{am}^n \tag{4.40}$$

is the damping belonging to the maximum stress.

The cross-section parameter can be written as

$$\alpha_c = \frac{1}{c^n A}\int_{-c}^{c} y^n t(y)dy \tag{4.41}$$

The length distribution parameter is

$$\alpha_L = \frac{1}{L}\int_0^L \left(\frac{\sigma_{az}}{\sigma_{am}}\right)^n dz \tag{4.42}$$

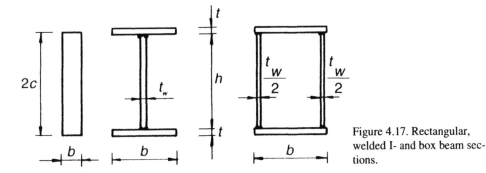

Figure 4.17. Rectangular, welded I- and box beam sections.

For the sake of comparison α_c is computed for three different cross-sections (Fig. 4.17).

For a rectangular section

$$\alpha_c = \frac{2}{c^n 2bc} \int_0^c y^n dy = \frac{1}{n+1} \tag{4.43}$$

for $n = 2.4$, $\alpha_c = 0.29$.

For welded I- and box section

$$\alpha_c = \frac{2}{(h/2)^n A} \left[t_w \int_0^{h/2} y^n dy + bt\left(\frac{h}{2}\right)^n \right] = 1 - \frac{n}{n+1} \frac{t_w h}{A} \tag{4.44}$$

For a welded I-beam, optimized for stress constraint, (see Chapter 8)

$$A = 2\beta h^2 \quad \text{and} \quad t_w h = \beta h^2$$

for box beam

$$A = 4\beta h^2 \quad \text{and} \quad t_w h = 2\beta h^2$$

so

$$\alpha_c = \frac{n+2}{2(n+1)}$$

for $n = 2.4$, $\alpha_c = 0.65$.

It can be seen that the cross-section parameter for the welded I-beam is significantly larger than that of rectangular cross-section one. It means, that the optimized welded sections are better in this sense, due to the higher stress level.

The length distribution parameter is the same for the three cross-sections

$$\alpha_L = \frac{1}{L} \int_0^L \left(\frac{\sigma_{az}}{\sigma_{am}}\right)^n dz = \frac{1}{L} \int_0^L \left(\frac{z}{L}\right)^n dz = \frac{1}{n+1} \tag{4.45}$$

for $n = 2.4$, $\alpha_L = 0.29$.

4.7 NONMATERIAL DAMPING

Friction or Coulomb damping is produced from the sliding and friction of two dry surfaces. The friction damping force F_c is equal to the product of the normal force between the two surfaces N and the coefficient of friction μ

$$F_c = \mu N \tag{4.46}$$

So the friction damping force depends on the normal force and on the coefficient of friction, but assumed to be independent of the relative velocity of motion between the two surfaces. The sign of the damping force is always opposite to that of the velocity. For a system with friction damping, the decay in response amplitude per cycle of vibration is a constant value 4 F_c/k where k is the spring constant. The motion will stop when the spring force is insufficient to overcome the static friction force.

The concept of friction damping is usually applied in structural joints. Damping force is introduced from slipping at the joint, and this results in energy dissipation at the joint. Coulomb damping can also be applied for the two-layer beam. The response will become nonlinear as soon as the midplane shear stress exceeds the yield limit. More detailed discussions and applications of Coulomb damping can be found in Plunkett (1981), and Earles (1966).

On a simple example we can show, how a friction damping works. A cantilever beam, with two layers forms a damper. The periodically varying bending force αF on the end and distributed compression forces p act on the two plates as shown in Figure 4.18.

There is friction due to the distributed forces between the two layers. We increase the bending force linearly. If the shear force is less than the static friction, $q = \tau b < q_0 = \mu\, pb$, there is no relative displacement between the layers. In this case the displacement is

$$y = \frac{\alpha F L^3}{24 EI} \tag{4.47}$$

where $I = bh^3 / 12$ is the moment of inertia.

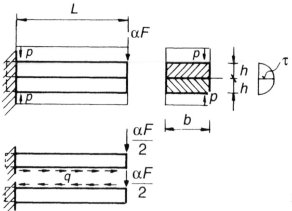

Figure 4.18. Friction damping of two compressed plates.

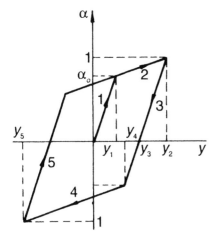

Figure 4.19. Hysteresis loop for friction damping.

This is represented by the 1st straight line on Figure 4.19. The limit point of this phase is

$$q = q_0 = \frac{3\alpha_0 F}{4h} \tag{4.48}$$

$$\alpha_0 F = \frac{4q_0 h}{3} \tag{4.49}$$

so the deflection is

$$y_1 = \frac{\alpha_0 F L^3}{24 EI} \tag{4.50}$$

If the shear force is greater than the static friction, $q = \tau\, b > q_0 = \mu\, pb$, in this second phase, there is a relative displacement between the layers and also energy dissipation. The external force αF is divided into two parts. The displacement can be determined from the force $\alpha(F/2)$ and from the specific shear force q_0, which causes uniformly distributed specific bending moments $m = q_0(h/2)$:

$$y = \frac{\alpha F L^3}{2*3EI} - \frac{m L^3}{3EI} = \frac{F L^3}{24 EI}(4\alpha - 3\alpha_0) \tag{4.51}$$

This is given by the 2nd line on Figure 4.19.
The limit point of this line is at $\alpha = 1$, at

$$y = y_1 = \frac{(4 - 3\alpha_0) F L^3}{24 EI} \tag{4.52}$$

In the 3rd phase the external bending force acts in the opposite direction and its absolute value is decreasing. This force is

$$(1 - \alpha) F$$

At the beginning there is no relative displacement between the two layers, so the relations are the same as in the 1st phase. The 3rd line is parallel with 1st one.

$$y = y_2 - \frac{(1-\alpha)FL^3}{24EI} = \frac{FL^3}{24EI}(3 - 3\alpha_0 + \alpha) \tag{4.53}$$

We need to compute y_3 to determine the area of the hysteresis loop:

$$y_3 = \frac{(1-\alpha_0)FL^3}{8EI} \tag{4.54}$$

The specific shear force in this phase is the following

$$q = q_0 - \frac{3(1-\alpha)F}{4h} \tag{4.55}$$

in limit state

$$q_0 - \frac{3(1-\alpha_1)F}{4h} = -q_0 \tag{4.56}$$

from this $\alpha_1 = 1 - 2\alpha_0$ and in the limit point

$$y_4 = \frac{FL^3}{24EI}(3 - 3\alpha_0 + \alpha_1) = \frac{FL^3}{24EI}(4 - 5\alpha_0) \tag{4.57}$$

In the 4th phase sliding proceeds in the opposite direction, so

$$y = y_4 - \frac{(\alpha_1 - \alpha)FL^3}{24EI} = \frac{FL^3}{24EI}(3\alpha_0 + 4\alpha) \tag{4.58}$$

and

$$y_5 = -\frac{(4 - 3\alpha_0)FL^3}{24EI} = -y_0 \tag{4.59}$$

The area of the hysteresis loop is

$$D_S = 4 y_3 \alpha_0 F = \frac{\alpha_0(1-\alpha_0)F^2 L^3}{2EI} = \frac{2\mu bhL^3(3Fp - 4\mu p^2 bh)}{9EI} \tag{4.60}$$

It can be seen, that the damping of this beam depends on the distributed compression force, on the area of the friction surface, and on the geometrical shape. If we plot the damping of the structure in the function of the distributed force we get a parabolic function (Fig. 4.20). The maximum value is

$$D_{S\,max} = \frac{F^2 L^3}{8EI} \tag{4.61}$$

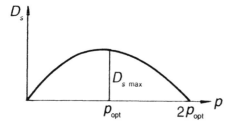

Figure 4.20. Structure damping in the function of compression.

in this case

$$p_{opt} = \frac{3F}{8\mu bh} \tag{4.62}$$

Friction damping does not occur when $p = 0$ and when there is no sliding, in the case of $p = 2\,p_{opt}$.

4.8 DAMPING OF WELDED STRUCTURES

In welded structures both material and friction damping occurs. Friction damping occurs due to compression between the plate elements, caused by residual welding deformations (Chapter 1) (Fig 4.21). To increase the material damping it is advisable to choose a suitable cross section as it was described above, i.e. the welded I-beam is better than the simple thick plate because the main part of the cross-section of an I-beam is loaded by larger stresses than that of a thick plate. The fatigue and damping behaviours of the structure are opposite. If we increase damping (friction damping) it will cause decrease of fatigue strength of the structure. The damping of welded I- or box beams with continuous longitudinal welds is smaller than that of beams with intermittent welds (Katzenschwanz 1967, Rangarajan & Rao 1968).

It is possible to increase the damping in a welded I-beam, by increasing the connected surfaces, which produce friction, or to form these surfaces close to the neutral axis, increasing the shear stress (Fig. 4.22).

An increase of the compression force between surfaces can increase damping too, e.g. in the case of prebending of the structural parts (Fig. 4.21c).

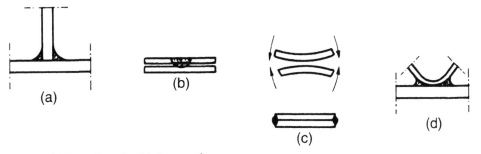

Figure 4.21. Damping of welded connections.

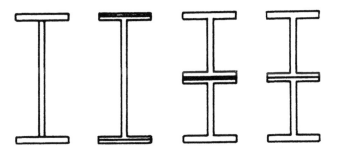

Figure 4.22. Welded I-beams with different friction surfaces.

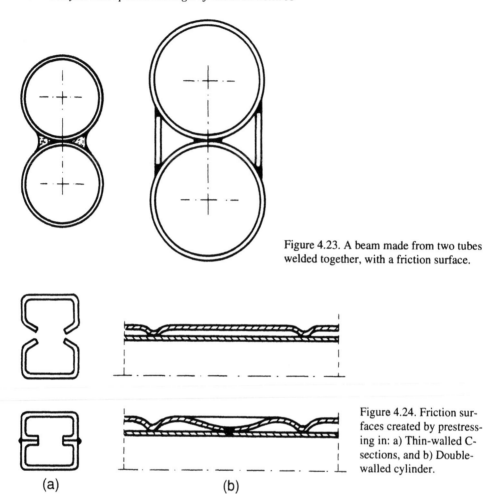

Figure 4.23. A beam made from two tubes welded together, with a friction surface.

Figure 4.24. Friction surfaces created by prestressing in: a) Thin-walled C-sections, and b) Double-walled cylinder.

(a)   (b)

If we weld two prebent thin plate together (Fig. 4.21d) the transversal shrinkage of welds will increase the friction damping (Kronenberg 1956). Welding tubular members together can cause also a friction surface with good damping (Fig. 4.23).

To increase the damping and decrease the vibration amplitudes in thin-walled cylinders of airplane motors, two surfaces should be welded together as shown in Figure 4.24b (Soyfer & Filekin 1958). The same increase of damping occurs when two thin-walled C-sections are welded together (Fig. 4.24a).

Overlapped joints are good for damping and bad for fatigue.

4.9   SANDWICH STRUCTURES

4.9.1   *Vibrations of three-layered damped beam structures*

The vibration damping of structures can be realized using high damping materials sandwiched between elastic layers to form laminated structures.

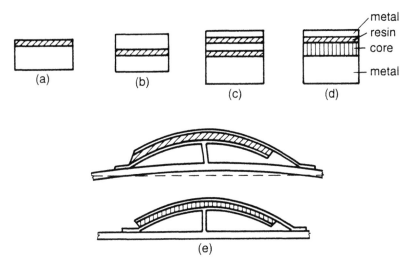

Figure 4.25. Sandwich beam with 2-5 layers.

When the beam or plate undergoes flexural vibration, the damped core is constrained and shear deformations occur in it. The vibration energy is dissipated by shear deformations. If the core is very soft, the shear strain is large but the stress required to cause it is small. The dissipated energy is then small. If the core is very stiff, shear strain is resisted, one of the face plates gets stretched while the other gets compressed, with the result that very little shear strain actually occurs, little energy is dissipated. In an intermediate range of core stiffness, the energy dissipated for a given flexural displacement reaches a maximum value so the beam damping has its highest value. Although many technical articles cover various aspects of this complicated physical phenomenon, the book authored by Nashif et al. (1985) is the only one that explicitly addresses relevant subjects in this vital technical area of vibration damping.

The study of vibratory characteristics of such composite structures has received much attention (Nashif et al. 1985, Kerwin 1965, Ross et al. 1962, Mead & Markus 1969, Yan & Dowell 1974, Rao 1990, Mead 1982, Ruzicka 1961a, b). The fundamental work in this field was first done by Ross et al. (1959), and Kerwin (1959). This chapter deals primarily with vibrations of three-layered damped sandwich beam structures (Fig. 4.25a-d).

Sandwich structures were applied first at airplanes during the II World War (Bert & Francis 1974). Later on they were used in machinery and building industry (Emerson 1974, Hammwill & Andrew 1974, Farah et al. 1977). Figure 4.25e shows the constrained layer damping, which can be very efficient.

The sandwich beam with outher layers of box section has the advantage that bending stiffness is great and the loss factor is also can be great. It was investigated by Yin et al. (1967) and the optimum design of this type of structure has been worked out (Farkas & Jármai 1982).

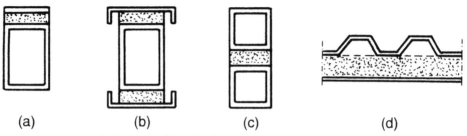

Figure 4.26. The sandwich beam with outher layers of box sections.

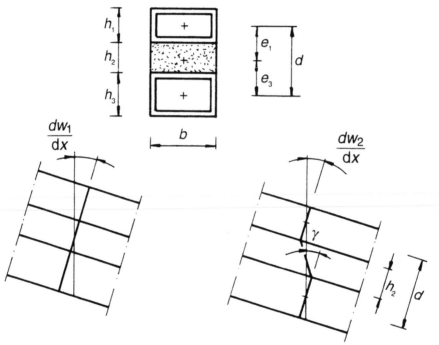

Figure 4.27. Static deformation of the three-layered sandwich beam.

4.9.2 *Static behaviour*

For a three layer sandwich beam the analytical calculations for static concentrated loading can be made according to Allen (1969), Stamm & Witte (1974), Grosskopf & Winkler (1973).

The beam can be seen in Figure 4.26.

For the analysis the Kerwin (1965) and Ungar (1962) models are necessary. Allen's (1969) assumptions are as follows:

– Stiffness of the core is negligible, the bending stresses occur only in the face plates,

– Shear stress in the core is uniform,

– In the core the cross stresses are negligible,

– Shear deformations of the face plates are negligible.

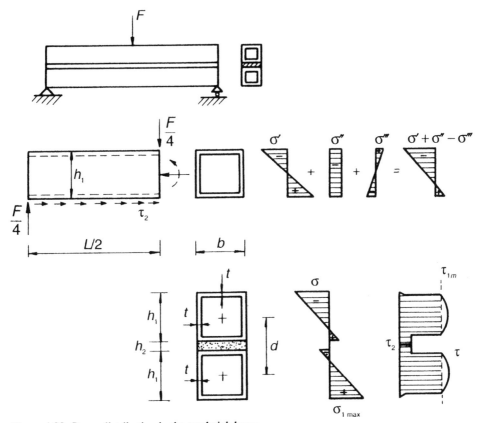

Figure 4.28. Stress distribution in the sandwich beam.

The static deformation can be divided into two parts: a) Primary deformation, the beam deforms like a homogeneous one, and b) Secondary deformation, there are relative displacement, sliding between the layers, the deformations of the different layers are separated (see Fig. 4.27).

The maximum deflection of the beam is expressed as

$$w_{max} = \frac{F_1 L^3}{48 B_f} + \frac{F_1 L}{4 B_{t2}} \left(1 - \frac{F_{f1}}{B_f}\right) S_1 \tag{4.63}$$

The maximum shear stress in core is

$$\tau_2 = -\frac{F_1}{2b(h_1 + h_2)} \left(1 - \frac{B_{f1}}{B_f}\right) S_2 \tag{4.64}$$

The maximum normal stresses in the face plates can be given as

$$\sigma_{max} = \frac{F_1 L}{4} \left[\left(h_1 + \frac{h_2}{2}\right) \frac{S_3}{B_f} + \frac{h_1}{2} \frac{1 - S_3}{B_{f1}}\right] \tag{4.65}$$

where the bending stiffness of the face plates

$$B_f = B_{f1} + B_{f2} \tag{4.66}$$

the bending stiffness of the core

$$B_{t2} = G_s \frac{b H_{13}^2}{h_2} \tag{4.67}$$

moment of inertia

$$I_1 = \frac{(h_1 - 2t)^3 t}{6} \tag{4.68}$$

$$B_{f1} = 2E_1 I_1, \quad B_{f2} = 2E_1 A_1 \left(\frac{h_1}{2} + \frac{h_2}{2}\right)^2, \quad H_{13} = \frac{h_1 + h_2}{2}$$

The cross sectional area is

$$A_1 = 2(h_1 - 2t)\,t + 2bt$$

the parameters:

$$S_1 = 1 - \frac{\sinh Q + \beta_1 (1 - \cosh Q)}{Q}, \quad S_2 = 1 - \sqrt{1 - \beta_1^2}, \quad S_3 = 1 - \frac{\beta_1}{Q}$$

$$Q = \frac{1}{2} \frac{B_{f1}}{L^2 B_{t2}} (1 - \frac{B_{f1}}{B_f})^{1/2}, \quad a^2 = \frac{B_f B_{t2}}{B_{f1}(B_f - B_{f1})}, \quad \varphi = aL_c \tag{4.69}$$

$$\beta_1 = \frac{\sinh Q - (1 - \cosh Q)\tanh \varphi}{\sinh Q \tanh \varphi + \cosh Q}$$

L_c is the length of the cantilever part of the beam.

The dimensions of cross section are shown in Figure 4.28. These equations were obtained by Allen for a central concentrated force. Grosskopf & Winkler (1973) have determined formulae for two concentrated forces and for distributed force too (Fig. 4.29).

The shear stress can be computed in the web plate part of the core according to Farkas & Jármai (1982)

$$\tau_1 = \frac{(F/2) - \tau_2 b h_2}{4t(h_1 - 2t)} \tag{4.70}$$

Measurements showed that most part of the shear stress occurs in the core and the web plates of the faces.

The stress distribution is given in Figure 4.28. For the static calculations it is necessary to determine the material parameters. On a tensile press machine one can measure the static shear modulus (see Fig. 4.30), and also the Young modulus of the metal face from the deflection of the simply supported beam.

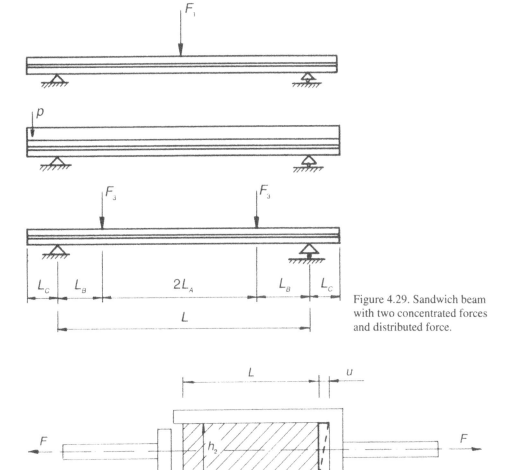

Figure 4.29. Sandwich beam with two concentrated forces and distributed force.

Figure 4.30. Static measurement of the shear modulus.

4.9.3 *Dynamic behaviour*

For a three-layered damped beam subjected to a time harmonic load of circular frequency of excitation the forced vibration equation can be described by Mead & Markus (1965).

The differential equation of sandwich beams according to Mead & Markus (1965) is the following:

$$\frac{\partial^6 w}{\partial x^6} - g_0(1+Y)\frac{\partial^4 w}{\partial x^4} = \frac{1}{B_{f1}}\left(\frac{\partial^4 p_0}{\partial x^4} - g_0 p_0\right) \tag{4.71}$$

where w is the displacement in the direction of vibration, the time dependent loading is

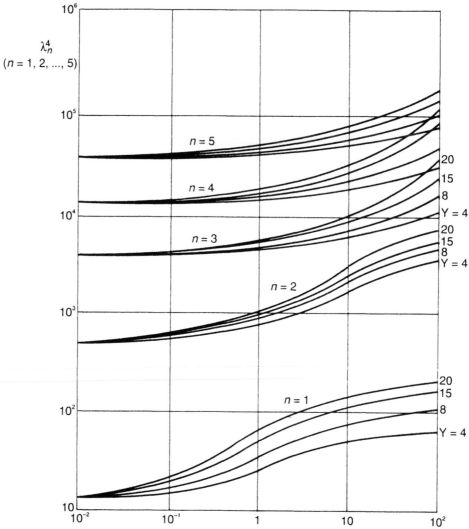

Figure 4.31. Eigenfrequencies of sandwich beams in the function of shear and geometrical parameters.

$$p_0 = -m\frac{\partial^2 w}{\partial t^2} + p(x,t)$$

for a symmetrical beam

$$g_1 = \frac{2G_d b}{A_1 E h_2}$$

the geometrical parameter

$$Y = \left[\frac{(h_1 + h_2)^2}{2B_{f1}}\right] A_1 E$$

G_d the dynamic shear modulus, E the Young-modulus.

To solve Equation (4.71), six boundary conditions, three at each end must be given for any particular three-layered beam system. Before all these conditions can be written down, expressions such as the face plate forces F_1 and F_3, the total bending moment M, the total shear force Q, and so on must be determined.

Similar to the classical beam structural system, the various boundary conditions of three-layered damped structures can be set by the prescribed values of displacements, rotations, moment, shear force, or a combination of these physical quantities at ends. Various boundary conditions in terms of displacement components are presented also by Rao (1978). End conditions given include free, free-riveted, pinned, pinned-riveted, clamped (unrestrained), clamped, sliding, and sliding-riveted ones.

The eigenfrequencies of a sandwich beam can be determined according to Markus et al. (1974) introducing shear (X) and geometrical parameters (Y) by Figure 4.31.

The loss factor of the beam can be determined according to Ungar (1962)

$$\eta = \frac{\eta_m XY}{1 + (2+Y)X + (1+Y)(1+\eta_m^2)X^2} \tag{4.72}$$

where the shear parameter is $X = g_0 r_k^2$, $r_k = CL/2\pi$, C depends on the boundary conditions. For a beam excited at the middle, free at both ends $C = 2$.

4.9.4 *Measurements on an experimental sandwich beam*

The measurements have been carried out in the laboratory of the University of Miskolc.

For a three layer sandwich beam the measured (subscript m) and the calculated (subscript c) static stresses and deflections can be seen in Table 4.1. The main data of the beam are: Length $L = 1500$, width/height of the face $b = h_1 = 40$, thickness of core $h_2 = 5$, thickness of the face section $t = 2$ (mm), concentrated force at the midspan $F = 7*10^4$ (N), $G_s = 2.36$ (MPa), $E = 7*10^4$ (MPa) (see Fig. 4.28).

Where τ_2 is the shear stress in the core, τ_{1c} and τ_{1m} are the calculated and the

Table 4.1. Measured and calculated static stresses and deflections.

F (N)	w_m (mm)	w_c (mm)	σ_m (MPa)	σ_c (MPa)	τ_2 (MPa)	τ_{1c} (MPa)	τ_{1m} (MPa)
350	11	14.2	14.36	14.44	0.04	0.59	0.58
613	22	24.9	23.78	25.27	0.07	1.03	0.98
924	38	37.7	35.90	38.11	0.11	1.56	1.51
1177	49	47.9	47.56	48.50	0.14	1.99	1.92
1429	65	58.3	54.74	58.89	0.17	2.42	2.39
1656	72	67.6	64.61	68.29	0.19	2.83	2.79
1888	84	77.0	74.49	77.92	0.22	3.21	3.29
2134	94	87.0	83.46	87.97	0.25	3.65	3.71
2365	105	96.4	94.23	97.48	0.28	4.02	4.36
2609	115	106.4	100.5	107.5	0.31	4.43	4.52
2839	125	115.8	109.4	117.0	0.33	4.82	4.99

Table 4.2. Eigenfrequencies of the sandwich beam.

Frequency	Measured (Hz)	Calculated (Hz)
f_1	110	110.3
f_2	515	541.6
f_3	1735	1670.3
f_4	2680	2867.5
f_5	4955	4858.6

measured shear stresses. It can be seen, that the measured and calculated values of deflection, normal and shear stresses are in good agreement in the measured range.

The dynamic measurements concern to the same specimen as used in static measurements. In this case the excitation is in the middle of the beam and the ends are free. In Equation (4.72) the core material loss factor η_m can be measured and calculated according to Equation (4.35). It is very important, that the stress level should be the same in both cases, at the beam and at the measurements according to Jones.

The core layer material loss factor is $\eta_m = 0.18$. The calculated shear parameter is $X = 1.898$, the geometrical parameter is $Y = 0.04363$. The calculated loss factor is $\eta_c = 0.04363$ according to Ungar (Eq. 4.72). The measured loss factor according to Oberst (1952) is $\eta_m = 0.04369$.

The measured and calculated eigenfrequencies of the measured beam can be found in Table 4.2. The agreement of the values is good.

CHAPTER 5

Fabrication costs

5.1 INTRODUCTION

The cost of a structure is the sum of the material, fabrication, transportation, erection and maintenance costs. The fabrication cost elements are the welding-, cutting-, preparation-, assembly-, tacking-, painting costs etc. It is very difficult to obtain such cost factors, which are valid all over the world, because there are great differences between the cost factors at the highly developed and at the developing countries, also there are great differences in the same country among factories, which are highly automated or not. If we choose times, as the basic data of fabrication phases, we can handle this problem. The fabrication time depends on the technological level of the country and the manufacturer, but it is much closer to the real process to calculate with (Malin 1985-1986). After computing the necessary time for a fabrication work element one can multiply it by a specific cost factor, which can represent the development level differences (Fern & Yeo 1990).

Although the whole production cost depends on many parameters and it is very difficult to express their effect mathematically, a simplified cost function can serve as a suitable tool for comparisons useful for designers and manufacturers (Pahl & Beelich 1982, Ott & Hubka 1985, Horikawa et al. 1992). The artificial intelligence is also applied for cost estimation (Ntuen & Mallik 1987).

The cost function can be expressed as

$$K = K_m + K_f = k_m \rho V + k_f \sum_i T_i \tag{5.1}$$

where K_m and K_f are the material and fabrication costs, respectively, k_m and k_f are the corresponding cost factors, ρ is the material density, V is the volume of the structure, T_i are the production times.

5.2 FABRICATION TIMES FOR WELDING

5.2.1 *Formulae proposed by Pahl & Beelich (1982)*

Equation (5.1) can be written in the following form

$$\frac{K}{k_m} = \rho V + \frac{k_f}{k_m}(T_1 + T_2 + T_3) \tag{5.2}$$

97

Table 5.1 Proposed values for the difficulty factor Θ_d. For skewed angle joints add 1-2 points

Structures	Welds	V-weld 60°	Fillet weld 90°
Planar	Long welds, flat position	1.0	2.0
Spatial	Short welds, plate, flat steel	1.5	2.5
Spatial	U-,L-profiles, tubes	2.0	3.0
Spatial	I-, T-profiles	2.5	4.0

where

$$T_1 = C_1 \Theta_d \sqrt{\kappa \rho V} \tag{5.3}$$

is the time for preparation, assembly and tacking, Θ_d is a difficulty factor, κ is the number of structural elements to be assembled.

Equation (5.3) can be derived as follows (Likhtarnikov 1968). For a plated structure consisting of κ elements the time of this part of the fabrication is proportional to the perimeter, for the ith element it is $T_i = c_1 P_i$. The mass of an element is proportional to the square of the perimeter $G_i = c_2 P_i^2$, thus $P_i = c_3 \sqrt{G_i}$ and $T_i = c_4 \sqrt{G_i}$. For the total structure, in average, it is $G = \kappa G_i$ and $T_1 = \kappa T_i = c_5 \kappa \sqrt{G/\kappa} = c_6 \sqrt{G \kappa}$.

Some proposed values for the difficulty factor are summarized in Table 5.1.

$$T_2 = \sum_i C_{2i} a_{wi}^{1.5} L_{wi} \tag{5.4}$$

is the time of welding, a_{wi} is the weld size, L_{wi} is the weld length, C_{2i} are constants given for different welding technologies. For manual-arc welding $C_2 = 0.8 \ast 10^{-3}$ and for CO_2-welding $C_2 = 0.5 \ast 10^{-3}$ min/mm$^{2.5}$.

$$T_3 = \sqrt{\Theta_d} \sum_i C_{3i} a_{wi}^{1.5} L_{wi} \tag{5.5}$$

is the time of additional fabrication actions such as changing the electrode, deslagging and chipping. $C_3 = 1.2 \ast 10^{-3}$ min/mm$^{2.5}$.

These formulae are used in (Farkas 1991) and in Sections 12.3.3, 12.3.4 and 12.5.

Ott & Hubka (1985) have proposed that $C_3 = (0.2-0.4)C_2$ in average $C_3 = 0.3C_2$. Thus, the modified formula for $T_2 + T_3$, neglecting $\sqrt{\Theta_d}$, is

$$T_2 + T_3 = 1.3 \sum C_{2i} a_{wi}^{1.5} L_{wi} \tag{5.6}$$

This formula is used in Section 8.3.

5.2.2 *The method based on COSTCOMP data*

The software COSTCOMP (1990) gives welding times and costs for different welding technologies (Bodt 1990). Using Equation (5.2) for T_1, the other times are calculated as

$$T_2 + T_3 = 1.3 \sum C_{2i} a_{wi}^{n} L_{wi} \tag{5.7}$$

This method is used in Sections 12.4, 13.3 and in Chapter 14.

The different welding technologies are as follows:
– SMAW (Shielded Metal Arc Welding),

– SMAW-HR (Shielded Metal Arc Welding, High Recovery),
– GMAW-C (Gas Metal Arc Welding with CO_2),
– GMAW-M (Gas Metal Arc Welding with mixgas),
– FCAW (Flux Cored Arc Welding),
– MCAW (Metal Cored Arc Welding),
– SSFCAW (Self Shielded Flux Cored Arc Welding),
– SAW (Submerged Arc Welding),
– GTAW (Gas Tungsten Arc Welding).

In the following figures and tables data are given for SMAW, GMAW-C and SAW techniques and for different weld types.

For GMAW-C welded I-butt double sided welds of size 2-8 mm $10^3 T_2 = 2(0.567 + 0.417a_w)$.

It should be noted that in values for SAW a multiplying factor of 1.7 is considered since in COSTCOMP different cost factors are given for various welding methods.

Table 5.2. Welding times T_2 (min) in function of weld size a_w (mm) for longitudinal fillet welds downhand position (see also Fig. 5.1).

Welding method	a_w (mm)	$10^3 T_2 = 10^3 C_2 a_w^n$
SMAW	2-5	$4.0\,a_w$
	5-15	$0.786\,a_w^2$
GMAW-C	2-5	$1.70\,a_w$
	5-15	$0.339\,a_w^2$
SAW	2-5	$1.190\,a_w$
	5-15	$0.236\,a_w^2$

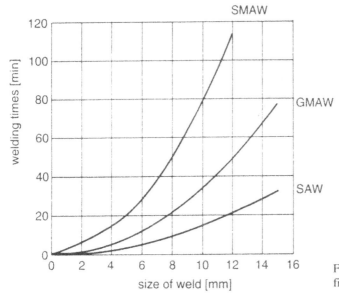

Figure 5.1. Welding times for fillet welds of size a_w

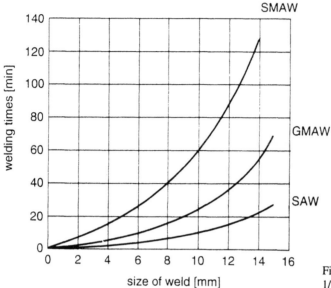

Figure 5.2. Welding times for 1/2 V-butt welds of size a_w.

Table 5.3. Welding times T_2 (min) in function of weld size a_w (mm) for longitudinal 1/2 V-butt welds downhand position (see also Fig. 5.2).

Welding method	a_w (mm)	$10^3 T_2 = 10^3 C_2 a_w^n$
SMAW	2-5	$3.86 \, a_w$
	5-15	$1.139 \, a_w^{1.758}$
GMAW-C	2-5	$1.26 \, a_w$
	5-15	$0.144 \, a_w^{2.348}$
SAW	2-5	$0.52 \, a_w$
	5-15	$0.178 \, a_w^2$

Table 5.4. Welding times T_2 (min) in function of weld size a_w (mm) for longitudinal V-butt welds downhand position (see also Fig. 5.3).

Welding method	a_w (mm)	$10^3 T_2 = 10^3 C_2 a_w^n$
SMAW	2-5	$2.88 \, a_w$
	5-15	$0.319 \, a_w^2$
GMAW-C	2-5	$0.72 \, a_w$
	5-15	$0.138 \, a_w^2$
SAW	2-5	$0.30 \, a_w$
	5-15	$0.096 \, a_w^2$

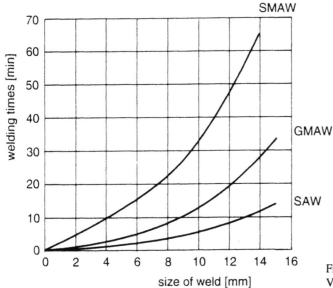

Figure 5.3. Welding times for V-butt welds of size a_w.

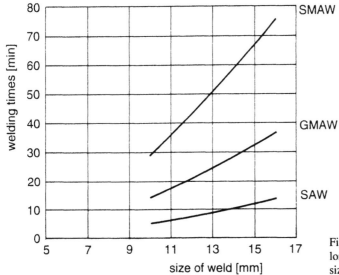

Figure 5.4. Welding times for longitudinal K-butt welds of size a_w

Table 5.5. Welding times T_2 (min) in function of weld size a_w (mm) for longitudinal K-butt welds downhand position (see also Fig. 5.4).

Welding method	a_w (mm)	$10^3 T_2 = 10^3 C_2 a_w^n$
SMAW	10-16	$1.4029 a_w^{1.25}$
GMAW-C	10-16	$0.129\, a_w^2$
SAW	10-16	$0.089\, a_w^2$

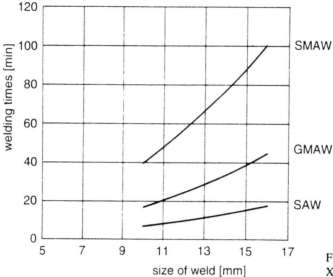

Figure 5.5. Welding times for X-butt welds of size a_w.

Table 5.6. Welding times T_2 (min) in function of weld size a_w (mm) for longitudinal X-butt welds downhand position (see also Fig. 5.5).

Welding method	a_w (mm)	$10^3 T_2 = 10^3 C_2 a_w^n$
SMAW	10 - 16	$0.488 a_w^2$
GMAW-C	10 - 16	$0.169\, a_w^2$
SAW	10 - 16	$0.116\, a_w^2$

5.3 SURFACE PREPARATION TIME

The surface preparation means the surface cleaning, painting costs, ground coat, top coat, sand-spraying, etc.

The surface cleaning time can be in the function of the surface area (A_s (m^2)) as follows:

$$T_S = \Theta_{ds} a_{sp} A_s \qquad (5.8)$$

where $a_{sp} = 8.33*10^{-3}$ (hour/m^2), Θ_{ds} is a difficulty factor.

5.4 PAINTING TIME

The painting means making the ground coat and the top coat.

The painting time can be in the function of the surface area (A_s (m^2)) as follows:

$$T_P = \Theta_{dp} (a_{gc} + a_{tc}) A_s \qquad (5.9)$$

where $a_{gc} = 8.33*10^{-3}$ (hour/m^2), $a_{tc} = 1.15*10^{-2}$ (hour/m^2), Θ_{dp} is a difficulty factor, $\Theta_{de}=1$, 2 or 3 means horizontal, vertical or overhead painting.

5.5 CUTTING AND EDGE GRINDING TIMES

The cutting and edge grinding can be made by hand grinding, hand flame cutting and machine flame cutting.

The cutting time can be in the function of the thickness and lengths (t (mm), L_e (m)) as follows:

$$T_G = \Theta_{dc}(a_{fc} + b_{fc}t^2)L_e \qquad\qquad (5.10)$$

where for hand grinding $a_{fc} = 4.12*10^{-2}$ (hour/m), $b_{fc} = 6.82*10^{-3}$ (hour/m/mm^2), for flame cutting $a_{fc} = 5.033*10^{-2}$ (hour/m), $b_{fc} = 2.47*10^{-4}$ (hour/m/mm^2), for machine flame cutting $a_{fc} = 3.45*10^{-2}$ (hour/m), $b_{fc} = 2.28*10^{-4}$ (hour/m/mm^2), Θ_{dc} is a difficulty factor, $\Theta_{dc} = 1, 2$ or 3.

Part 2
Optimum design

CHAPTER 6

Mathematical methods for structural synthesis

6.1 HISTORICAL BACKGROUND

People in their everyday life always make optimization on a conscious or a subconscious way 'to reach the best, which is possible with the resources available'. The consciousness makes the act more efficient. They have always targets to reach and constraints to control them. The birth of optimization methods as mathematical techniques can be dated back to the days of Newton, Lagrange and Cauchy. The further development in optimization was possible by the developments of differential calculus by Newton and Leibnitz, the variational calculus by Bernoulli, Euler, Lagrange and Weierstrass, the introduction of unknown multipliers by Lagrange. The concept of multiobjective optimization was formulated 100 years ago by Pareto in 1896.

The first written analytical work published on structural optimization was made by Maxwell in 1890, followed by the well known work of Michell in 1904. These works provided theoretical weight minima of trusses, using highly idealised models, but the analytical way of solution of the structural optimization problem is still usable.

During the Second World War and in the late 1940's and the early 1950's the development of optimization concerned to the minimum weight design of aircraft structural components: Columns, stiffened panels, subject to compressive loads and to buckling. Digital computers appear in the early 1950's and gave a strong impulse to the application of linear programming techniques. The applications were focused primarily on steel frame structures.

In the late 1950's and 1960's the applications of structural optimization on lightweight structures concentrated to the aircraft and space industries. This time a some new optimization techniques have been developed by works of Rosenbrock, Box and Powell. The great development of this period is that the finite element method, which is a powerful tool for analysis of complex structures, has been invented by Zienkiewich and applied by many others for structural analysis.

Modern structural optimization can be dated from the paper of Schmit in 1960, who drew up the role of structural optimization, the hierarchy of analysis and synthesis, the use of mathematical programming techniques to solve the nonlinear inequality constrained problems. The importance of this work is that it proposed a new philosophy of engineering design, the structural synthesis, which clarifies the methodology of optimization.

6.2   DESIGN VARIABLES, OBJECTIVE FUNCTIONS, CONSTRAINTS AND PREASSIGNED PARAMETERS

The objective function (more functions at multiobjective optimization), the design variables, the preassigned parameters and the constraints describe an optimization problem.

6.2.1   *Design variables and preassigned parameters*

The quantities, which describe a structural system can be divided into two groups: Preassigned parameters and design variables. The difference between them is that the members of the first group are fixed during the design, the second group is the design variables, which are varied by the optimization algorithm. These parameters can control the geometry of the structures. It is the designer choice, which quantities will be fixed or varied. They can be cross-sectional areas, member sizes, thicknesses, length of structural elements, mechanical or physical properties of the material, number of elements in a structure (topology), shape of the structure, etc.

For example, in the case of a simple beam the quantities are as follows: 1. Span length of the beam, 2. Sizes and area of the cross section, 3. Characteristics of applied materials, 4. Loadings, bending moments, 5. Shape of the beam, 6. Type of supports, end conditions, 7. Number of supports etc. Some of them can be design variable, all the others should be preassigned quantities.

Cross-sectional design variables
Size, or dimension variables are the simplest and the most natural design ones. The cross-section area of tension and compression members, the moment of inertia of bent members, or the plate thicknesses can be design variables of this kind. In simple cases a single design variable (i.e. area) is sufficient to describe the cross section, but for a more detailed design several variables may be necessary. For example, if we consider the overall buckling of members, the moment of inertia or the radius of gyration would be also important as design variable. It should emphasise that less variables for the same problem mean considerable advantage in the solution from the optimization point of view. From the analytical point of view, the result can be opposite.

Material design variables
The Young modulus, yield stress, material density, thermal conductivity, specific heat coefficient etc. can be material design variables. These properties has a discrete character, i.e. a choice is to be made only from a discrete set of variables. In most cases the optimization procedure is nondiscrete, so these discrete variables complicate the optimization problem. In this case it is advisable to use discrete optimization techniques like Backtrack. For a few number of available materials non-discrete technique would probably be more efficient to perform the optimization separately for each material.

Geometrical variables
Geometrical variables are the span length of a beam, the coordinates of joints in a

truss or in a frame. Although many practical structures have geometry which is selected before optimization, geometrical variables can be treated by most optimization methods. In general, the geometry of the structure is represented by continuous variables.

Topological variables
Topology means the structural layout, number of supports, number of elements etc. These can be of discrete or continuous type. In truss systems the topology can be optimized automatically if we allow members to reach zero cross section size. The uneconomical members can be eliminated during the optimization process. Integer topological variables can be the number of spans of a bridge, the number of columns supporting a roof system, or the number of elements in a grillage system.

6.2.2 *Constraints*

Behaviour means those quantities that are the results of an analysis, such as forces, stresses, displacements, eigenfrequencies, loss factors etc. These behaviour quantities form usually the constraints. A set of values for the design variables represents a design of the structure. If a design meets all the requirements, it will be called feasible design. The restrictions that must be satisfied in order to produce a feasible design are called constraints. There are two kinds of constraints, explicit and implicit ones.

Explicit constraints
Explicit constraints which restrict the range of design variables may be called size constraints or technological constraints. These constraints may be derived from various considerations such as functionality, fabrication, or aesthetics. Thus, a size constraint is a specified limitation, upper or lower bounds on a design variable. Examples of such constraints include minimum slope of a portal frame structure, minimum thickness of a plate, minimum or maximum ratio of a box section height and width, etc.

Implicit constraints
Constraints derived from behaviour requirements are called behavioural constraints. Limitations on the maximum stresses, displacements, or local and overall buckling strength, eigenfrequency, damping are typical examples of behavioural constraints. The behaviour constraints can be regarded as implicit variables. The behavioural constraints are often given by formulae presented in design codes or specifications. Other part of the behavioural constraints are computed by numerical technique such as FEM. In any case the constraints can be evaluated by analytical technique. From a mathematical point of view, all behavioural constraints may usually be expressed as a set of inequalities.

The constraints may be linear or nonlinear functions of the design variables. These functions may be explicit or implicit in the feasible region X and may be evaluated by analytical or numerical techniques. However, except for special classes of optimization problems, it is important that these functions should be continuous and have continuous first derivatives in X.

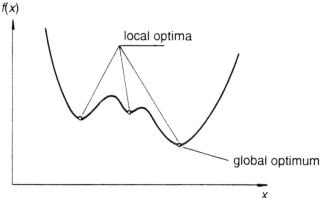

Figure 6.1. Optima for one variable.

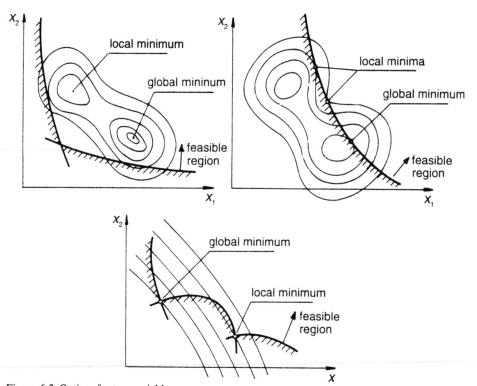

Figure 6.2. Optima for two variables.

Design space, feasible region

We may regard each design variable as one dimension in a design space and any particular set of variables as a point in this space. In cases with two variables the design space reduces to a plane problem. In the general case of N variables, we have an N-dimensional hyperspace.

Considering only the inequality constraints the set of values of the design variables that satisfy the equation $g_j(\mathbf{x}) = 0$ forms a surface in the design space. These are boundary points. This surface cuts the space into two regions: 1. Where $g_j(\mathbf{x}) < 0$ (these are interior points), and 2. Where $g_j(\mathbf{x}) > 0$ (these are the exterior points). The set of all feasible designs forms the feasible region. The solution of the constrained optimization problem in most cases lies on the surface. The solution can be local or global optimum (see Figs 6.1 and 6.2).

Any vector $\mathbf{x}$ that satisfies both the equality and inequality constraints is called a feasible point or vector. The set of all points which satisfy the constraints constitutes the feasible domain of $f(\mathbf{x})$ and will be represented by X; any point not in X is termed nonfeasible.

Convexity, concavity
It is very important to determine, under what condition a local optimum is also a global one. It depends on the form of the feasible region, determined by the con-

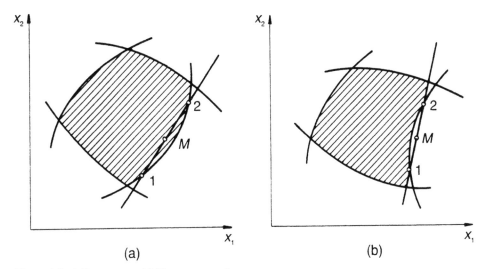

(a)

(b)

Figure 6.3. a) Convex sets, b) Non-convex sets.

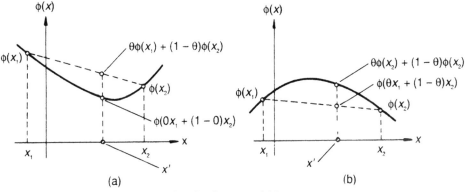

(a)

(b)

Figure 6.4. Convex and non-convex function in one variable.

straints. For a convex region the local optimum is a global one, otherwise there are several local optima (Figs 6.3a and b). Convex is a region, if between any two interior points all points are also interior, otherwise the region is non-convex.

The set F (the feasible region) is convex, if for any points x_1, x_2 in the set the line segment joining these points is also in the set. Mathematically the Φ function is convex, if $\Phi(\Theta x_1 + (1 + \Theta) x_2) \leq \Theta \Phi(x_1) + (1 - \Theta) \Phi(x_2)$ over the feasible domain. Θ is a scalar with the range $0 \leq \Theta \leq 1$. The sets shown in Figure 6.4a are convex, those in Figure 6.4b are not. They called non-convex, or concave. No analytical method is to classify a problem as being convex or non-convex (Yu 1974).

6.2.3 *Objective function*

In most practical cases an infinite number of feasible designs exists. In order to find the best one, it is necessary to form a function of the variables to use it for comparison of design alternatives. The objective function (also termed the cost, or merit function) is the function whose least, or greatest value is sought in an optimization procedure. It is usually a nonlinear function of the variables $\mathbf{x}$, and it may represent the mass, the cost of the structure, or any other function, which extremum can give a possible and useful solution of the problem. The minimization of $f(\mathbf{x})$ is equivalent with the maximization of $-f(\mathbf{x})$.

The selection of an objective function can be one of the most important decisions in the whole optimum design process. If we choose several objectives to be minimized, we reach the area of multiobjective optimization (described in detail in Section 6.12) and the greatest decision is to find in this case the relative importance of the different objective functions. Mass is the most commonly used objective function due to the fact that it is readily quantified, although most optimization methods are not limited to mass minimization. The minimum mass is usually not the cheapest. Cost is of wider practical importance than mass, but it is often difficult to obtain sufficient data for the construction of a real cost function. A general cost function may include the cost of materials, fabrication, welding, painting, maintenance, etc. (Chapter 5).

6.3 DIVISIONS IN OPTIMIZATION TECHNIQUES

The different single-objective optimization techniques make the designer able to determine the optimum sizes of structures, to get the best solution among several alternatives. The efficiency of these mathematical programming techniques is different. A large number of algorithms has been proposed for the nonlinear programming solution Himmelblau (1972), Vanderplaats (1984), Schittkowski et al. (1994). Each technique has its own advantages and disadvantages, no one algorithm is suitable for all purposes. The choice of a particular algorithm for any situation depends on the problem formulation and the user.

The general formulation of a single-criterion nonlinear programming problem is the following:

$$\text{minimize } f(\mathbf{x}) \quad \mathbf{x} = \{x_1, x_2, ..., x_N\} \tag{6.1}$$

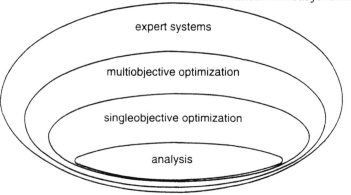

Figure 6.5. The hierarchy of different design stages.

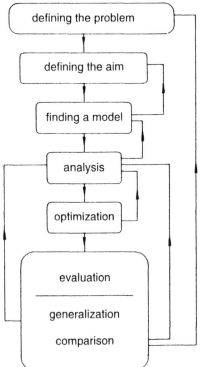

Figure 6.6. Logical structure of optimum design.

$$\text{subject to } g_j(\mathbf{x}) \le 0 \qquad j = 1, 2, ..., P \tag{6.2}$$

$$h_i(\mathbf{x}) = 0 \qquad i = P+1, ..., P+M \tag{6.3}$$

f(x) is a multivariable nonlinear function, $g_j(\mathbf{x})$ and $h_i(\mathbf{x})$ are nonlinear inequality and equality constraints respectively.

If we consider the system analysis, the single- and multiobjective optimization and expert systems (described in Chapter 7) the hierarchy is as in Figure 6.5.

The logical structure of the optimum design can be seen in Figure 6.6. It follows

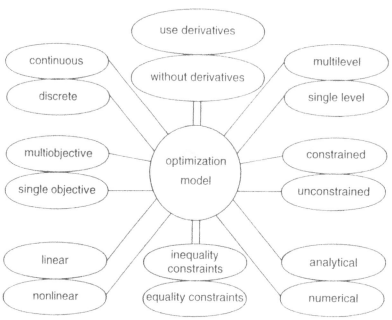

Figure 6.7. Different optimization models.

the logical thinking of a human. There are close connections between the different levels. If there are problems finding the optimum, we should go back to the analysis stage. If there are problems in the analysis stage it can be necessary to go back to the modelling stage. A good solution in the optimization is achieved by several loops in the scheme.

The optimization models can be very different from each other. Figure 6.7. shows the main alternatives.

Analytical and numerical

Analytical methods use the mathematical theory of differential calculus and the calculus of variations. These methods seek the extremum of a function $f(x)$ by finding the values of x that cause the derivatives of $f(x)$ with respect to x to vanish. When the extremum of $f(x)$ is sought in the presence of constraints, techniques such as Lagrange multipliers and constrained variation are used. For the application of analytical methods, the problem to be optimized must be described in mathematical terms, so that the functions and variables can be manipulated by known rules. For large, highly nonlinear problems, analytical methods prove unsatisfactory.

Numerical methods usually employ mathematical programming. Numerical methods use past information to generate better solutions to the optimization problem by means of iterative procedures. Numerical methods can be used to solve problems that cannot be solved analytically, because of this efficiency in practical problems, we deal with numerical methods of nonlinear programming. The iterative process of numerical optimization can be seen on Figure 6.8.

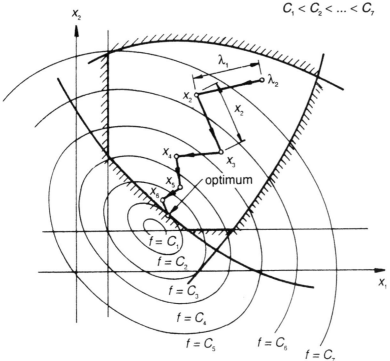

Figure 6.8. Iterative process of numerical optimization.

Unconstrained and constrained

In early generation of the optimization techniques, the aim was only the function minimization of maximization without any constraints. Some techniques, for example SUMT (Section 6.5.1), transforms the constrained problem into an unconstrained one, using penalty functions. For most practical problems, with several constraints, constrained optimization techniques are applicable.

Single- and multivariable

The differential calculus by Newton and Leibnitz, and the variational calculus by Bernoulli, Euler, Lagrange and Weierstrass create the basis of single variable optimization. For practical problems there are usually more variables. Most of the constrained multivariable optimization techniques are efficient, if the number of unknowns variables is not too much (2-20-50). For more variables (some thousands) special techniques were developed (Rozvany 1989).

Single- and multiobjective

The objective can be the mass, cost, strength, reliability, etc. If all objectives are relevant but not equally important (i.e., no dominant criterion can be defined), multicriteria optimization techniques are used to generate the set of Pareto optima (or nondominated solutions) and to select the satisfying solution, i.e., one that achieves the best compromise between all conflicting criteria (see Section 6.12).

The basis of the multiobjective optimization methods is the transformation of the vector optimization problem into a sequence of single objective optimization problems. Alternatively, the most important criterion is selected as the only objective function and limits are imposed on the other objective functions.

Discrete and nondiscrete

Most optimization techniques use nondiscrete, continuous unknowns, variables and they can be very robust and quick in this sense. Discrete optimization is a smaller stream in this field, but also a promising one. The main advantage of discrete optimization is, that the result is closer to the realistic solution (Karczewski et al. 1991). Several hybrid techniques connect the two streams.

Structure free and structure dependent techniques

There are some optimization techniques, which depend on the problem, solved. If one changes something at the problem, or try to use the optimization technique for another one, he should adopt the technique for the modified or new problem. One example is the optimality criterion method (Section 6.7). Other techniques can be regarded as universal techniques in the sense, that the core of the procedure is uniform, only the problem description: Objective function, inequality constraints, size constraints are different for different problems.

Single- and multilevel optimization

Multilevel optimization can be applied to solve large scale optimization problems using decomposition methods. Large scale systems can be decomposed into a few subsystems that can be optimized separately. The coordination level leads to the solution of the decomposed system, which converges to the solution of the original system (Kirsch 1975, Lin 1976, Sobieszczanski-Sobieski et al. 1985).

Many civil engineering structures can be decomposed into subsystems. For example, a hall structure can be divided as follows: Roofing panels, trusses, columns, walls and foundations. In the optimization process, it is necessary to choose the local design variables concerned with particular subsystems, and global design variables common to at least two or more subsystems. The particular subsystems are optimized independently of one another with respect to the local variables. Next, coordination is performed according to the global objective function with respect to the coordination (global) variables.

6.4 METHODS WITHOUT DERIVATIVES

Optimization techniques, which require the evaluation of function values only during the search, called methods without derivatives (zeroth-order methods). These methods are usually reliable and easy to program. Often can deal effectively with nonconvex functions. The price to pay for this generality is that these methods often require thousands of function evaluations to achieve the optimum. Thus these methods can be considered as most useful for problems, in which the function evaluation is not computationally expensive and we can rerun the programs from different points to avoid local optima.

6.4.1 *Complex method*

This method is a constrained minimization technique which uses random search so it does not require derivatives. Although the complex method was designed by Box (1961) to be applied to nonlinear programming problems with inequality constraints, we include the method in this section because of its use of random search directions. The method evolved from the simplex method of Spendley et al. (1962). Vertices are deleted and added as in the simplex method, but no attempt is made to preserve the regular figure so characteristic of the simplex method. The difficulty with the methods of Spendley et al. (1962) and Nelder & Mead (1964) on repetitively encountering a constraint is that it is necessary to withdraw the nonfeasible vertex until it becomes feasible. After many such withdrawals the polyhedron will collapse into $(N-1)$ or fewer dimensions, and the search will be quite slow. Furthermore, if the constraint ceases to be active, the collapsed polyhedron cannot readily expand back into the full N-dimensional space again. To avoid these difficulties, Box selected a polyhedron with more than $(N+1)$ vertices, which he termed a complex.

Using random numbers the so-called 'complex' is generated from the upper and lower bounds of variables:

$$\text{the explicit constraints} \quad x_i^L \leq x_i \leq x_i^U \qquad i = 1, 2, ..., N \qquad (6.4)$$

$$\text{the implicit constraints} \quad x_i^L \leq x_i \leq x_i^U \qquad i = N+1, ..., N+M \qquad (6.5)$$

where M is the number of implicit constraints, x_i^L the lower, x_i^U the upper limit for the variables.

Generation of starting complex
In the first iteration cycle (IT = 0) an original complex is generated. The complex contains $K \geq N+1$ feasible points or vertices in an N-dimensional design space. It is assumed that at least one initial feasible point exists, that is the starting point. The remaining $(K-1)$ points are generated randomly. For clarity a two-subscript notation is used, the first subscript indicating the co-ordinate of the point and the second one indicating the point number. Thus, the jth point is generated from random numbers r_{ij} and from the upper and lower bound on the ith independent variable as follows:

$$x_{ij} = x_i^L + r_{ij}(x_i^U - x_i^L) \qquad i = 1, 2,..., N \quad j = 2, 3,..., K \qquad (6.6)$$

The random numbers have a uniform distribution over the interval 0-1.

The point generated in the foregoing fashion will satisfy the explicit constraints but not necessarily the implicit constraints. If an implicit constraint is violated, the trial point is moved halfway towards the centroid of the previous points as follows:

$$x_{iC} = \frac{1}{k-1} \sum_{j=1}^{k-1} x_{ij} \qquad (6.7)$$

and take a new point

$$x_{ij}^N = \frac{x_{iC} + x_{ij}^W}{2} \qquad (6.8)$$

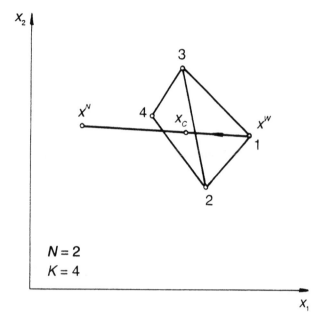

Figure 6.9. Complex in two dimensions.

in which x_{ij}^N is the ith co-ordinate of this new point. Superscripts N and W mean the new and the worst points. Subscript C means the centroid point. The process is repeated a few times. If the design space is convex, halving will eventually lead to a feasible point (Fig. 6.9).

Search procedure

Since the points are generated in a random fashion, they are located throughout the feasible design space. The function values are evaluated at each of these points. The points with the worst, and the best function values are determined. For function minimization the 'worst' means the greatest value, the 'best' means the smallest one. The convergence criterion checked:

$$f_{max} - f_{min} < \beta \tag{6.9}$$

The generally used value of β is $(10^{-3} - 10^{-4}) f_{max}$. If it is not fulfilled, the worst point x^W, for which $f(x^W) = f_{max}$, is rejected to the centroid and replaced with a new one

$$x_{ik}^N = \alpha(x_{iC} - x_{ik}^W) + x_{iC} \quad i = 1, 2,..., N \tag{6.10}$$

where

$$x_{iC} = \frac{1}{k-1} \sum_{j=1}^{k} (x_{ij} - x_{ij}^W) \quad i = 1, 2,..., N \tag{6.11}$$

is the centroid of the remaining points.

This new point is first examined to see whether it satisfies the explicit constraints. If not, it is moved a small distance δ inside the violated limit. If so, a further check is

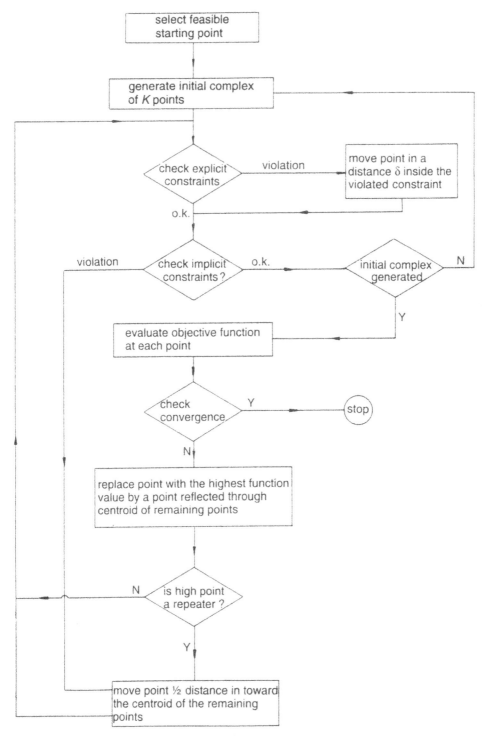

Figure 6.10. Flow chart of the complex method.

made to see whether any of the implicit constraints are violated. If the function value is less than the worst value and all the implicit constraints are satisfied, then this point is accepted as an improvement.

We continue with the next iteration cycle (IT = 1). However, if either the function value of the new point is equivalent or larger than the worst value or any of the implicit constraints are violated, the point is replaced by selecting another point half-way towards the centroid from its present position according to Equation (6.8). This point is checked for feasibility.

If all constraints are satisfied, but the point remains the worst one, i.e. $f(x_{ij}^N) = f_{max}$ a new point must be calculated according to Equation (6.8).

If $f(x_{ij}^N) = f_{max}$, we continue with step Equation (6.8) until some predetermined criterion (convergence, iteration number) stops the process. Figure 6.10 shows the flow chart of the Complex method.

Box recommended from empirical studies that the reflection coefficient $\alpha = 1.3$ can be used and that ($K = 2N$) vertices be used in the search. The reflection coefficient in the reflection stage of the polyhedron and the use of more than ($N + 1$) vertices are features of the procedure designed in order to continually enlarge the Complex and to compensate for moves halfway toward the centroid, to prevent the polyhedron from flattening near constraints. The search will continue until the polyhedron is reduced essentially to the centroid to within a given precision.

Numerical results demonstrated that the rate of convergence of the complex method depends on the character of the initial polyhedron. Box concluded that Rosenbrock's method is more efficient than the simplex or complex methods for unconstrained problems, and that the number of functional evaluations increased twice as rapidly for the Simplex and Complex methods as they did for Rosenbrock's method as the dimensionality (N) of $f(\mathbf{x})$ increased.

Modifications of the original algorithm
Some modifications, including those suggested by some authors Ghani (1972) were implemented. The following briefly describes the main modifications of the complex algorithm and some salient features of the computer program.

Expansion step
If on reflection of the worst point (Eq. 6.10) none of the explicit or the implicit constraints is violated and the function value of the point is better than the best point value, this direction of search is considered to be promising. In this case an expansion step is taken by simply reflecting the point again in Equation (6.10), and the results compared.

If unsuccessful, the point is restored to the previous position. This procedure allows a rapid movement of the complex towards the optimum.

Limitation of halving steps
The halving steps bring the new point closer to the centroid in a geometric progression. The centroid was always a feasible point. Instead of continuously halving the distance between the wrong reflected point and the centroid, it is halved only five times, and then the centroid is used.

Variation in size of complex (K), convergence and reflection coefficients (β, α)
Since the result may be improved by using an other value of K, it is useful to repeat the computation with other K-values. Greater K means a greater point set with more information being more reliable, but needs more computation. The program varies the value of K automatically if it is needed. The decreasing of the convergence coefficient (β) improves the result, but increases the computer time remarkably.

The iteration number (IT) is reducible with the variation of value of reflecting coefficient (α) between certain bounds. The best range of α is between 1.3-1.7 (Jármai 1982).

To start the method, a feasible starting point is necessary x_{i1}, otherwise the method is unmoveable.

The procedure is robust, it gives global optima, but if the number of unknowns (N) and the size of Complex (K) are large, it becomes very slow.

6.4.2 *Flexible tolerance method*

This method is a constrained random search technique. The Flexible Tolerance algorithm (Himmelblau 1982) improves the value of the objective function by using information provided by feasible points, as well as certain nonfeasible points termed near-feasible points. The near-feasibility limits are gradually made more restrictive as the search proceeds toward the solution, until in the limit only feasible **x** vectors are accepted.

With this strategy Equations (6.1)-(6.3) can be replaced by a simpler problem, having the same solution:

$$\text{minimize } f(x)$$
$$\text{subject to } \Phi^{(k)} - T(x) \geq 0 \tag{6.12}$$

where $\Phi^{(k)}$ is the value of the flexible tolerance criterion for feasibility on the kth stage of the search, and $T(x)$ is a positive functional of all the equality and/or inequality constraints of Equation (6.3), used as a measure of the extent of constraint violation.

The tolerance criterion $\Phi^{(k)}$ is selected to be a positive decreasing function of the vertices of the flexible polyhedron in the design space. The function $\Phi^{(k)}$ is a tolerance criterion for constraint violation during the entire search, and also serves as a criterion for termination of the search. Many alternative definitions of $\Phi^{(k)}$ are possible, but the one incorporated into the algorithm to be described is

$$\Phi^{(k)} = \min\left[\Phi^{(k-1)}, \frac{M+1}{r+1} \sum_{i=1}^{r+1} \left\|x_i^{(k)} - x_{r+2}^{(k)}\right\|\right] \tag{6.13}$$

$$\Phi^{(0)} = 2(M+1)q \tag{6.14}$$

where q is the size of initial polyhedron, M is the number of equality constraints, $x_i^{(k)}$ is the ith vertex of polyhedron in the design space, $r = (N - M)$ is the number of degrees of freedom of $f(\mathbf{x})$ in the original problem, $x_{r+2}^{(k)}$ is the vertex corresponding to centroid, with $N = r$, $k = 0, 1,...$ is an index referring to number of completed stages of search, $\Phi^{(k-1)}$ is the value of tolerance criterion on $(k-1)$st stage of search.

The second term in the braces of expression is $\Phi^{(k)}$

$$\Theta^{(k)} = \frac{M+1}{r+1}\sum_{i=1}^{r+1}\left\|x_i^{(k)} - x_{r+2}^{(k)}\right\| = \frac{M+1}{r+1}\left[\sum_{i=1}^{r+1}\sum_{j=1}^{N}(x_{ij}^{(k)} - x_{r+2,j}^{(k)})^2\right]^{1/2} \qquad (6.15)$$

where $x_{ij}^{(k)}, j = 1,..., N$ are the co-ordinates of the ith vertex of the flexible polyhedron in the design space. The value of $\Theta^{(k)}$ will depend on the size of the polyhedron, which may remain unchanged, expand, or contract, depending on which one of the four operations is used to carry out the transition from $x_i^{(k)}$ to $x_i^{(k+1)}$. Thus $\Theta^{(k)}$ behaves as a positive decreasing function of x, although $\Theta^{(k)}$ may increase or decrease during the progress of the search, and as the solution of the problem is approached, both $\Theta^{(k)}$ and $\Phi^{(k)}$ approach zero.

The criterion for constraint violation T(x)
Consider now a functional of the equality and inequality constraints of the original problem.

$$T(x) = +\left[\sum_{i=1}^{M}h_i^2(x) + \sum_{i=M+1}^{P}U_i g_i^2(x)\right]^{1/2} \qquad (6.16)$$

where U_i is the Heaviside operator such that $U_i = 0$ for $g_i(x) \leq 0$ and $U_i = 1$ for $g_i(x) \geq 0$. Therefore $T(x)$ is defined as the positive square root of the sum of the squared values of all the violated equality and/or inequality constraints of the original problem.

$T(\mathbf{x})$ is a convex function with a global minimum $T(\mathbf{x}) = 0$ for all feasible $\mathbf{x}$ vectors.

If $T(x^{(k)}) = 0$, $x^{(k)}$ is feasible; if $T(x^{(k)}) > 0$, $x^{(k)}$ is nonfeasible. A small value of $T(x^{(k)})$ implies that $x^{(k)}$ is relatively near to the feasible region, and a large value for $T(x^{(k)})$ implies that $x^{(k)}$ is relatively far from the feasible region.

The concept of near-feasibility
Near-feasible $\mathbf{x}$ vectors are those points that are not feasible, but almost feasible, in the sense given below. The $x^{(k)}$ vector is
   – Feasible, if $T(x^{(k)}) = 0$
   – Near-feasible, if $0 < T(x^{(k)}) < \Phi^{(k)}$
   – Nonfeasible, if $T(x^{(k)}) > \Phi^{(k)}$
The region of near-feasibility is defined as

$$\Phi^{(k)} - T(x) \geq 0 \qquad (6.17)$$

On any transition from $x^{(k)}$ to $x^{(k+1)}$, the move is said to be feasible if $T(x^{(k)}) = 0$, near-feasible if $0 < T(x^{(k)}) < \Phi^{(k)}$, and nonfeasible if $T(x^{(k)}) > \Phi^{(k)}$.

It is very important to choose a good size of initial polyhedron, which is difficult, when the difference between the values of unknowns is great. If the upper and lower bounds on x are known $(x_i^U - x_i^L)$, a reasonable estimate is as follows:

$$q = \min\left\{\frac{0.2}{N}\sum_{i=1}^{N}(x_i^U - x_i^L)\right\} \qquad (6.18)$$

One advantage of the flexible tolerance strategy is that the extent of the violation of the constraints is progressively decreased as the search moves toward the solution of problem (Eq. 6.13) because the equality and/or the inequality constraints are loosely satisfied in the early stages of the search, and more tightly satisfied only as the search approaches the solution of problem (Eq. 6.13), the overall computation effort required in the optimization is considerably reduced.

Another advantage of the flexible tolerance strategy is that $\Phi^{(k)}$ can be conveniently used as a criterion for termination of the search. For all practical purposes it is sufficient to continue the search until $\Phi^{(k)}$ becomes smaller than some arbitrarily selected positive number ε. In the final stages of the search, $\Phi^{(k)}$ is also a measure of the average distance from each vertex $x^{(k)}$ to the centroid $x_{r+2}^{(k)}$ of the polyhedron. In the developed flexible tolerance technique, Himmelblau implemented the Nelder-Mead method for the minimization of $T(x)$ (Eq. 6.16), but any other techniques are usable.

6.4.3 *Hillclimb method*

This method is a direct search one without derivatives. Rosenbrock's (1960) method is an iterative procedure that bears some correspondence to the exploratory search of Hooke and Jeeves (1961) in that small steps are taken during the search in orthogonal coordinates. However, instead of continually searching the co-ordinates corresponding to the directions of the independent variables, an improvement can be made after one cycle of co-ordinate search by lining the search directions up into an orthogonal system, with the overall step on the previous stage as the first building block for the new search coordinates. Rosenbrock's method locates $x^{(k+1)}$ by successive unidimensional searches from an initial point $x^{(k)}$ along a set of orthonormal directions.

The method is executed as follows:
Minimize the objective function $f(x_i) \to$ min.
Design constraints are:

$$\text{explicit } x_i^L \le x_i \le x_i^U \quad (i = 1, 2,..., N) \tag{6.19}$$

$$\text{implicit } g_j(x_i) \ge 0 \quad (j = 1, 2,.., M)$$

1. Before starting the minimization process, define a set of 'initial' step lengths S_i, to be taken along the search directions M_i, $i = 1, 2,..., N$. The starting point must satisfy the constraints and should not lie in the boundary zones.

2. After each function evaluation, the following steps are carried out: Define by f^0 the current best objective function value for a point where the constraints are satisfied, and $f(\mathbf{x})$ where in addition to this the boundary zones are not violated. f^0 and $f(\mathbf{x})$ are initially set equal to the objective function value at the starting point.

3. The first variable x_1 is stepped a distance S_1 parallel to the axis and the function evaluated. If the current point objective function value, f, is worse (greater or less) than f^0 or if the constraints are violated, the trial point is a failure and S_1 decreased by a factor β, $0 < \beta \le 1$, and the direction of movement reversed. If the move is termed a success, S_1 increased by a factor $\alpha, \alpha \ge 1$. The new point is retained, and a success is recorded. The values of α and β are usually taken as 3,0 and 0,5 respectively.

4. Continue the search sequentially stepping the variables, x_i, a distance S_i parallel to the axis. The same acceleration or deceleration and reversal procedure is followed for all variables, until at least one step has been successful and one step has failed in each of the N directions. Perturbations are continued sequentially in the search directions until a success is followed by a failure in every direction, at which time the kth stage is terminated. Since an equal value of a function counts as a success, a success is eventually reached in each direction as the multipliers of reduce the magnitude of the step length. The final point obtained becomes the initial point for the succeeding stage $x^{(k+1)} = x^{(k)}$. The normalized direction $S_i^{(k+1)}$ is chosen parallel to $x_0^{(k+1)} - x_0^{(k)}$, and the remaining directions are chosen orthonormal to each other and to $S_i^{(k+1)}$.

5. Compute the new set of directions $M_{i,j}^{(k)}$ rotating the axes by the following equations. In general, the orthogonal search directions can be expressed as combinations of all the co-ordinates of the independent variables as follows:

$$M_{i,j}^{(k+1)} = \frac{D_{i,j}^{(k)}}{\left[\sum_{l=1}^{n} (D_{i,j}^{(k)})2 \right]^{1/2}} \tag{6.20}$$

where

$$D_{i,1}^{(k)} = A_{i,1}^{(k)} \tag{6.21}$$

$$D_{i,1}^{(k)} = A_{i,1}^{(k)} - \sum_{l=1}^{j-1} \left[\left(\sum_{n=1}^{j} M_{n,j}^{(k+1)} A_{n,j}^{(k)} \right) M_{i,j}^{(k+1)} \right] \tag{6.22}$$

$$j = 2, 3, ..., N$$

$$A_{i,j}^{(k)} = \sum_{l=j}^{N} d_i^{(k)} M_{i,l}^{(k)} \tag{6.23}$$

$$i = 1, ..., N \quad j = 1, ..., N$$

d_i -sum of distances moved in the i direction since last rotation of axes.

6. Search is made in each of the **x** directions using the new co-ordinate axes. In each **x** direction the variables are stepped a distance S_i parallel to the axis and the function is evaluated.

$$\text{new } x_i^{(k)} = \text{old } x_i^{(k)} + S_j^{(k)} * M_{i,j}^{(k)} \tag{6.25}$$

7. If the current point lies within a boundary zone, the objective function is modified as follows:

$$f(\text{new}) = f(\text{old}) - (f(\text{old}) - f^*)(3\lambda - 4\lambda^2 + 2\lambda^3) \tag{6.26}$$

where the boundary zones are defined as follows:

$$\lambda = \frac{\text{distance into boundary zone}}{\text{width of boundary zone}}$$

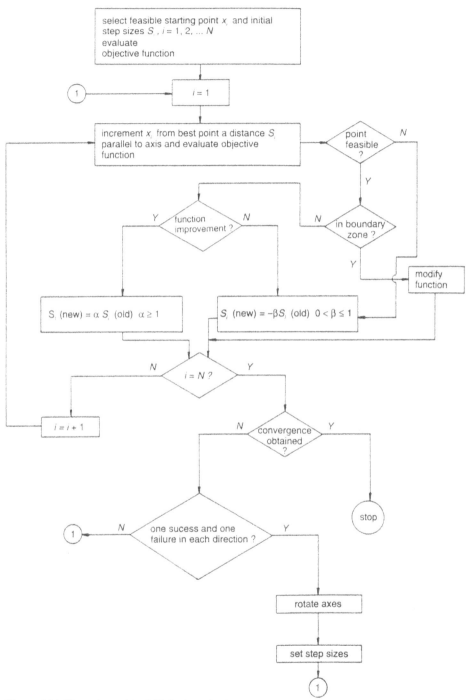

Figure 6.11. Flow chart of the hillclimb method.

lower zone:

$$\lambda = \frac{x_i^L + (x_i^U - x_i^L) * 10^{-4} - x_i}{(x_i^U - x_i^L) * 10^{-4}} \tag{6.27}$$

upper zone:

$$\lambda = \frac{x_i - (x_i^U - (x_i^U - x_i^L) * 10^{-4})}{(x_i^U - x_i^L) * 10^{-4}} \tag{6.28}$$

At the inner edge of the zone, $\lambda = 0$, i.e., the function is unaltered (f(new) = f(old)). At the constraints, $\lambda = 1$, and thus f(new) = f^*.

For a function which improves as the constraint is approached, the modified function has an optimum in the boundary zone.

8. f^* is set equal to f^0 if an improvement in the objective function has been obtained without violating the boundary zones or constraints.

9. The search procedure to find the continuous values of the variables is terminated when the convergence criterion is satisfied.

10. The procedure was modified by a secondary search to find the discrete values of the variables. In details it is written in Section 6.9.4. A flow chart illustrating the above procedure is given in Figure 6.11.

The procedure stops if the convergence criterion or the iteration limit is reached. The procedure is very quick, but it gives usually local optima, so it is advisable to use more starting points. The Turbo/Borland C version of Hillclimb technique can be found in Appendix C.

6.5 METHODS WITH FIRST DERIVATIVES

Methods with first derivatives (first order methods), which utilize gradient information, are usually more efficient than zero-order methods. The price paid for this efficiency is that gradient information must be supplied, either by finite-difference computations or analytically. These methods are not efficient, if the function has discontinuous first derivatives. However in most cases first order methods can be expected to perform better than zeroth order methods.

6.5.1 *Penalty methods: SUMT, exterior, interior penalties*

Penalty methods belong to the first attempts to solve constrained optimization problems satisfactorily. The basic idea is to construct a sequence of unconstrained optimization problems and to solve them by any standard minimization method, so that the minimizers of the unconstrained problems converge to the solution of the constrained one. (SUMT = Sequential Unconstrained Minimization Technique.) This type of optimization methods use two different types of penalty function: Ones for inequality constraints and ones for equality constraints. Equality constraints cannot be implemented as two inequality constraints, because then there is no feasible region.

The procedure was developed by Fiacco & McCormick (1968). The technique

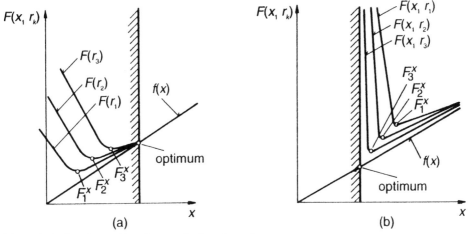

Figure 6.12. Exterior and interior penalty function values.

uses the problem constraints and the original objective function to form an uncon-strained objective function which is minimized by any appropriate unconstrained, multivariable technique.

Exterior penalty function
The exterior penalty function associates a penalty with a violation of a constraint. The term 'exterior' refers to the fact, that penalties are applied only in the exterior of the feasible domain. The most common exterior penalty function is proportional to the square of violation. The advantage of this method is, that the procedure can be started from an infeasible point. Another advantage is that only active constraints af-fect the objective function at the optimum. The disadvantage of the technique is, that one cannot have a feasible solution until the optimum is reached. Another drawback is that the penalty parameter must approach infinity at the optimum design.

$$F(x, r_k) = f(x) + r_k \left\{ \sum_{j=1}^{M} \left\{ \max[0, g_j(x)] \right\}^2 + \sum_{k=1}^{l} [h_k(x)]^2 \right\} \qquad (6.29)$$

where the limit

$$\lim_{r_k \to \infty} F_{\min} = f_{\min}$$

The first part of the penalty denotes the positive part, or the maximum of the $(0, g_j(x))$ region. Illustration of the exterior penalty function method can be seen on Figure 6.12a. F^* means the minimum function value.

Interior penalty function
In the interior penalty function, or barrier function method all intermediate solutions lie in the feasible region and converge to the solution from the interior side of the feasible domain. This penalty function is the inverse of the constraint function when the constraint is inactive. The advantage of the method, that every iterations are fea-

sible, one can stop the procedure at any time. The constraints become critical only near the optimum. Disadvantage of the method is that all constraints affect the objective regardless of whether they are active at the optimum or not. It has the serious drawback that infeasible design results in large negative penalties. This means that the optimization must start in the feasible region and must never evaluate an infeasible design.

To construct the unconstrained problems, so-called penalty terms are added to the objective function which penalize $f(\mathbf{x})$ whenever the feasible region is left. A factor r_k controls the degree of penalizing f. Proceeding from a sequence $f(r_k)$ with $r_k \to 0$ for $k = 0, 1, 2,...$ penalty functions can be defined, for example, by

$$F(\mathbf{x}, r_k) = f(\mathbf{x}) - r_k \sum_j \frac{1}{g_j(\mathbf{x})} \quad \text{or} \quad F(\mathbf{x}, r_k) = f(\mathbf{x}) - r_k \sum_j \ln g_j(\mathbf{x}) \qquad (6.30)$$

where the limit

$$\lim_{r_k \to \infty} F_{\min} = f_{\min}$$

Illustration of the interior penalty function method can be seen on Figure 6.12b. F^* means the minimum function value.

Extended interior penalty function
This penalty function is a merging of the interior and exterior penalty function methods already described. A new parameter ε is introduced to qualitatively control where the switch between the two function occurs. One version of the interior extended penalty function is defined as follows:

$$P(x) = -\frac{r}{g(x)} \qquad g(x) \le -\varepsilon \qquad (6.31)$$

$$P(x) = \frac{r}{\varepsilon} \left[\left(\frac{g(x)}{\varepsilon} \right)^2 + \frac{3g(x)}{\varepsilon} + 3 \right] \qquad g(x) > -\varepsilon \qquad (6.32)$$

This penalty function is constructed so that the first and second derivatives of the penalty are continuous at the transition location, if the constraint is smooth. For large violations, the penalty function approaches the exterior penalty function method. This method has some of the advantages of both the interior and exterior methods. The requirement for feasibility is removed. The disadvantage of this method is that as the transition parameter is reduced, violations of the constraints result in extremely large values of the penalty function.

The unconstrained nonlinear programming problems are solved by any common technique, e.g. a quasi-Newton search direction combined with a line search. However, the line search must be performed quite accurately due to the steep, narrow valleys created by the penalty terms. In our computer program the Davidon-Fletcher-Powell method is built into SUMT using interior penalty function. The list of the program can be found in ANSI C in Appendix B.

There exists a large variety of other proposals and combinations of them (e.g. Fiacco & McCormick 1968). The main disadvantage of penalty type methods is that

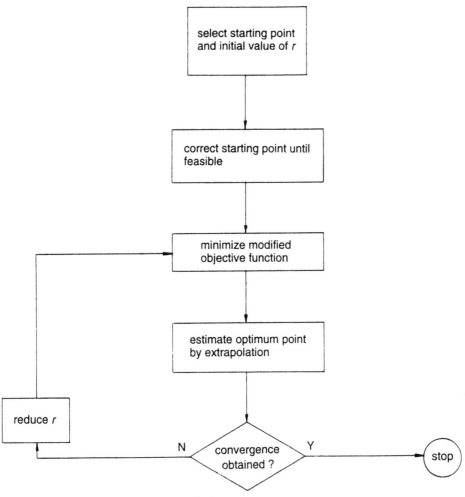

Figure 6.13. Flow chart of the SUMT method.

the condition number of the Hessian matrix of the penalty function increases when the parameter r_k becomes too large.

An interior penalty function algorithm proceeds as follows:
1. A modified objective function is formulated consisting of the original function and penalty functions with the form

$$F(x, r_k) = f(x) + r_k \sum_{j=1}^{P} \frac{1}{g_j(x)} + r_k^{-1/2} \sum_{i=P+1}^{P+M} h_i^2(x) \qquad (6.33)$$

where r_k is a positive constant. As the algorithm progresses, r_k is re-evaluated to form a monotonically decreasing sequence $r_1 > r_2 > ... > 0$. As r_k becomes small, under suitable conditions F approaches f and the problem is solved.

2. Select a starting point (feasible or nonfeasible) and an initial value for r_k.

3. Determine the minimum of the modified objective function for the current value of r_k using an appropriate technique.

4. Estimate the optimum solution using extrapolation formulae.

5. Select a new value for r_k and repeat the procedure until the convergence criterion is satisfied. The flow chart of this method can be seen in Figure 6.13. The value of r_k is decreased during the procedure to 10^{-4} or so.

The penalty function of the inequality constraints can be another function in Equation (6.30), e.g instead of reciprocal $\sum_{j=1}^{r} 1/g_j(x)$ a logarithmic function $\sum_{j=1}^{r} \ln g_j(x)$. The efficiency of the penalty function depends on the problems.

6.5.2 *Davidon-Fletcher-Powell method*

The variable metric method of Davidon (1959) was extended by Fletcher & Powell (1963). This method is one of the best general-purpose unconstrained optimization techniques making usmmmme of the derivatives that are currently available.

1. Starts with an initial point **x** and $N*N$ size positive definite symmetric matrix H. This H matrix is usually taken as the identity matrix I. The iteration number is $i = 1$.

2. The method computes the gradient of the function $f(x)$ at the initial point and sets

$$S_i = -H_i \nabla f_i \tag{6.34}$$

3. Find the optimum length λ_i^*, in the direction S_i and set

$$x_{i+1} = x_i + \lambda_i^* S_i \tag{6.35}$$

where H_i is taken as the identity matrix.

4. Find the new point x_{i+1}, test for optimality and if x_{i+1} is optimum, terminate the iterative process, otherwise continue the calculation.

5. Update the H matrix as

$$H_{i+1} = H_i + M_i + N_i \tag{6.36}$$

$$M_i = \lambda_i^* \frac{S_i S_i^T}{S_i^T Q_i} \tag{6.37}$$

$$N_i = -\frac{(H_i Q_i)(H_i Q_i)^T}{Q_i^T H_i Q_i} \tag{6.38}$$

and $Q_i = \nabla f(x_{i+1}) - \nabla f(x_i)$

6. Set the new iteration number $i = i + 1$ and go to step 2.

The interior penalty function method is used in the algorithm to be able to handle constraints (Eq. 6.30).

The cubic interpolation method is used for finding the minimizing step length λ_i^* in four stages. Sometimes there is an overflow at $F(x, r_k)$ function, because $g_j(x)$ is close to zero near the optimum, so the convergence criterion is very important.

6.6 METHODS WITH SECOND DERIVATIVES

The second-derivative methods, among which the best-known is Newton's method, are originated from the quadratic approximation of $f(x)$. They make use of second-order information, that is an information obtained from the second partial derivatives of $f(x)$ with respect to the independent variables.

6.6.1 *Newton's method*

A classical second order method is the Newton's method. This technique begins with the second order Taylor series expansion. The direction of search s for Newton's method is chosen as follows:

If $(x - x^{(k)})$ is replaced by $\Delta x^{(k)} = x^{(k+1)} - x^{(k)}$, the quadratic approximation of $f(x)$ in terms of $\Delta x^{(k)}$ is

$$\begin{aligned} f(x) &\approx f(x^{(k)}) + \nabla^T f(x^{(k)})(x - x^{(k)}) \\ &+ \frac{1}{2}(x - x^{(k)})^T \nabla^2 f(x^{(k)})(x - x^{(k)}) \end{aligned} \tag{6.39}$$

The minimum of $f(x)$ in the direction of $\Delta x^{(k)}$ is obtained by differentiating $f(x)$ with respect to each of the components of Δx and equating the resulting expressions to zero to give

$$\Delta x^{(k)} = -\left[\nabla^2 f(x^{(k)})\right]^{-1} \nabla f(x^{(k)}) \tag{6.40}$$

where

$$\left[\nabla^2 f(x^{(k)})\right]^{-1}$$

is the inverse of Hessian matrix $H(x(k))$ defined in Section 6.12 (the matrix of second partial derivatives of $f(x)$ with respect to x evaluated at $x(k)$).

Note that in Equation (6.40) a matrix inversion is required, and one must be very cautious to use a technique that guarantees a positive definite inverse, as will be explained subsequently. Many standard digital computer programs for matrix inversion are unsatisfactory in this respect. Also note that analytical second partial derivatives must be evaluated or approximated, which may not be practical in some instances. The criterion to guarantee convergence in Newton's method, assuming the function $f(x)$ is twice differentiable, is that the inverse of the Hessian matrix of the objective function should be positive definite.

The minimum of $f(x)$ in the direction of S^k is obtained by differentiating $f(x)$ with respect to each of the components of x and equating the resulting expressions to zero.

The new vector of design variables is the following

$$x^{(k)} = x^{(k-1)} + \alpha_k^* S^{(k)} \tag{6.41}$$

where k is the iteration number, $S^{(k)}$ is the search direction, α_k^* is a scalar multiplier determining the amount of change in x for this iteration.

From Equation (6.40) we get

$$S^{(k)} = -\left[H(x^{(k)})\right]^{-1} \nabla f(x^{(k)})$$ (6.42)

where

$$\left[H(x^{(k)})\right]^{-1}$$

is the inverse of Hessian matrix $H(x^{(k)})$.

Therefore Equation (6.42) provides a search direction to use in a one-dimensional search.

$H(x) > 0$ and if the objective function can be approximated reasonably well by a quadratic function, which is just the region in which steepest descent methods perform the poorest. Far from the minimum, steepest descent methods may be superior. One can conclude that a suitable combination of steepest descent and Newton's method should exhibit superior performance to either method alone.

We should provide not only function values and gradient information, but second-derivative matrix H as well. If the function being minimized is a true quadratic in the design variables, the use of this search direction with a move $\alpha = 1$ will provide the solution in only one iteration.

Principal difficulty with Newton's method is that the H matrix may be singular, or at least not positive definite as it is required to guarantee a solution for a minimum $f(x)$. The H matrix will be singular any time the objective is linear in one or more of the design variables. If the Hessian matrix has negative eigenvalues, this identifies a nonconvex problem. Another difficulty is, that the move predicted by the Newton's method may be so great as to cause oscillation in the solution. For this reason it is advisable to provide reasonable move limits at each iteration in order to avoid ill-conditioning.

In case we can calculate the matrix of second derivatives easily, Newton's method is almost always the best approach.

6.6.2 *Sequential quadratic programming (SQP)*

Sequential quadratic programming or SQP methods are the standard general purpose algorithms for solving smooth nonlinear optimization problems under the following assumptions:
  – The problem is not too big,
  – The functions and gradients can be evaluated with sufficiently high precision,
  – The problem is smooth and well-scaled.

The mathematical convergence and the numerical performance properties of SQP methods are very well understood now and have been published in so many papers that only a few can be mentioned here: e.g. Thanedar (1986) or Qian (1984) for a review. Theoretical convergence has been investigated by Powell (1978a, b), Schittkowski (1983), for example, and the numerical comparative studies of Schittkowski (1980) and Hock & Schittkowski (1981) show their superiority over other mathematical programming algorithms under the above assumptions.

The key idea is to approximate also second-order information to obtain a fast final convergence speed. Thus we define a quadratic approximation of the Lagrangian function $L(x, u)$ and an approximation of the Hessian matrix by a so-called quasi-Newton matrix.

Feasible sequential quadratic programming (FSQP)

FSQP is a set of FORTRAN subroutines for the minimization of the maximum of a set of smooth objective functions (possibly a single one) subject to general smooth constraints. If the initial guess provided by the user is infeasible for some inequality constraints or some linear equality constraint, the program first generates a feasible point for these constraints; subsequently the successive iterates generated by FSQP all satisfy these constraints. Nonlinear equality constraints are turned into inequality constraints (to be satisfied by all iterates) and the maximum of the objective functions is replaced by an exact penalty function which penalizes nonlinear equality constraint violations only (Zhou & Tits 1991, 1992, 1993).

The user has the option of either requiring that the (modified) objective function decreases at each iteration after feasibility for nonlinear inequality and linear constraints have been reached (monotone line search), or requiring a decrease within at most four iterations (nonmonotone line search). The user must provide subroutines that define the objective functions and constraint functions and may either provide subroutines to compute the gradients of these functions or require that FSQP estimate them by forward finite differences. FSQP implements two algorithms based on SQP, modified so as to generate feasible iterates. In the first one (monotone line search), a certain Armijo type arc search is used with the property that the step of one is eventually accepted, a requirement for superlinear convergence. In the second one the same effect is achieved by means of a (nonmonotone) search along a straight line.

The merit function used in both searches is the maximum of the objective functions if there is no nonlinear equality constraint. If the initial guess provided by the user is infeasible for nonlinear inequality constraints and linear constraints, FSQP first generates a point satisfying all these constraints by iterating on the problem of minimizing the maximum of these constraints. Then, using Mayne-Polak's scheme nonlinear equality constraints are turned into inequality constraints.

The resulting optimization problem therefore involves only linear constraints and nonlinear inequality constraints. Subsequently, the successive iterates generated by FSQP all satisfy these constraints. The user has the option of either requiring that the exact penalty function (the maximum value of the objective functions if without nonlinear equality constraints) decreases at each iteration after feasibility for original nonlinear inequality and linear constraints have been reached, or requiring a decrease within at most three iterations. He must provide subroutines that define the objective functions and constraint functions and may either provide subroutines to compute the gradients of these functions or require that FSQP estimate them by forward finite differences.

Thus, FSQP solves the original problem with nonlinear equality constraints by solving a modified optimization problem with only linear constraints and nonlinear inequality constraints. For the transformed problem, it implements algorithms that are described and analysed in refinements.

These algorithms are based on a SQP iteration, modified so as to generate feasible iterates. An Armijo-type line search is used to generate an initial feasible point when required. After obtaining feasibility, either 1. An Armijo-type line search may be used, yielding a monotone decrease of the objective function at each iteration, or 2. A nonmonotone line search and analysed, may be selected, forcing a decrease of the objective function within at most four iterations. In the monotone line search scheme, the SQP direction is first tilted if nonlinear constraints are present to yield a feasible direction, then possibly 'bent'' to ensure that close to a solution the step of one is accepted, a requirement for superlinear convergence.

The nonmonotone line search scheme achieves superlinear convergence with no bending of the search direction, thus avoiding function evaluations at auxiliary points and subsequent solution of an additional quadratic program. After turning nonlinear equality constraints into inequality constraints, these algorithms are used directly to solve the modified problems.

6.7 OPTIMALITY CRITERIA METHODS

Optimality criteria (OC) methods based on Kuhn-Tucker necessary conditions of optimality. The advantage of these techniques, that they are computationally very efficient. The disadvantages that they depend on the behaviour of the structure and the convergence is not always guaranteed (Khot & Berke 1984).

To solve the general formulation of a single-criterion nonlinear programming problem according to Equations (6.1) and (6.2), i.e.

$$\text{minimize } f(x) \quad x_1, x_2, ..., x_N$$

$$\text{subject to } g_j(x) \leq 0, \quad j = 1, 2, ..., P$$

where *f(x)* is a multivariable nonlinear function, $g_j(x)$ is nonlinear inequality constraints, we use the Lagrange multipliers.

We convert the inequality constraints by introducing Y_j parameters into equality ones:

$$g_j + Y_j^2 = 0$$

The Lagrange function is a follows:

$$L(x, \lambda_j, Y_j) = f(x) + \sum_{j=1}^{P} \lambda_j \left[g_j(x) + Y_j^2 \right] \tag{6.43}$$

To find the local minima of this function, the necessary conditions are as follows:

$$\frac{\partial L}{\partial x} = \nabla f(x) + \sum_{j=1}^{P} \lambda_j \nabla g_j(x) = 0 \tag{6.44}$$

$$\frac{\partial L}{\partial x} = g_j(x) + Y_j^2 = 0 \tag{6.45}$$

$$\frac{\partial L}{\partial x} = 2\lambda_j Y_j = 0 \tag{6.46}$$

Equations (6.44) and (6.45) show, if $g_j = 0$, i.e. the constraint is active, then $Y_j = 0$ and $\lambda_j \geq 0$. If the constraint is not active, then $g_j < 0$, $Y_j \neq 0$, and $\lambda_j \geq 0$. In conclu-

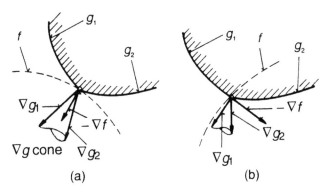

Figure 6.14. The Kuhn-Tucker optimality conditions if there is an optimum (a), or not (b).

sion, if the constraint is active, then $Y_j = 0$, so the Y_j elements are negligible and instead of Equations (6.44) and (6.45) one can use the following constraints

$$\lambda_j \geq 0 \quad \text{and} \quad \lambda_j g_j = 0 \tag{6.47}$$

The Kuhn-Tucker necessary conditions of optimality are as follows

$$\nabla f(x) = -\sum_{j=1}^{P} \lambda_j \nabla g_j(x)$$
$$\lambda_j \geq 0, \quad \lambda_j g_j = 0 \tag{6.48}$$

The geometrical meaning of the first condition is, that it is possible to get the gradient of the objective function from the linear combination of the gradients of constraints, i.e. in the optimum the gradient of the objective function is in the cone of the gradients of constraints (Fig. 6.14). In other words, *the contour of the objective function touches the feasible region in the optimum point*. This point can be global optimum, if the feasible region in convex. In the case of concave feasible region, point can be only local optimum.

If all constraints are active, the condition (Eq. 6.47) and the equations $g_j = 0$ give $n + p$ equalities for $n + p$ unknowns of x_i and λ_j.

Several solution methods have been proposed, because the solution depends on the type of objective function and constraints. Optimization of truss structures are the subject of main interest so far, although plates and shells have also been considered, connecting the OC methods to finite element techniques (Kiusalaas & 1977), Rozvany (1989).

6.8 DISCRETE OPTIMIZATION TECHNIQUES

In practical design, cross-sectional variables may be restricted to some discrete values. Such are the rolled steel members, which are produced in distinct sizes with unevenly spaced cross-sectional properties. In such cases the design variable is permitted to take on only one of a discrete set of available values. However, as discrete variables increase the computational time, the cross-sectional design variables are usually assumed to be continuous.

6.8.1 *Backtrack method*

The backtrack method is a combinatorial programming technique, solves nonlinear constrained function minimization problems by a systematic search procedure. The advantage of the technique is, that it uses only discrete variables, so the solution is usable. The general description of backtrack can be found in the works of Walker (1960), Golomb & Baumert (1965) and Bitner & Reingold (1975). This method was applied to welded girder design by Lewis (1968) and Annamalai (1970). Farkas & Szabó (1980) have used it for the minimum cost design of hybrid I-beams. An estimation procedure for efficiency of backtrack programming was proposed by Knuth (1975). In the book of Farkas (1984) the following problems have been solved by backtrack method: Welded I-beam subject to bending, compression struts of square hollow section, a tubular truss, hybrid I-beams with one welded splices on the flanges, welded box beams subject to bending and shear, a crane runway girder of asymmetrical I-section, closed press frames of welded box section.

The general formulation of a single-criterion nonlinear programming problem is the following:

$$\text{minimize } f(x) \qquad x_1, x_2, ..., x_N$$
$$\text{subject to } g_j(x) \leq 0 \qquad j = 1, 2, ..., P$$
$$h_i(x) = 0 \qquad i = P+1, ..., P+M$$

f(x) is a multivariable nonlinear function, *g_j(x)* and *h_i(x)* are nonlinear inequality and equality constraints. The equality constraints should be transfered to inequality ones to handle them by the program:

$$h_i(x) - \varepsilon \leq 0 \quad i = P+1, ..., P+M$$
$$h_i(x) - \varepsilon \geq 0 \tag{6.49}$$

ε is a given small number.

The algorithm is suitable to find optimum of those problems which are characterized by monotonically increasing or decreasing objective functions. Thus, the optimum solution can be found by decreasing the variables. Originally the procedure can find the minimum of the problem, starting from the maximum values.

6.8.1.1 *Interval halving procedure*
To find the optimum for a single variable many single variable search techniques are available. An efficient and suitable search method is the interval halving procedure to reduce the search time.

We assume that the objective function is monotonously decreasing, if the variables decrease. At the line search, when only one variable is changing, the aim is to find the minimum feasible value of the variable, starting from the maximum value.

The starting point, i.e. the maximum value, should satisfy the constraints. Second step is the investigation of the minimum value. If it satisfies the constraints, then the solution is found. If not, the region is divided into two subregions with the middle value. If the constraints are satisfied with the middle value, then the upper region is feasible, all points satisfy the constraints. In this case we should investigate the lower region, to find the border between the feasible and unfeasible regions.

Sign [means feasibility, sign means { unfeasibility. The halving procedure works as follows:

Assume, that the variable is a thickness given by the following series of discrete values:

$$6 \quad 8 \quad 10 \quad 12 \quad 15 \quad 18 \quad 20 \quad 25 \quad 30 \text{ mm}$$
$$\{..]$$

Furthermore assume that the maximum value is feasible, the minimum is unfeasible. If the middle value is feasible, the region to be investigated is as follows:

$$6 \quad 8 \quad 10 \quad 12 \quad 15$$
$$\{..]$$

In the upper part of the region one cannot find any solution, so it is possible only in the lower part. We can leave the upper region without any further calculations. Continuing with the middle point of the lower region, if it is unfeasible, then the remaining region is only one quarter of the original one, after two checkings.

$$10 \quad 12 \quad 15$$
$$\{.......................]$$

If the middle point is feasible, then it gives the solution.

$$12$$
$$]$$

The ratio of the number of total discrete points and checked discrete values is 9/4. If we have 1025 discrete values, then this ratio is much better, at the first halving step we can leave 512 discrete numbers without further investigations. The halving procedure stops, if the step length is less, that the distance between two discrete points. The step length should not be uniform between every discrete values, but for practical reasons we usually use a uniform value. The number of discrete values should be $2^k + 1$, where k is an integer number.

In the case of a completely general series the latter can be completed with the maximum values as follows:

Basic:	1	2	3	4	5	6	7	8	9
Completed:	4	6	8	10	12	14	16	16	16

At the backtrack method the variables are in a vector form $x = \{x_i\}^{\mathrm{T}}$ $(i = 1,..., n)$ for which the objective function $f(x)$ will be a minimum and which will also satisfy the design constraints $g(x) \geq 0$ $(j = 1,..., P)$. For the variables, series of discrete values are given in an increasing order. In special cases the series may be determined by and by the constant steps Δx_k between them. The flow chart for the backtrack method is given in Figure 6.15.

First a partial search is carried out for each variable and if all variations have been investigated, a backtrack is made and a new partial search is performed on the previous variable. If this variable is the first one: No variations have to be investigated (a number of backtracks have been made), then the process stops. The main phases of the calculation are as follows.

1. With a set of constant values of $x_{i,t}$ $(i = 2,..., n)$ the minimum $x_{i,m}$ value satisfy-

ing the design constraints is searched for. The interval halving method can be employed. This method can be employed if the constraints and the objective function are monotonous from the sense of variables.

2. As in the case of the first phase, the halving process is now used with constant values, and the minimum $x_{1,m}$ value, satisfying the design constraints is then determined.

3. The least value $x_{n,m}$ is calculated from the equation relating to the objective function $f(x)$

$$f(x_{1,m}, \ldots, x_{n,m}) = f_0$$

where f is the value of the cost function calculated by inserting the maximum **x**-values. Regarding the $x_{n,m}$ value, three cases may occur as follows.

3a. If we decrease x_{n-1} step-by step till it satisfies the constraints or till $x_{n,\min}$, the minimum values are reached. If all variations of the x_n value have been investigated, then the program jumps to the x_{n-1} and decreases it step-by step till x satisfies the constraints or till $x_{n-1,\min}$ are reached.

3b. If $x_{n,m} < x_{n,1}$, we backtrack to x_{n-1}.

3c. If $x_{n,m}$ does not satisfy the constraints, we backtrack to $x_{n-1,m}$. If the constraints are satisfied, we continue the calculation according to 3a.

The number of all possible variations is $\Pi_{i=1}^{n} ti$ where t_i is the number of discrete sizes for one variable. However, the method investigates only a relatively small number of these. Since the efficiency of the method depends on many factors (number of unknowns, series of discrete values, position of the optimum values in the series, complexity of the cost function and/or that of the design constraints), it is difficult to predict the run time. The main disadvantage of the method is, that the runtime increases exponentially, if we increase the number of unknowns.

We have made the program in C language modifying the procedure in the sense, that originally the program depended on the number of variables. All variables were computed by the halving procedure except the last one, which was computed from the objective function. The modified version is independent from the number of variables. The list of the program in ANSI C can be found in Appendix C. Advantage of the method is, that it gives discrete values, usually finds global minimum. The disadvantage of the method that it is useful only for few variables because of the long computation time.

6.8.1.2 *An example of using backtrack for solving a combinatorial problem*
A simple example to show the procedure in details is to place four queens on a four by four chess table not to beat each other.

The objective of the problem is to maximize the number of queens. The constraints are that the queens beat each other if they are at the same row, or column, or every row means one variable, one queen, because it is not possible to place two queens in the same row.

We can place the first queen, can find the place of the second queen, but no place for the third queen (Fig. 6.16a).

If there is no place for the third variable, we jump back to the second one and look for another place for it. Finding the next place, we can place the third variable too. Unfortunately there is no place for the fourth queen (Fig. 6.16b).

We have investigated all variations at the third variable, so we should jump back to the second variable, but there are no new possibilities, so should backtrack to the first variable and replace it. This case we can get a solution, which means, that we could place the maximum queens on the chess table, it is four and they don't beat each other, so satisfy the constraints (Fig. 6.16c).

There are more solutions like the following (Fig. 6.16d).

The discrete value of the method is that to place the queens on the chess table is possible only in the squares, it means that for a technical variable also a limited number of discrete values are given.

6.8.1.3 *An example of using backtrack for optimum design of a welded I-beam*

The problem is finding the minimum mass solution of a welded I-beam, which is simply supported, subject to bending moment and normal force (Fig. 6.17).

The unknowns are the height of web plate $h = x_1$, the thickness of web $t_w = x_2$, the area of the flange $A_f = x_3$.

The objective function is the mass of the structure. The span length is given, the material is also known (steel Fe 360), so the minimum mass is proportional to the minimum area of cross-section.

$$f(x) = x_1 x_2 + 2x_3 \tag{6.50}$$

The design constraints are as follows:

$g_1(x)$ is the normal stress constraint

$$\sigma_b + \sigma_c = \frac{M_b}{W_x} + \frac{N}{A} \le \frac{f_y}{\gamma_{M1}} \tag{6.51}$$

where M_b is the bending moment, N is the compression force, W_x is the section modulus, A is the cross-section area, f_y is the yield stress, γ_{M1} the partial safety factor.

$g_2(x)$ is the local buckling constraint approximation

$$\frac{h}{t_w} = 145.4 \sqrt{\frac{\left[1 + \left(\sigma_c / \sigma_b\right)\right]^2}{1 + 173\left(\sigma_c / \sigma_b\right)^2}} \tag{6.52}$$

Note, that in Eurocode 3 another calculation method is prescibed.

Data are as follows:

$$M_b = 320 \text{ kNm}, \quad N = 128 \text{ kN}, \quad f_y = 240 \text{ MPa}, \quad \gamma_{M1} = 1.2$$

Table 6.1. Upper and lower limits of the variables.

	Upper	Lower	Step length
h	740	660	20
t_w	9	5	1
A_f	2200	1400	100

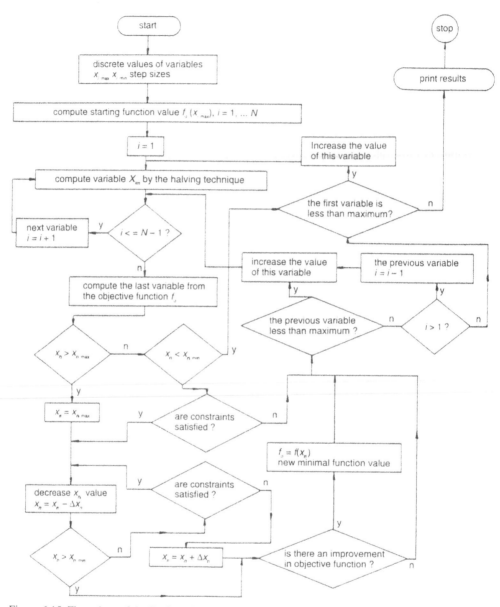

Figure 6.15. Flow chart of the Backtrack method.

The upper and lower limits and step length is shown in Table 6.1. The numerical calculation is given in Table 6.2.

During the procedure the minimum function value and its variable values are saved and at the end of the procedure the final results come from these saved values. The search is not a direct one. If the second values give the solution it will continue the search till there are no possibilities for backtracking, jumping back to the previous variable and try to reduce the objective function value.

Table 6.2. A numerical example of the backtrack method.

x_1	x_2	x_3	f	g_1	g_2	Remarks
740	9	2200	11060	+	+	$f_0 = 11060$ mm^2
660	9	2200	10340	+	+	x_{1min} feasible, halving procedure for x_2
660	5	2200	7700	+	−	
660	7	2200	9020	+	+	
660	6	2200	8360	+	+	
660	6	2200	8360	+	+	$x_3 = (11060 - 660*6)/2 = 3550$
660	6	2100	8160	+	+	$3550 > x_{3max} f_0 = 7960$
660	6	2000	7960	+	+	backtrack to x_2
660	6	1900	7760	−	+	
660	7	1600	7820	−	+	$x_3 = (7960 - 660*7)/2 = 1670$
660	8	1300				$1670 < x_{3max}$ backtrack to x_2
680	9	2200	10520	+	+	Halving procedure for x_2
680	5	2200	7800	+	−	
680	7	2200	9160	+	+	
680	6	2200	8480	+	+	
680	6	1900	7880	+	+	$x_3 = (7960 - 680*6)/2 = 1940$
680	6	1800		−	+	$f_0 = 7880$ backtrack to x_2
680	7	1500	7680	−	+	$x_3 = (7880 - 680*7)/2 = 1560$
680	8	1200	7760			$1560 < x_{3min}$ backtrack to x_1
700	9	2200	10700	+	+	Halving procedure for x_2
700	5	2200	7900	+	−	
700	7	2200	9300	+	+	
700	6	2200	8600	+	+	
700	6	1800	7800	+	+	$x_3 = (7880 - 700*6)/2 = 1840$
700	6	1700	7600	−	+	$f_0 = 7800$ backtrack to x_2
700	7	1400	7700	−	+	$x_3 = (7800 - 700*7)/2 = 1450$
700	8	1100				$1450 < x_{3min}$ backtrack to x_1
720	9	2200	10880	+	+	Halving procedure for x_2
720	5	2200	8000	+	−	
720	7	2200	9440	+	+	
720	6	2200	8720	+	−	
720	7	1300				$x_3 = (7800 - 720*7)/2 = 1380$
						$1380 < x_{3min}$ backtrack to x_1
740	9	2200	11060	+	+	Halving procedure for x_2
740	5	2200	8100	+	−	
740	7	2200	9580	+	+	
740	6	2200	8840	+	−	
740	7	1300				$x_3 = (7800 - 740*7)/2 = 1310$
						$1310 < x_{3min}$ no backtrack
						Results are $x_1 = 700$, $x_2 = 6$, $x_3 = 1800$

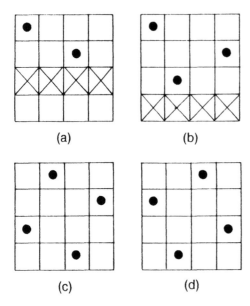

(a)   (b)

(c)   (d)

Figure 6.16. a) Place the first and second queens, b) Place the first and second and third queens, c and d) Solution for four queens.

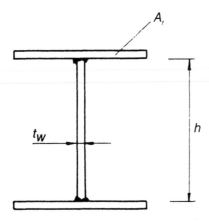

Figure 6.17. Section of welded I-beam.

6.8.2 *Discretization after continuous optimization*

To make the search more practicable it is advisable to use discrete member sizes. After continuous optimization a secondary search is necessary to find discrete optimum sizes in such a way, that not only the explicit and implicit constraints satisfied are but the merit function takes its minimum as well. It is assumed that the optimum discrete sizes are near to the optimal continuous ones (Jármai 1982).

Starting from the optimum continuous values, the secondary search chooses the nearest discrete sizes for each continuous size from the series of discrete values. The number of chosen discrete sizes for one continuous size can be two, three or more. The possible variations can be obtained using binary, ternary or larger systems. In our numerical example we use the binary system, two discrete sizes, upper and

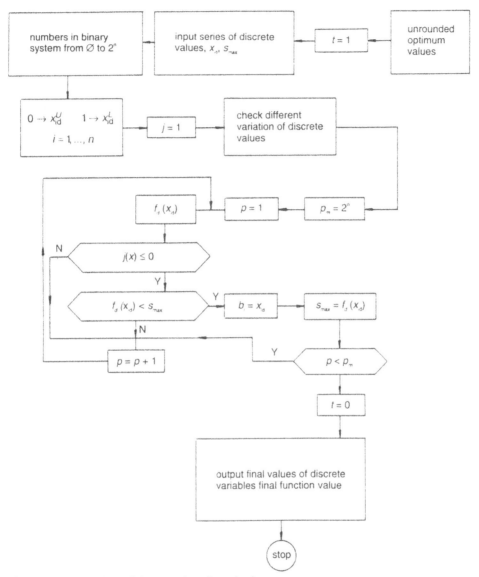

Figure 6.18. Flow chart of the secondary discretization.

lower, belonging to one continuous value. In a binary system the figure zero means the upper discrete size, the figure one means the lower one. The first $2n$ number in binary system gives the all possible variations. Each variation is tested, whether the explicit and implicit constraints are satisfied, and the optimal values minimizing the merit function are determined. The flow chart of the secondary discretization can be seen in Figure 6.18.

The unrounded optimum values of fourth variable are as follows: ♣ ♦ ♥ ♠

$$_1\text{Lower}\ldots\ldots\ldots\clubsuit\ldots\ldots\ldots\ldots{}_1\text{Upper}$$
$$_1\text{Lower}\ldots\ldots\blacklozenge\ldots\ldots\ldots\ldots{}_1\text{Upper}$$
$$_1\text{Lower}\ldots\ldots\ldots\ldots\heartsuit\ldots\ldots{}_1\text{Upper}$$
$$_1\text{Lower}\ldots\spadesuit\ldots\ldots\ldots\ldots{}_1\text{Upper}$$

The number 0000 means the lower discrete values of all variables, the number 1111 means the upper discrete values of all variables. The other numbers in the binary system are the variants of the possible discrete solution. That one of the tested variants is the solution, which gives the minimum objective function value.

6.9 SENSITIVITY ANALYSIS

In any design problem the optimum solution must be evaluated in terms of the sensitivity of the value of the criterion and the degree of satisfaction of the constraints to small perturbations in the starting point of the vector of in the independent variables, the optimal value of the same vector, and, especially, the values of the coefficients in the criterion and constraints. A significant degree of sensitivity in critical aspects of the design should lead to reevaluation of the criterion, process models, and/or source of the coefficients. Certainly if the coefficients are secured from databases or experiments in which random effects play a role, a high degree of sensitivity for a few variables can be very troublesome. Some of the interior penalty function codes avoid the introduction of sensitivity to slight variations in the problem but these codes inherently exhibit poor performance in solving problems.

The opposite side of the coin is that the criterion may be quite insensitive near the extremum to changes in the values of the variables. Consequently, premature termination may take place or, alternatively, the computation time and function evaluations may become excessive as the precision of the termination criteria are increased.

6.10 APPROXIMATION TECHNIQUES

Some of the mathematical programming procedures require linear or quadratic approximations of $f(x)$, $g(x)$, and $h(x)$. A linear, or first-order approximation of the objective function, $f(x)$, for example, can be made by a truncated Taylor series about $x^{(k)}$.

$$f(x) \approx f(x^{(k)}) + \nabla f(x^{(k)})(x - x^{(k)}) \tag{6.53}$$

A quadratic approximation of $f(x)$ can be made by neglecting the third- and higher-order terms in the Taylor series.

$$f(x) \approx f(x^{(k)}) + \nabla^T f(x^{(k)})(x - x^{(k)})$$
$$+ \frac{1}{2}(x - x^{(k)})^T \nabla^2 f(x^{(k)})(x - x^{(k)}) \tag{6.54}$$

where $\nabla^2 f(x^{(k)})$ is the Hessian matrix of $f(x)$, $H(x)$, that is, the square matrix of the second partial derivatives of $f(x)$ evaluated at $x^{(k)}$.

$$\nabla^2 f(x^{(k)}) = H(x^{(k)}) = \begin{bmatrix} \dfrac{\partial^2 f(x^{(k)})}{\partial x_1^2} & \cdots & \dfrac{\partial^2 f(x^{(k)})}{\partial x_1 \partial x_n} \\ \dfrac{\partial^2 f(x^{(k)})}{\partial x_n \partial x_1} & \cdots & \dfrac{\partial^2 f(x^{(k)})}{\partial x_n^2} \end{bmatrix} \qquad (6.55)$$

6.11 MULTIOBJECTIVE OPTIMIZATION

The first work on multiobjective (multicriterion, vector) optimization was presented by Pareto (1896). After at least fifty years, problems of multiobjective optimization were again considered by von Neumann & Morgenstern (1947). A relatively modern formulation of the multiobjective optimization problem was presented by Zadeh (1963). However, wider interest in optimization theory, operations research, and control theory was not taken until the late 1960s, and since then numerous papers have been published on the subject. Most are concerned with the theory and applications of multiobjective decision making from a general viewpoint, and few applications to engineering design can be found. Comprehensive bibliographies on multicriterion decision-making and related areas have been written by Stadler (1986, 1988), Cohon (1978) presenting the foundations of modern multiobjective optimization.

Almost all decisions are multiobjective. In complex engineering design problems there often exist several non-commensurable objectives which must be considered. Such a situation is formulated as a multiobjective optimization problem in which the designer's goal is to minimize and/or maximize several objective functions simultaneously to get compromise between them (Hwang & Masud 1979).

For decades engineers have employed a single measure, like costs, weights, or benefits, to determine an optimum.

For example design a simple beam with two supports at the end, the cost of the beam can be an objective, to be minimized, but also the rigidity should be maximal, or with another measure the deflection should be minimum. These two objectives: Cost and deflection are in conflict. A singleobjective solution cannot be a reasonable one. A compromise is necessary to reduce these two conflicting values as much as possible. Multiobjective programming techniques are the tools doing this. Multiobjective analysis and optimization represent a general philosophy of design. It puts the designer a more useful position of providing decision makers a set of good alternative solutions rather than a single optimal solution (Lounis & Cohn 1995).

The most popular criteria used in structural optimization are: Minimum mass or cost, maximum stiffness, minimum displacement at specific structural points, maximum frequency of free vibration and so on. These criteria are very often in conflict. In such cases, it is necessary to formulate a multicriteria optimization problem and look for the set of compromise solutions in the objective space. Next, the so-called preferred solution should be chosen taking into account an additional criterion or using a so-called global criterion like the utility function, distance function or hierarchical method (see Eschenauer et al. 1990, Jendo 1990, Koski 1984, Azarm & Eschenauer 1993).

A multicriteria optimization problem can be formulated as follows :
Find **x** such that

$$\mathbf{f(x^*)} = \text{opt } \mathbf{f(x)} \tag{6.56}$$

such that

$$g_j(\mathbf{x}) \geq 0 \quad j = 1,..., P$$
$$h_i(\mathbf{x}) = 0 \quad i = P,..., P + Q \tag{6.57}$$

where **x** is the vector of decision variables defined in n-dimensional Euclidean space and **f(x)** is a vector function defined in *r*-dimensional Euclidean space. $g_j(x)$ and $h_i(x)$ are inequality and equality constraints.

The solutions of this problem are the Pareto optima. The definition of this optimum is based upon the intuitive conviction that the point x^* is chosen as the optimal, if no objective can be improved without worsening at least one other objective.

6.12 DESCRIPTION OF THE METHODS OF MULTIOBJECTIVE OPTIMIZATION

6.12.1 *Method of objective weighting*

6.12.1.1 *Weighting objectives method*
The pure weighting method means to add all the objective functions together using different weighting coefficients for each. It means, that we transform our multicriteria optimization problem to a scalar one by creating one function of the form:

$$\mathbf{f(x)} = \sum_{i=1}^{r} w_i f_i(\mathbf{x}) \quad \text{where} \quad w_i \geq 0 \quad \text{and} \quad \sum_{i=1}^{r} w_i = 1 \tag{6.58}$$

If we change the weighting coefficients result of solving this model can vary significantly, and depends greatly from the nominal value of the different objective functions.

6.12.1.2 *Normalized objectives method*
The normalized objectives method solves the problem of the pure weighting method e.g. at the pure weighting method, the weighting coefficients do not reflect proportionally the relative importance of the objective, because of the great difference on the nominal value of the objective functions. At the normalized weighting method w_i reflect closely the importance of objectives.

$$\mathbf{f(x)} = \sum_{i=1}^{r} w_i f_i(\mathbf{x}) / f_i^0 \quad \text{where} \quad w_i \geq 0 \quad \text{and} \quad \sum_{i=1}^{r} w_i = 1 \tag{6.59}$$

The condition $f_i^0 \neq 0$ is assumed.

6.12.2 *Method of distance functions*

Let f^0 be the ideal solution that simultaneously yields minimum values for all criteria. Such a solution does not exist but is introduced in compromise programming as a target or a goal to approach, although impossible to reach (perfection is impossible).

In compromise programming, the 'best' or satisfying solution is defined as one that minimizes the distance from the set of nondominated solutions to the ideal solution. The criterion used in compromise programming is the minimization of the normalized deviation from the ideal solution f^0 measured by the family of L_p metrics defined in several different forms.

This family of L_p metrics indicates how close the satisfying solution is to the ideal solution, and represents the feasible set. In this chapter, the satisfying solutions are determined for two particular values of p, namely, $p = 2$ and $p = \infty$ (which correspond to the minimization of the Euclidean and maximum distances, respectively), and are given below. For the case $p = \infty$, the largest deviation is the criterion of comparison and is referred to as min-max criterion.

6.12.2.1 *Global criterion method type I*
Global criterion method means that a function which describes a global criterion is a measure of closeness the solution to the ideal vector of f^0. The common form of this function is:

$$f(x) = \sum_{i=1}^{r} [(f_i^0 - f_i(x)) / f_i^0]^P \qquad P = 1, 2, 3,... \qquad (6.60)$$

It is suggested to use $P = 2$, but other values of P such as 1, 3, 4, etc. can be used. Naturally the solution obtained will differ greatly according to the value of P chosen, $P = 1$ means a linear correlation, $P = 2$ a quadratic one, etc.

6.12.2.2 *Global criterion method type II*
The deviations in the absolute sense are as follows:

$$L_P(f) = \left[\sum_{i=1}^{k} \left| f_i^0 - f_i(x) \right|^P \right]^{1/P} \qquad 1 \leq P \leq \infty \qquad (6.61)$$

if

$$P = 1 \qquad L_P(f) = \sum_{i=1}^{k} \left| f_i^0 - f_i(x) \right| \qquad (6.62)$$

if

$$P = 2 \qquad L_P(f) = \left[\sum_{i=1}^{k} \left| f_i^0 - f_i(x) \right|^2 \right]^{1/2} \qquad \text{Euclidean metric} \qquad (6.63)$$

if

$$P = \infty \qquad L_P(f) = \max \left| f_i^0 - f_i(x) \right| \qquad \text{Chebysev metric } i = 1,..., k \qquad (6.64)$$

6.12.2.3 *Global criterion method type III*

Instead of deviations in the absolute sense it is recommended to use relative deviations such as

$$L_P(f) = \left[\sum_{i=1}^{r} \left| \frac{f_i^0 - f_i(x)}{f_i^0} \right|^P \right]^{1/P} \qquad 1 \leq P \leq \infty \tag{6.65}$$

In this case the *P* has a larger set.

6.12.3 *Min-max method*

At the min-max method the maximum loss of the collective objective will be minimized. The min-max optimum compares relative deviations from the separately reached minima. The relative deviation can be calculated from

$$z_i'(x) = \frac{|f_i(x) - f_i^0|}{|f_i^0|} \quad \text{or} \quad z_i''(x) = \frac{|f_i(x) - f_i^0|}{|f_i(x)|} \tag{6.66}$$

If we know the extremes of the objective functions which can be obtained by solving the optimization problems for each criterion separately, the desirable solution is the one which gives the smallest values of the increments of all the objective functions. The point x^* may be called the best compromise solution considering all the objective functions simultaneously and on equal terms of importance.

$$z_i(x) = \max \left\{ z_i'(x), \ z_i''(x) \right\} \quad i \in I \tag{6.67}$$

$$\mu(x^*) = \min \max \left\{ z_i(x) \right\} \qquad x \in X \ \ i \in I \tag{6.68}$$

where *X* is the feasible region.

6.12.4 *Constrained method*

The basis of these methods is the transformation of the vector optimization problem into a sequence of single objective optimization problems by retaining one selected objective as the primary criterion to be optimized and treating the remaining criteria as some predetermined constraints. These constants are then altered within their defined ranges, and the subset of Pareto optima is systematically generated. This approach has gained wide acceptance because it is more practical and rational than the weighting objectives method, if there is a dominant objective.

Minimize

$$f_L(x) \tag{6.69}$$

subject to the constraints

$$g_j(x) \geq 0 \qquad j = 1,..., M$$
$$f_i(x) \geq h_i \qquad i = 2,..., Q \tag{6.70}$$

and side bounds on the design variables as

$$x_i^L \leq x_i \leq x_i^U \quad i = 1,..., N \tag{6.71}$$

where h_i are the parametrically varied target levels of Q objective functions. Each vector, or constraint set, H of the various h_i, will produce one Pareto solution. As in the weighting method, many different combinations of values for each h_i, must be examined in turn to generate the entire Pareto set. The constraint method provides direct control of the generation of members of the Pareto set and generally provides an efficient method for defining the shape of the Pareto set. It should be noted that the constraint method does not recognize a weak Pareto optima.

After transformation of the vector optimization problem into a scalar optimization problem, the latter may be solved using some appropriate mathematical programming techniques.

6.12.5 *Hybrid methods*

6.12.5.1 *Weighting global criterion method*
The weighting global criterion method is made, by introducing weighting parameters, one could get a great number of Pareto optima with Equation (6.65) (Jármai 1989). If we choose $P = 2$, which means the Euclidean distance between Pareto optimum and ideal solution Jármai (1989a). The coordinates of this distance are weighted by the parameters as follows:

$$L_P(f) = \left[\sum_{i=1}^{r} w_i \left| \frac{f_i^0 - f_i(x)}{f_i^0} \right|^2 \right]^{1/2} \tag{6.72}$$

where P is the dimension of the function space, x indicates the design variables and X the constraint set, r is the number of objective functions, f^0 is the optimum of the ith objective function, and w_i are the weighting factors.

The solution obtained by minimizing Eq. (6.72) differs greatly depending on the value of P chosen.

6.12.5.2 *Weighting min-max method*
The weighting min-max method one gets combining the min-max approach with the weighting method, a desired representation of Pareto optimal solutions can be obtained

$$z_i(x) = \max \left\{ w_i z_i'(x), \ w_i z_i''(x) \right\} \quad i \in I \tag{6.73}$$

The weighting coefficients w_i reflect exactly the priority of the criteria, the relative importance of it. We can get a distributed subset of Pareto optimal solutions.

6.12.6 *Selection of the 'best' solution*

Once a subset of Pareto optima has been generated, the designer has to make an important decision concerning the selection of the best solution from this subset. The selection is not obvious when several conflicting criteria are considered but may be made subjectively by giving preference to one criterion over the others.

CHAPTER 7

Expert systems

7.1 ARTIFICIAL INTELLIGENCE

In 1950 Alan Turing gave the following definition for machine intelligence: A machine can be considered intelligent only if a human being communicating with the machine from a distance through a teletype cannot recognize whether he is communicating with a machine or a human being. No computer at present can pass this so-called Turing test. In a broad sense, artificial intelligence (AI) is a branch of computer science concerned with making computers act like human beings.

The emerging fields of AI and knowledge engineering offer means to carry out qualitative reasoning on computers. Advanced programs that can solve a variety of new problems based on stored knowledge without being reprogrammed, are called knowledge-based systems. If their level of competence approaches that of human experts, they become expert systems, which is the popular name for all knowledge systems.

AI techniques provide powerful symbolic computation and reasoning facilities that accommodate intuitive knowledge used by experienced designer. AI techniques, knowledge engineering in particular, can be used in conjunction with numerical programs to serve as an interface between the alternatives, constraints and the designer. AI should be used in the following context:

– To track the available design alternatives and relevant constraints and to infer candidate modifications in order to improve the design,

– To observe the relationship – intuitive or numerical – between specifications and decision variables, and give advice on how to formulate the problem for optimization, in particular, to identify the limiting constraints and specifications.

Computer programs using AI techniques to assist people in solving difficult problems involving knowledge, heuristics, and decision-making are called Expert Systems (ES), intelligent systems, or smart systems. An ES is an 'intelligent' interactive computer program that can play the role of a human expert by using heuristic knowledge or rules of thumb. The heuristics are usually accumulated by a human expert over a number of years (Dym & Lewitt 1991).

During the early days of AI, there was a considerable hope of simulating the learning ability of human beings by machines. The topic was dropped partly due to hardware limitations, but with the recent development of powerful supercomputers and new parallel architectures, machine learning is again being pursued actively by

AI researchers (Forsyth & Rada 1986). The promising area of neural networks shows the possibilities of learning (Hajela & Berke 1991).

In the sixties, AI researchers try to model the human thinking process by developing general problem solvers and general-purpose programs for solving broad classes of problem. In the seventies, the main focus concentrated to the knowledge representation and efficient search techniques. In the mid-seventies AI researchers realized that the problem-solving ability of humans lies basically in their knowledge of a particular domain and not so much in the inferential mechanism they use: i.e. the computer program should contain a large domain-specific knowledge base in order to be able to solve complex real word problems and thus become intelligent (Waterman 1986).

7.2 EXPERT SYSTEMS

Civil and mechanical engineering system design is a complex process that involves creative thinking, use of intuitive knowledge and quantitative analysis. The past several decades have shown the development of digital computers and of various, mostly quantitative, computational aids such as finite element techniques, optimization programs, dynamic simulators, etc.

Although these programs greatly enhance the designer's abilities to carry out quantitative analyses, they are limited to the quantitative aspects of design.

AI techniques, such as ES-s are the best utilized in identifying and evaluating design alternatives and their relevant constraints while leaving the important design decisions to the human designer. The emerging field of ES and knowledge engineering offer means to carry out qualitative reasoning on computers. There were some attends to connect the ES-s and structural optimization (Adeli 1988). Two of them are an ES for finding the optimum geometry of steel bridges (Balasubramanyan 1990) and for cranes (Hanna 1996).

We are all experts in many things of life – no one could survive without having a great deal of this day-to-day expertise. The essence of intuitive decision making is that it is done rapidly by the unconscious mind, using a vast store of previous experience and knowledge, with no formulation of reasons, logic or background knowledge. It is common that there will be intermediate conclusions, and after these are reached, the expert may ask for additional evidence before proceeding to the next stage.

A typical engineering example would be an ES for material selection. In the first stage the expert might select a broad class of material, say steel. Additional questions would then be asked about aspects of the application, and the expert would then go on to decide what kind of steel grade, what heat treatment, and so on. It thus uses hierarchical concepts, going from the more general to the more specific.

It is very difficult to extract rules which only exist in the unconscious mind, and actually in an unknown format. Procedures are available which could be considered as a kind of machine learning, in which the rules are generated by an analysis of examples for which an expert has given solutions.

Hayward (1985) described the difficulties in defining ES-s including the exploratory nature of the ES development techniques. The main problem is the translation of the expert's knowledge into a computable form. While some classic ES-s such as DENDRAL solve the problem of combinatorial explosion using heuristics, some other early ES-s such as MYCIN, PROSPECTOR, and XCON do not handle a large search problem but use specialized knowledge in their respective domains.

There is also a fundamental difference between the fully implemented ES-s which are mainly based on heuristic knowledge and the ES-s for most engineering applications. Any significant engineering ES probably must rely on well-established theories and principles requiring numerical computations in addition to heuristics. Thus, interaction of numerical data processing and AI technologies becomes an imperative consideration for developing ES-s in most engineering domains (Gero 1987).

There are some side benefits from developing ES-s and performing research in this area:
– Understanding the problem-solving ability and the reasoning process of human experts,
– Accumulation and codification of knowledge in a particular domain,
– Better articulation, representation of knowledge.

In Chapter 6 both single- multiobjective optimization techniques were described. The connection between single- and multiobjective optimization made it possible in the structural optimization to form a decision support system (DSS): i.e. producing a great number of optima to choose the solution from. In the multiobjective optimization several so called weighting coefficients serve for the designer to give relative importance of the objective functions (Jármai 1989, 1990). The DSS and the ES are close together, but it is necessary to build an inference engine. The key concept in our approach is to give the user control of important design decisions.

Therefore, our approach in applying AI to engineering design is to use AI techniques for keeping track of all design alternatives and constraints, for evaluating the performance of the proposed design by means of a numerical model, and for helping to formulate the optimization problem.

Using the ES the human designer evaluates the information and advices given by the computer, assesses whether significant constraints or alternatives have been overlooked, or violated, decides on alternatives, and makes relevant design decisions.

7.3 COMPARISON OF CONVENTIONAL PROGRAMS AND EXPERT SYSTEMS

The following differences may be found between traditional computer programs and ES-s (Adeli 1986):
– ES-s are knowledge-intensive programs,
– The output of an ES can be qualitative rather than quantitative,
– ES-s can use heuristics in a specific domain of knowledge in order to improve the efficiency of search,
– the knowledge base used in an ES is usually separated and independent from the methods for applying the knowledge to the current problem,
– ES-s can provide advice, answer questions, and justify their conclusions,
– ES-s are usually highly interactive.

A program that is procedural in nature can programmed in a conventional computer language like Basic, Fortran, C, or Pascal. For ES Lisp, Prolog languages and also the different own languages of expert shells are very useful.

7.4 ARCHITECTURE OF AN EXPERT SYSTEM

An ES has three main components as follows:

1. *Knowledge base.* This is the collection of information available in a particular domain. The knowledge base may consist of well-established and documented definitions, facts, and rules, as well as judgemental information, and heuristics,

2. *Inference engine* (inference mechanism or reasoning mechanism). It controls the reasoning strategy of the ES by making assertions, hypotheses, and conclusions. In rule-based systems the inference mechanism determines the order on which rules should be invoked and resolves any conflict among the rules when several rules are satisfied,

3. *User interface.* An important part of the ES for making the communication more easy between the user and the computer. An intelligent user interface includes explanation facility, answers to questions and justifies answers (e.g. why a specific conclusion or recommendation is made), help facility, which help and guides the user how to use the system effectively and easily, uses windows, graphics, helps to form the knowledge base, it can have debugging facility, etc.

7.5 ADVANTAGES OF EXPERT SYSTEMS (Andriole 1985, Adeli 1986)

– The knowledge base can be separated from the inference mechanism, it means that at a general system with one inference mechanism an ES can be developed for different types of applications simply by changing the knowledge base.

– Knowledge is more explicit, accessible, and expandable. There is similarity between ES-s and the human reasoning process. The human mind absorbs new information without disturbing the knowledge already stored in the brain or affecting the way in which it processes information. The same argument applies to modify information already stored in the brain.

– The knowledge base can be gradually and incrementally developed over an extended period of time. The modularity of the system allows continuous expansion and refinement of the knowledge base. In rule-based systems, for example, any rule can be deleted or new rules can be added independently.

– An ES can explain its behaviour through an explanation facility.

– An ES can check the consistency of its knowledge entities or rules and point out the faulty ones through a debugging facility.

– An ES uses a systematic approach for finding the answer to the problem.

– The use of ES-s results in a more effective, accurate, and consistent distribution of expertise.

7.6 CAPABILITIES OF EXPERT SYSTEMS

Depending on the application, an ES can perform eleven type of projects as follows:

1. *Selection.* Choose the most appropriate analysis methodology, tool, chemical process catalyst, tunnelling method, weld test procedure, and so on, from a finite set of alternatives,

2. *Interpretation.* A system describes a situation based on its interpretation of incoming data,

3. *Prediction.* Select the most likely future values from a known set of possible values for some important attributes of a system. Application would be e.g. an ES for weather forecasting,

4. *Diagnosis.* Locates or identifies malfunction within an electronic, biological or other system. Most diagnosis systems also select a repair or recommendation strategy to correct the cause of the failure or malfunction,

5. *Design.* In design problems there is no well-established list of all possible outcomes. Instead, the system creates a recommendation, depending upon observations and user needs. In the more complex design problems there can be incompatible constraints. This is often called a conflict resolution system,

6. *Planning.* A system to explore possible future actions to produce a series of plausible steps leading to a desired goal,

7. *Monitoring.* Application is one in which an ES examines real-time data from a large industrial system and watches for developing malfunctions,

8. *Debugging.* The system is an interactive aid to a designer to test the design to determine if it would work as planned, and if not, the system would isolate the fault and suggest a design change to correct it,

9. *Repair.* The system combines the activities of diagnostic and planning systems to create and execute plans for repairing faulty systems,

10. *Instruction.* The system plays the role of teacher to a student,

11. *Control.* The system is one which is actually planned in control of a complex process such as air traffic control.

7.7 INFERENCE ENGINE

The majority of knowledge-based ES-s developed so far is based on the production system paradigm. The inference engine stands between the user and the knowledge base. The inference mechanism or the control strategy is the heart of an ES. The built in inference engine applies the inference and control strategy, guides the knowledge system, as it uses the facts and rules stored in its knowledge base and the information it acquires from the user. The inference mechanism fires rules according to its built-in reasoning process. The two main inference mechanisms are forward-chaining and backward-chaining.

– Backward-chaining (consequent reasoning and goal-driven strategy),

– Forward-chaining (antecedent reasoning and data-driven control strategy),

– A combination of backward-chaining and forward-chaining (the hybrid approach).

Backward-chaining starts with the goal and backs through the rules looking for facts with establish the goal. Forward-chaining is the reverse of backward-chaining. It begins with facts and proceeds to fire rules to see where they lead. Backward-chaining chooses among pre-existing goals, solutions. Forward-chaining often assembles new solutions. This makes it a useful technique for problems that don't have a fixed set of possible answers.

Forward-chaining systems are not creative the same way that people are. When people come up with unique solutions to problems, they often bring ideas from other disciplines. ES-s do not have this ability, their knowledge is limited to a single, narrow domain. However they can be creative in many modest ways, constructing unique, new solutions to fit the resources and constraints of a specific problem. If the goal state is not known and has to be constructed or the number of possible outcomes is large then the forward-chaining mechanism is often recommended. Complex planning problems can be tackled by this approach.

In backward-chaining systems, the inference engine has a great deal of authority. It dictates most of the system's flow, setting up subgoals, applying rules and generating queries. The knowledge engineer's main responsibility is to put rules into the system, trusting the inference engine to use them wisely in order to solve an overall goal. If the values of goal state are known and their number is small then backward-chaining seems to be quite efficient. Backward-chaining is often employed in diagnostic ES-s.

In forward-chaining the developer has a greater responsibility for looking after issues of control. This is because forward-chaining inference engine behaves more unpredictably than backward one and thus requires more supervision.

Hybrid approach can be utilized through the use of the 'blackboard' environment. The blackboard will keep track of the simultaneous application of forward and backward reasoning chains.

7.8 STEPS FOR THE DEVELOPMENT OF EXPERT SYSTEMS

The major steps in development of an ES are (Harmon & King 1985):
 – Selection of an ES programming language, environment or shell,
 – Selection of AI techniques for representation and inference mechanism,
 – Analysis, acquisition, and conceptualization of the knowledge to be included in the knowledge base,
 – Formalization and development of the knowledge base,
 – Development of a prototype system using the knowledge base and AI tools,
 – Evaluation, review, and expansion of the ES-s,
 – Refinement of the user interface, ie. help facilities, etc.,
 – Maintenance and updating of the system.

The traditional way of developing an ES assumes that a knowledge engineer co-operates with one or a group of domain expert(s) knowledgeable about the particular domain (Hayes-Roth et al. 1983). The knowledge engineer must first make himself or herself familiar with the application domain and then try to acquire, compile, organize and formalize the domain expert's knowledge mostly through interviews.

The conventional knowledge engineer with a particular engineering domain may not be achieved easily. Explaining the knowledge of an engineering domain to a knowledge engineer can be difficult. Thus, it appears that most of the engineering ES-s will be developed by engineers knowledgeable in their domain as well as ES development tools and techniques. These engineers will substantiate and complement the knowledge base of the ES by consulting other highly knowledgeable domain experts.

7.9 KNOWLEDGE REPRESENTATION

The two fundamentally different approaches to knowledge representation are the procedural and declarative representation. Procedural representation is commonly used in traditional algorithmic programming and very efficient. In this type of representation, knowledge is context-dependent and embedded in the code. The disadvantage of this technique is, that it is difficult to modify the knowledge.

In declarative representation, knowledge is encoded as data and is therefore more understandable, easier to modify, and context independent. While semantics in procedural representation are distributed over the code, in declarative representation they are collected in one place.

The advantages of declarative knowledge representation are the following:
- Ease of comprehension,
- Ease of modification,
- Context independence,
- Semantic transparency.

These characteristics are essential in ES-s. Thus, ES-s usually use declarative knowledge representation. In engineering applications where substantial numerical computation is involved, a hybrid procedural-declarative knowledge representation appears to be the best solution.

Several approaches for declarative representation of knowledge are available in the AI literature. The major ones are as follows:
- Formal methods based on the predicate calculus and mathematical logic,
- Semantic networks,
- Semantic triples (object-attribute-value triplets),
- Rule-based or production systems,
- Frames consisting of generic data structures in predefined information categories called slots.

7.9.1 *Semantic networks, object-oriented systems (Garret 1990)*

Semantic networks consist of a collection of nodes for representation of concepts, objects, events, etc., and links for connecting the nodes and characterizing their interrelationship. An advantage of this representation method is its flexibility which means new nodes and links may be added whenever needed. Another characteristic of semantic networks is inheritance. That is, each node can inherit the characteristics of its connected nodes.

7.9.2 *Object-attribute-value triplets*

Semantic triples or object-attribute-value (O-A-V) triplet represent a special case of semantic networks in which there are only three types of nodes, i.e. objects, attributes. and values. Objects can be physical or conceptual entities. Attributes are general properties or features of objects. The value determines the particular character of an attribute in a specified situation.

7.9.3 *Production rules*

The production system has been the most favourable representation approach for building computer ES-s. A production system is a collection of rules, which consist of an IF part and a THEN part or antecendent-consequent or situation-action parts.

RULE N
IF [{antecendent 1} ....{antecendent n}]
THEN [{consequent 1 with certainty c_1}.......................................
        {consequent n with certainty c_n}]

The rule number is a unique number for identifying the rule. The value of this number does not specify the order of application of the rule. Each rule represents an independent piece of knowledge. Antecedents can be considered as patterns and consequents as conclusions or actions to be taken. The antecedent part of a rule is matched with the content of the working memory. When all the conditions in the antecedent part are satisfied the rule is fired. Certainty factors will be discussed in Section 7.9.5.

7.9.4 *Frames*

Frame systems are suitable for more complex representation of knowledge. A frame consists of a number of attributes, called slots, in which different characteristics and features of an object or a piece of information are described. Slots can store values. They may contain default values, pointers to other frames, or procedures. The procedures may determine the values of slots. This is called procedural attachment.

7.9.5 *Certainty factors*

Various methods have been used to deal with uncertain or incomplete information in the knowledge base. Certainty factors were first used in MYCIN, a medical ES. Certainty factors are attached to rules in rule-based systems. Each rule can contain a certainty factor usually in the range −1 to 1 (or 0-1 or 0-100). Certainty factors indicate the level of confidence in a piece of information. They are simply informal measures of confidence and not probabilities. Simple rules are used to combine the uncertainties in various pieces of information. In MYCIN, for example, the following equations are used (Buchanan & Shortlife 1984):

$$c_f(a,b) = \begin{cases} c_f(a)+c_f(b)-c_f(a)c_f(b) & \text{if} \quad c_f(a),c_f(b)>0 \\ \dfrac{c_f(a)+c_f(b)}{1-\min\left[\left|c_f(a)\right|,\left|c_f(b)\right|\right]} & \text{if} \quad c_f(a) \text{ or } c_f(b)<0 \\ -c_f(-a,-b) & \text{if} \quad c_f(a) \text{ and } c_f(b)<0 \end{cases} \quad (7.1)$$

The range of certainty factors in the above equations is for total belief to 1 for total disbelief to 0. There is a similarity between the above rule and the probability of joint events in the probability theory. The certainty factor is increased with the number of pieces of evidence.

7.10 KNOWLEDGE ACQUISITION

In an engineering problem, knowledge can be represented by well-established equations, graphs, tables of data, algorithmic analysis procedures, and experiential knowledge. Various types of experiential knowledge may be recognized such as the steps to be followed for solving a problem, use of the right equation or data, and use of the proper application programs. Considering the increasing availability of powerful ES development environments or shells, knowledge acquisition and encoding is appearing as the most time-consuming part of developing an ES.

In developing an ES for an engineering problem, knowledge may be acquired through various means:
– Technical literature (books, manuals, papers, etc.),
– Interviews with domain experts. This can be in the form of questions and answers and example problem-solving sessions. The two methods complement each other. In the latter case, the expert will be asked to solve example cases,
– Questionnaires sent to experts,
– Experimentation. If the necessary knowledge for solving a particular problem is obtained it can be used in a knowledge base for solving similar problems in the future.

Knowledge acquisition through machine experimentation appears to be useful in certain complex domains of engineering such as structural optimization.

7.11 EXPERT SYSTEM SHELLS AND ES-S FOR STRUCTURAL DESIGN

A great number of expert shells exist to help the developing of an ES. Some of them remained research tools, some of them made for workstations and PC-s (Harmon & Sawyer 1990, Adeli 1988).
– *Early research tools.* EMYCIN (diagnosis of deseases), KAS (Knowledge Acquition System for diagnosis and classification).
– *For workstations.* EXPERT (written in FORTRAN), ROSIE (written in InterLisp), S1 (written in Lisp), KEE (Knowledge Engineering Enviroment in Lisp), ART (Automated Reasoning Tool, Inference Corporation in Lisp), KEE (Intellicorp), ESE.
– *For PC-s.* EXPERT-EASE (Expert Software International in Pascal), M1 (Technoledge Inc. in Prolog), RuleMaster (Radian Corporation), INSIGHT 2+ (Level Five Research in Pascal), 1st CLASS (written in Pascal), Personal Consultant Easy (Texas Instruments), Level 5 Object (Level Five Research), Kappa PC, Intelligence Compiler (Intelligence Ware Inc.), Symbologic Adept (Symbologic Corporation), GURU (Micro Data Base Systems), Goldworks III (Goldworks Inc.).
– *ES-s for structural design.* SACON (interact MARC finite element program),

SSPG (for stiffened plate girders), CDA, SPEX, GEPSE (General Engineering Problem Solving Environment in C), BTEXPERT (optimum design of bridge trusses), HI-RISE (high rise building design).

7.12 OVERVIEW ON PERSONAL CONSULTANT

Personal Consultant is an EMYCIN-like program developed by Texas Instruments to run on PC-s. Facts are represented as object-attribute-value triplets with accompanying confidence factors. Production rules represent heuristic knowledge. Personal Consultant can build systems of up to about 400 rules. A rule tests the value of an O-A-V fact and concludes about other facts. The inference engine is a simple back-chainer. As with EMYCIN, one states a goal attribute and whenever a consultation system runs, the goal provides the starting point for backward-chaining.

Control is governed primarily by the order of clauses in the rules. Uncertain information is marked by confidence factors ranging from 0 to 100. Personal Consultant accepts an unknown as an answer to its questions and continues to reason with available information. There are explanation facilities in the program as well as trace functions for knowledge base debugging. Personal Consultant uses questions to prompt the designer to enter the initial information into a knowledge base. The tool provides several programming aids for debugging.

Personal Consultant is implemented in IQLISP. Sources of data can be other language programs or procedures such as FORTRAN, C, C++, data bases such as DBASE, LOTUS. The program has some graphics functions as well (DR HALO). The tool uses an Abbreviated Rule Language, (ARL) to write the rules (Personal Consultant Easy 1987). The second author could use Personal Consultant Easy during his stay in Chalmers University of Technology, Gothenburg and acknowledges this help.

7.13 OVERVIEW OF LEVEL 5 OBJECT

LO5 is an object-oriented ES development and delivery environment. It provides an interactive, windows-based user interface integrated with Production Rule Language (PRL), the development language used to create L5O knowledge bases. The PRL Syntax Section provides syntax diagrams to follow logically when writing a knowledge base. System classes are automatically built by L5O when a new knowledge base is created, thereby providing built-in logic and object tools. The developer can use system classes in their default states or customize them. In this way, the developer can control devices, files, database interactions and the inferencing and windowing environments.

The most remarkable tools of LO5 are:
– Object oriented programming (OOP),
– Relational database handling (RDB),
– Computer aided software engineering (CASE),
– Graphical development system.

The most remarkable tools of LO5 for IBM compatible PCs are:
– Microsoft Windows,
– Programming with an object-oriented language (Borland C++),
– Direct connection with dBase,
– Direct connection with the fourth generation FOCUS data handling system, offers means to carry out qualitative reasoning on computers.

Using LO5 there are two ways of developing programs: they can be generated either by word processors or in the developing environment. Taking these capabilities into account, L5O was found suitable for development of ES-s for structural engineering.

As an application our aim was to develop an ES, which is able to find the optimum version of *the welded main girder of overhead travelling crane* due to different loading, steel grades and design codes (Figs 7.1 and 7.2). The main girder can be an I-beam (the application of the I-beam is limited to hanging crane types), box beam, can be a stiffened or an unstiffened type. Another example is to find the optimum version of *belt-conveyor bridges* due to different geometry, loading, steel grades and design codes. The truss structures can be constructed with four or three chords. The belt-conveyor can be placed on or in the bridge. Instead of a truss structure a tube or a stiffened shell can be used as the main girder.

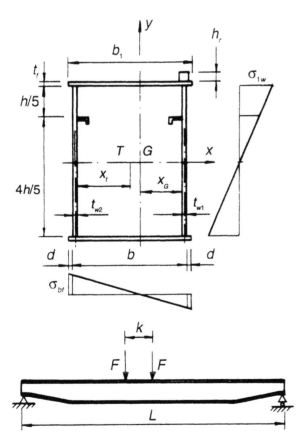

Figure 7.1. Cross section of the welded main girder of an overhead travelling crane.

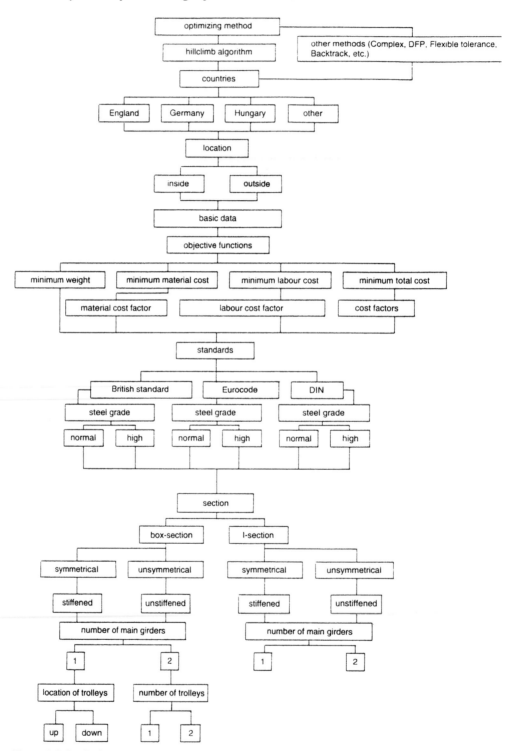

Figure 7.2. Logical structure of an overhead travelling crane girder design.

7.14   APPLICATION OF AN ES FOR THE OPTIMUM DESIGN OF THE MAIN GIRDERS OF OVERHEAD TRAVELLING CRANES

The aim was to develop an ES, which is able to find the optimum sizes of the welded box girder of the crane due to different geometry, loading, steel grades and design codes. The model can be seen in Figure 7.1. The logical structure of ES, the different variants can be seen in Figure 7.2 (Jármai et al. 1995).

The total number of variants is about 60,000 and it can be increased if we take into account other aspects and constraints in a modular way. The decision support system, which was connected to the expert one, contains five various single-objective and seven various multiobjective optimization techniques (Chapter 6). These techniques are able to solve nonlinear optimization problems with practical nonlinear inequality constraints. It could contain finite element procedures to compute the mechanical behaviour of the structures.

7.14.1   *Objective functions*

Material cost of the girder,

$$K_m = k_m \rho V \text{ (kg)} \tag{7.2}$$

where ρ is the material density, V is the volume of the girder, k_m is the specific material cost.

– Labour cost contains welding cost and surface preparation cost,

$$K_l = K_w + K_s \tag{7.3}$$

– Welding cost,

$$K_w = k_w (a_w^2 / \sqrt{2}) L_w \rho Q_d \text{ (\$)} \tag{7.4}$$

where a_w is the effective size of weldment, $a_w^2 / \sqrt{2}$ is the area of the weld, L_w is the length of weldment, Θ_d is the difficulty factor of welding, which depends on the position of welding.

Note that this type of cost calculation is different from the method described in Chapter 5. In Equation (7.4) the cost is proportional to the weld volume.

– Surface preparation and painting costs,

$$K_s = k_s (2 b L + 2 h L) \text{ (\$)} \tag{7.5}$$

where b and h are width and height of the girder, k_s is the specific cost of manufacturing.

– Total cost contains material and labour costs,

$$K_t = K_w + K_l \tag{7.6}$$

7.14.2   *Design constraints*

Constraint on the static stress at midspan due to biaxial bending according to BS 2573 and 5400 (1983) is described by

$$M_x / W_y + M_y / W_y \leq \alpha_d \sigma_{\text{adm}} \tag{7.7}$$

where M_x, M_y are the bending moments,

$$M_x = \frac{L^2}{8}(1.05A + p_r + p_s)g + \frac{F}{2L}\left(L - \frac{k}{2}\right)^2 \tag{7.8}$$

$$M_y = 0.3\frac{z_b}{z_t}\frac{L^2}{8}(1.05A + p_r + p_s)g + \frac{G_t}{8L}\left(L - \frac{h}{2}\right)^2 + M_w \tag{7.9}$$

the wheel load is

$$F = \frac{\psi H + G_t}{4} \tag{7.10}$$

W_x, W_y are section moduli, σ_{adm} is the admissible static stress, α_d is the duty factor, G_t is the mass of trolley, ψ is the dynamic parameter, k is the distance between the trolley axes, M_w is the bending moment caused by the wind, $g = 9.81$ m/s², z_b and z_t are the numbers of the braked and the total wheels of trolley, p_r and p_s the specific mass of the rail and the sidewalk, H is the hook load, A is the cross section area.
    – Constraint on fatigue stress is as follows

$$M_{xf} / W_x + M_y / W_y \leq \Delta\sigma_N \tag{7.11}$$

where M_{xf} contains the live load multiplied by the impact factor and the spectrum factor. $\Delta\sigma_N$ is the fatigue stress range.
    – Local flange buckling constraint is

$$\sigma_{1f} / (\sigma_{\text{adm}}K_{1f}) + \{(\sigma_{bf} / (\sigma_{\text{adm}}K_{bf})\}^2 \leq 1 \tag{7.12}$$

where $\sigma_{1f} = M_x / W_x$, $\sigma_{bf} = M_y / W_y$, the K factors depend on the slenderness of the plate

$$\lambda_f = (b/t_f)\sqrt{f_{yf} / 355}$$

where f_{yf} is the yield stress of the flange plate.
    – Local web buckling constraint can be formulated according to BS 2573 and 5400, see also Jármai (1990).
    – Deflection constraint due to wheel load can be expressed as

$$w_{\text{max}} \leq L / (800 - 1000) \tag{7.13}$$

where L is the span length.

7.14.3 *Main data of an example solved by Personal Consultant*

Hook load is $H = 240$ kN, length is $L = 25$ m, mass of trolley is $G_t = 30$kN, distance between the trolley axes is $k = 2.5$ m, height of rail is $h_r = 50$ mm, mass of the rail is $p_r = 80$ kg/m, the Young modulus is $E = 2.06$ GPa, $z_b/z_t = 2$, class of the crane is A7, steel grade is Fe 430, $f_y = 255$ MPa, $\Delta\sigma_N = 167$ MPa, admissible deflection is $L/800$, stiffeners are 120*80*8 mm angle profiles.

In the ES one part of the rules are concerning to the selection of the crane (see Fig. 7.2).

The second part is concerning to the selection of optimization techniques. The weighting factors at the multiobjective optimization system and the uncertainty parameters at the expert system for the various objective functions are the same. The range is from 0 to 100%. It means the relative importance of the objective function.

The third part of the rules are concerning to the results of the optimization, to choose the smallest objective function value, where the ratio of web height and flange width and the ratio of the two web thicknesses are acceptable. First ratio should be near to the golden ratio, the second ratio has technological reasons.

$$0.4 \leq \ b/h \ \leq \ 0.8 \tag{7.14}$$

$$t_{w1}/t_{w2} \ \leq \ 1.5 \tag{7.15}$$

The multi-objective optimization is performed using the min-max method (Chapter 6). The weighting coefficients are given according to Table 7.1.

The dimensions of the minimum cost crane girder chosen by the ES from the calculated variations according to the value of the total cost (Eq. 7.6).

The discrete value ranges of the variables are as follows: For h and b the step sizes were 20 and 10 mm respectively, for the thicknesses step size was 1 mm. Further development can be the installation of the new Eurocodes in the analysis to build the system in Borland C++, to use the Object Oriented Programming (OOP).

The main differences using the Personal Consultant Easy and the L5O expert shells were, that in Easy all values for the computation should be given in advance, so the program goes on a given way bordering by the rules, but LO5 asks for the unknowns during the computation, it knows what to ask for, easier to jump from one level to another on the rules' tree and the optimization part is built into the expert shell. It means that the second ES is much close to the original aim of artificial intelligence.

The numerical example shows that it is important to include the optimum design in an ES to make it possible to select the most suitable structural versions.

Table 7.1. Questions and possible answers in the expert system for overhead travelling crane design.

Question	Possible answers	Answer
In which country the crane is made for	England, Germany, Hungary, etc.	England
Location of the crane	Inside, outside	Inside
What kind of standard do you wish to use	British Standard, DIN, Eurocode, etc.	British Standard
What is the steel grade	Normal steel ($f_y = 230\text{-}240$ MPa) Higher strength steel ($f_y = 330\text{-}340$)	Normal steel
How many main girders at the crane	One, two	Two

Table 7.1. Continued.

Question	Possible answers	Answer
How many trolleys of the crane	One, two	One
Place of the trolley(s)	Up, down	Up
What is the cross section of the main girder	Welded box, or I-section	Welded box
Other section features	Symmetrical, unsymmetrical	Symmetrical
Other section features	Stiffened web plate, unstiffened web plate	Stiffened web plate, stiffener is at the $h/5$ on the web from the upper flange
Objective function	Minimum mass, material cost, labour cost, total cost	Minimum total cost
Material, welding, surface preparation specific costs	Positive numbers	Material, welding, surface preparation specific costs, material $k_m = 1$ ($/kg), welding $k_w = 10$ ($/kg), surface preparation $k_s = 100$ ($/m^2) respectively
Weighting parameter for every objective function	Between 0 and 1 for each	$w_1 = 0.3$, $w_2 = 0.3$, $w_3 = 0.4$, for total cost = 0.4 , material cost = 0.3, labour cost = 0.3
Singe-objective optimization technique	Hillclimb procedure, Complex method, Davidon-Fletcher-Powell method, Direct random search technique, Flexible tolerance method (see Chapter 6)	Hillclimb technique
Multi-objective optimization technique	Min-max, weighting min-max, pure weighting, normalized weighting, global type I, global type II, weighting global (see Chapter 6)	All multiobjective optimization techniques are used
Parameters for the multiobjective optimization	Relative interest of the objective function	Weighting parameters for every objective function are used here
Weldability of the box section	b/h	between 0.4 and 0.8

Table 7.2. Optimum solutions of the overhead travelling crane design, dimensions in mm.

Objective function	Eq. (7.2)	Eq. (7.5)	Eq. (7.6)	Eq. (7.3)	Multi-objective
Web height is	$h = 1320$	1520	1260	1140	1260
Main web thickness is	$t_{w1} = 6$	6	6	7	6
Secondary web thickness is	$t_{w2} = 5$	5	5	5	5
Width of the flange is	$b = 650$	550	700	750	790
Thickness of the flange is	$t_f = 18$	17	18	19	16
Total cost of the structure is ($)	$K_t = 16745$	17251	16677	16834	16737

CHAPTER 8

Statically determinate beams subjected to bending and shear

8.1 OPTIMUM DESIGN NEGLECTING SHEAR

8.1.1 *Analytical method*

We treat the problem for welded I-sections, but, as it will be shown later, the derived formulae are valid also for box-sections, only small changes should be performed in the calculations.

Note that we treat here only homogeneous sections. Hybrid sections consisting of two types of steels (flanges of higher-strength steel) have been investigated in the book (Farkas 1984).

The cross-sectional area is (Fig. 8.1)

$$A = ht_w + 2bt_f \tag{8.1}$$

The approximate formula for the moment of inertia is

$$I_x = h^3 t_w / 12 + 2bt_f (h/2)^2 \tag{8.2}$$

The elastic section modulus is

$$W_{xe} \cong 2I_x / h = h^2 t_w / 6 + bt_f h \tag{8.3}$$

The plastic section modulus is expressed by

$$W_{xp} \cong h^2 t_w / 4 + bt_f h \tag{8.4}$$

We need also the section modulus of flanges

$$W_{xf} \cong bt_f h \tag{8.5}$$

From Equation (8.1) one obtains

$$bt_f = A/2 - ht_w / 2 \tag{8.6}$$

Substituting Equation (8.6) into Equations (8.3) and (8.4) as well as Equation (8.5) a generalized formula for the section modulus can be written as

$$W_x = Ah/2 - \varsigma h^2 t_w / 2 \tag{8.7}$$

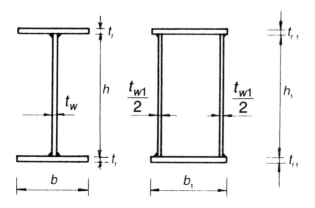

Figure 8.1. Dimensions of I- and box-sections.

where $\varsigma_e = 2/3$, $\varsigma_p = 1/2$, $\varsigma_f = 1$.

The stress constraint can be expressed as

$$W_x \geq W_0 = M / f_{y1} \tag{8.8}$$

where W_0 is the required section modulus, $f_{y1} = f_y / \gamma_{M1}$, f_y is the yield stress, $\gamma_{M1} = 1.1$ is the partial safety factor.

The local web buckling constraint is expressed by

$$t_w \geq \beta h \tag{8.9}$$

where the limiting web slenderness has the following values
– For elastic design

$$1 / \beta_e = 124\varepsilon \tag{8.10a}$$

– For plastic design

$$1 / \beta_p = 72\varepsilon \tag{8.10b}$$

where $\varepsilon = \sqrt{235 / f_y}$ (f_y in MPa).

Treating Equation (8.9) as active, from Equation (8.7) we get

$$A = 2W_0 / h + \varsigma\beta h^2 \tag{8.11}$$

and the optimal web height to minimize the cross-sectional area can be calculated from the condition $dA/dh = 0$

$$h_{\text{opt}} = \sqrt[3]{\frac{W_0}{\varsigma\beta}} \tag{8.12}$$

Expressing W_0 from Equation (8.12) and substituting it into Equation (8.11) we get

$$A_{\text{min}} = 3\varsigma\beta h_{\text{opt}}^2 = \sqrt[3]{27\varsigma\beta W_0^2} \tag{8.13}$$

This formula can be used for comparison of elastic and plastic minimum cross-sectional areas

$$\frac{A_{min.p}}{A_{min.e}} = 3\sqrt{\frac{\varsigma_p \beta_p}{\varsigma_e \beta_e}} = 1.089 \tag{8.14}$$

i.e. the plastic design results in cross-sections by 8.9% larger than the elastic design. The plastic design is advantageous only in the optimum design of statically indeterminate structures.

The other optimum dimensions can be calculated from Equation (8.6)

$$bt_f = \delta b^2 = 3\varsigma\beta h^2 / 2 - \beta h^2 / 2 = \beta h^2 (3\varsigma - 1) / 2 \tag{8.15}$$

$$b_{opt} = h_{opt} \sqrt{\beta / (2\delta)} \sqrt{3\varsigma - 1}, \quad \delta = t_f / b, \quad (1/\delta)_L = 28\varepsilon \tag{8.16a}$$

For the compression flange of a box beam the limiting plate slenderness is

$$(1/\delta_1)_L = 42\varepsilon \tag{8.16b}$$

Note that, for a box beam (Fig. 8.1) the local web buckling constraint is

$$t_{w1} \geq 2\beta h_1 \tag{8.17}$$

Comparing Equation (8.17) with Equation (8.9) it can be concluded that the formulae derived for I-beams are valid also for box beams when we take 2β instead of β.

8.1.2 *Graphoanalytical method*

In many cases, in addition to the stress constraint, also the deflection limitation should be considered. In these cases the graphoanalytical method is very advantageous.

The problem to minimize the objective function with four variables, Equation (8.1) can be solved by reducing the number of variables to two. Calculations show that the local buckling constraints for web and compression flange (Eqs 8.9 and 8.16) are always active, thus, the objective function can be written in the form

$$A = \beta h^2 + 2\delta b^2 \tag{8.18}$$

We treat here only the elastic design, i.e. $\varsigma_e = 2/3$.

In order to have linear contours of the objective function, we linearize Equation (8.18) taking

$$x_1 = h^2 \quad \text{and} \quad x_2 = b^2 \tag{8.19}$$

to obtain

$$A = \beta x_1 + 2\delta x_2 \tag{8.20}$$

The new form of the stress constraint Equation (8.8) is

$$W_x = \beta h^3 / 6 + \delta b^2 h = \beta x_1^{3/2} x_2 \geq W_0 \tag{8.21}$$

The equation of the limit curve for the stress constraint is

$$x_2^{(\sigma)} = \frac{W_0}{\delta x_1^{1/2}} - \frac{\beta}{6\delta} x_1 \tag{8.22}$$

– *The deflection constraint*, neglecting the effect of shear deformations, is given by

$$w_{max} = C_w / I_x \leq c^* L \tag{8.23}$$

or in other form

$$I_x \geq I_0 = \frac{C_w}{c^* L} \tag{8.24}$$

C_w is a constant depending on load, span length L and modulus of elasticity E. For instance, for a simply supported, uniformly loaded beam with constant cross-section

$$C_w = \frac{5pL^4}{384E} \tag{8.25}$$

where p is the load intensity, c^* is the allowable deflection ratio, e.g. for simply supported beams in buildings, according to EC3, $c^* = 1/300$.

The deflection constraint Equation (8.24) can be expressed as

$$I_x = \frac{\beta h^4}{12} + \frac{\delta b^2 h^2}{2} = \frac{\beta x_1^2}{12} + \frac{\delta}{2} x_1 x_2 \geq I_0 \tag{8.26}$$

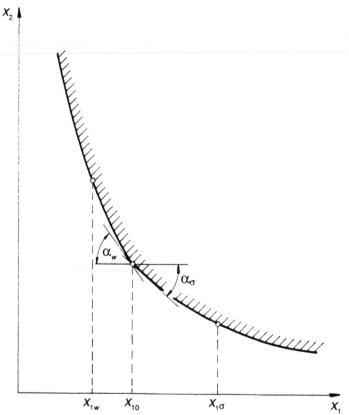

Figure 8.2. Graphoanalytical optimization of welded I- and box-sections.

and the limit curve is defined by

$$x_2^{(w)} = \frac{2I_0}{\delta x_1} - \frac{\beta}{6\delta} x_1 \qquad (8.27)$$

The abscissa x_{10} of the point of intersection of the two limit curves is (Fig. 8.2)

$$x_{10} = 4I_0^2 / W_0^2 \qquad (8.28)$$

The inclinations of the tangents for the limit curves at the point x_{10} are as follows:

$$\left| \tan \alpha_w \right| = \left| \frac{dx_2^{(w)}}{dx_1} \right|_{x_1 = x_{10}} = \frac{W_0^4}{8\delta I_0^3} + \frac{\beta}{6\delta} \qquad (8.29)$$

Similarly, from Equation (8.21) one obtains

$$\left| \tan \alpha_\sigma \right| = \frac{W_0^4}{16\delta I_0^3} + \frac{\beta}{6\delta} \qquad (8.30)$$

The inclination of the objective function, Equation (8.20) is

$$\left| \tan \alpha_A \right| = \frac{\beta}{2\delta} \qquad (8.31)$$

The stress constraint is active if $\left| \tan \alpha_A \right| \leq \left| \tan \alpha_\sigma \right|$, i.e. when

$$16\beta I_0^3 \leq 3W_0^4 \qquad (8.32)$$

On the other hand, the deflection constraint is active if $\left| \tan \alpha_A \right| \geq \left| \tan \alpha_w \right|$, i.e. when

$$8\beta I_0^3 \geq 3W_0^4 \qquad (8.33)$$

Finally, in the case when

$$8\beta I_0^3 \leq 3W_0^4 \leq 16\beta I_0^3 \qquad (8.34)$$

the optimum is given by the point of intersection.

If the stress constraint is active, the optimum h-value may be obtained from the condition that the contour of the objective function touches the feasible region at the optimum point, i.e.

$$\left| \tan \alpha_A \right| = \left| \frac{dx_2^{(\sigma)}}{dx_1} \right| \qquad (8.35)$$

Equation (8.35) yields

$$x_{1.\text{opt}}^{1/2} = h_\sigma = \sqrt[3]{\frac{3W_0}{2\beta}} \qquad (8.36)$$

which is the same as obtained by analytical method (Eq. 8.12).

If the deflection constraint is active, the local buckling constraints should be modified considering that the maximum stress is smaller than f_{y1}. An approximate

Table 8.1. Characteristics of cross-sections optimized with stress constraint.

I-section	Box-section
$h_\sigma = \sqrt[3]{1.5 W_0 / \beta}$	$h_{1\sigma} = \sqrt[3]{0.75 W_0 / \beta}$
$t_{w\sigma} = \beta h_\sigma$	$t_{w1\sigma} / 2 = \beta h_{1\sigma}$
$A_\sigma = 2\beta h_\sigma^2 = \sqrt[3]{18\beta W_0^2}$	$A_{1\sigma} = 4\beta h_{1\sigma}^2 = \sqrt[3]{36\beta W_0^2}$
$b_\sigma = h_\sigma \sqrt{\beta / (2\delta)}$	$b_{1\sigma} = h_{1\sigma} \sqrt{\beta / \delta_1}$
$t_{f\sigma} = \delta b_\sigma$	$t_{f1\sigma} = \delta_1 b_{1\sigma}$
$I_{x\sigma} = \beta h_\sigma^4 / 3$	$I_{x1\sigma} = 2\beta h_{1\sigma}^4 / 3$
$W_{x\sigma} = 2\beta h_\sigma^3 / 3$	$W_{x1\sigma} = 4\beta h_{1\sigma}^3 / 3$

Table 8.2. Characteristics of cross-sections optimized with deflection constraint.

I-section	Box-section
$h_w = \sqrt[4]{6 I_0 / \beta}$	$h_{1w} = \sqrt[4]{3 I_0 / \beta}$
$t_{w.w} = \beta h_w$	$t_{w1.w} / 2 = \beta h_{1w}$
$A_w = 4\beta h_w^2 / 3 = \sqrt{32\beta I_0 / 3}$	$A_{1w} = 8\beta h_{1w}^2 / 3 = \sqrt{64\beta I_0 / 3}$
$b_w = h_w \sqrt{\beta / (6\delta)}$	$b_{1w} = h_{1w} \sqrt{\beta / (3\delta_1)}$
$t_{fw} = \delta b_w$	$t_{f1.w} = \delta_1 b_{1w}$
$I_{xw} = \beta h_w^4 / 6$	$I_{x1w} = \beta h_{1w}^4 / 3$
$W_{xw} = \beta h_w^3 / 3$	$W_{x1w} = 2\beta h_{1w}^3 / 3$

solution can be obtained when we calculate with the unmodified local buckling constraints. Then, from the condition

$$\left| \tan \alpha_A \right| = \left| \frac{dx_2^{(w)}}{dx_1} \right| \tag{8.37}$$

one obtains

$$x_{1.opt}^{1/2} = h_w = \sqrt[4]{\frac{6 I_0}{\beta}} \tag{8.38}$$

Characteristics of the optimum cross-sections for elastic design are summarized in Tables 8.1 and 8.2.

When the deflection constraint is active, the limiting plate slendernesses in the local buckling constraints can be modified calculating with the maximum stress instead of yield stress (Chapter 3). This calculation has been treated in the book (Farkas 1984).

– *Numerical example of an I-beam.* Data: Steel Fe 360, f_y = 235 MPa, f_{y1} = 235/1.1 = 213.64 MPa. $M = 3*10^9$ Nmm, $W_0 = M/f_{y1} = 1.40426*10^7$ mm³, β = 1/124. Two cases of the deflection constraint should be considered: Case a) $I_0 = 5*10^9$ mm⁴, and Case b) $I_0 = 15*10^9$ mm⁴.

Case a): Since

$$I_0 = 5*10^9 < \sqrt[3]{3W_0^4/(16\beta)} = 9.6695*10^9 \text{ mm}^4$$

the stress constraint is active, so using Equation (8.36) or Table 8.1 one obtains $h_\sigma = 1377$ mm and $A_\sigma = 30590$ mm^2.

For purposes of comparison, it is interesting to check the case when both stress and deflection constraints are active. Then $h = 2I_0/W_0 = 712$ mm and

$$A = \frac{W_0^2}{I_0} + \frac{8\beta I_0^2}{3W_0^2} = 42165 \text{ mm}^2$$

this solution is not optimum and the difference is 38%.

Case b): Since

$$I_0 = 15*10^9 > \sqrt[3]{3W_0^4/(8\beta)} = 12.18*10^9 \text{ mm}^4$$

the deflection constraint is active, and the approximate solution, according to Equation (8.38) and Table 8.2 is $h_w = 1828$ mm and $A_w = 35921$ mm^2.

8.1.3 *Some comparisons*

By means of the simple formulae derived above it is possible to compare the characteristics of various types of sections and beams, which can be useful for designers. In the following only elastic sections optimized for stress constraint will be used. Note that we have compared sections obtained by elastic and plastic design in Section 8.1.1.

8.1.3.1 *Cross-sectional areas of I- and box-sections*
From Table 8.1 it can be seen that

$$\frac{A_1 - A}{A} = \sqrt[3]{2} - 1 = 0.26$$

so that the area of a box-section is 26% larger than that of an I-section. Thus the latter is more economical for pure bending when a larger torsional stiffness is not required and when the lateral buckling is not governing.

8.1.3.2 *I-beams made of steels Fe 360 or 510*
According to Table 8.1

$$A = \sqrt[3]{18\beta(M/f_{y1})^2}$$

and from Equation (3.55)

$$\beta = \sqrt{\frac{12(1-v^2)f_{y1}}{k_\sigma \pi^2 E}}$$

thus

$$A \sim (f_{y1})^{-1/2}$$

In the case of static loads

$$1 - \frac{A_{510}}{A_{360}} = 1 - \sqrt{\frac{f_{y1.360}}{f_{y1.510}}} = 0.18$$

thus, the mass of an I-beam made of steel Fe 510 is 18% less than that corresponding to steel Fe 360. Note that the deflection of a beam made of steel Fe 510 is larger. On the basis of the formula

$$I_x = \frac{\beta h_\sigma^4}{3} = \frac{\beta}{3} \left(\frac{3M}{2\beta f_{y1}} \right)^{4/3}$$

one obtains that $I_x \sim (f_{y1})^{-3/2}$. Thus,

$$\frac{w_{510}}{w_{360}} = \left(\frac{355}{235} \right)^{3/2} = 1.86$$

i.e. the deflection is 86% larger.

Note that larger savings can be achieved by hybrid sections (Farkas 1984).

8.1.3.3 *I-beam made of steel 360 or Al-alloy*

Formulae used in Section 8.1.3.2 show that $A \sim (f_{y1})^{-1/2}$ and $A \sim E^{-1/6}$ so that the ratio of masses is

$$1 - \frac{G_{al}}{G_s} = 1 - \frac{\rho_{al}}{\rho_s} \sqrt{\frac{f_{ys}}{f_{y.al}}} \sqrt[6]{\frac{E_s}{E_{al}}}$$

With values of $f_{ys} = 235$ MPa, for an Al-alloy of 6082-T2 according to BS 8118 $f_{y.al} = 240$ MPa, $\rho_s = 7850, \rho_{al} = 2750$ kg/m^3, and $E_{al}/E_s = 1/3$ we get

$$1 - G_{al} / G_s = 0.57$$

i.e. the mass of an I-beam made of Al-alloy is 57% less than that of made of steel Fe 360. The deflections are, however, larger. Considering that

$$\beta \sim E^{-1/2} f_y^{1/2} \text{ and } I_x \sim E^{1/6} f_y^{-3/2}$$

we get

$$\frac{w_{al}}{w_s} = \frac{E_s I_{xs}}{E_{al} I_{x.al}} = \left(\frac{E}{E_{al}} \right)^{7/6} \left(\frac{f_{y.al}}{f_{y.s}} \right)^{3/2} = 3.26$$

i.e. the deflections of an I-beam made of Al-alloy are 226% larger.

8.1.3.4 *Torsional stiffnesses of I- and box-sections*

The torsional inertia of an I-section can be calculated using Equation (2.30)

$$I_t = \alpha_0 \left(h t_w^3 + 2 b t_f^3 \right) / 3 = \frac{\alpha_0}{3} h^4 \beta^2 \left(\beta + \frac{\delta}{2} \right)$$

With the values $\alpha_0 = 1.5, \beta = 1/124, \delta = 1/28$ one obtains $I_t = 8.43*10^{-7} h^4$.

The torsional inertia of a box-section, according to Equation (2.37), is

$$I_{t1} = \frac{4b_1^2 h_1^2}{2b_1 / t_{f1} + 4h_1 / t_{w1}} = \frac{2\beta h_1^4}{1 + \delta_1 / \beta} = 4.08*10^{-3} h_1^4$$

Since $h_1 = 2^{-1/3} h$, we get $I_{t1}/I_t = 1920$, i.e. the torsional stiffness of a box-section is much larger than that of an I-section.

8.1.3.5 *Normal stresses in I- and box-beams due to warping torsion*

In the case of a simply supported I-beam (Fig. 8.3) the maximum warping torsional moment is expressed by (see Chapter 2, Eq. 2.65)

$$B_{\omega max} = Fe \frac{\tanh \alpha l}{\alpha}, \quad \alpha^2 = \frac{GI_t}{EI_\omega}$$

where the warping section constant is

$$I_\omega = \frac{h^2 b^3 t_f}{24} = \delta \left(\frac{\beta}{2\delta} \right)^2 h^6 = 1.90*10^{-5} h^6$$

With $G/E = 0.8/2.1$ and $I_t = 8.43*10^{-7} h^4$ one obtains $\alpha^2 = 1.6902*10^{-2} h^{-2}$ and $\alpha l = 0.1300 l / h$. Assuming that $\alpha l < 0.5$ (i.e. $h > 0.26 l$), the following approximations may be used: $\tanh \alpha l \approx \alpha l$ and $B_{\omega max} \cong Fel$.

The maximum normal warping stress is given by

$$\sigma_{\omega max} = \frac{B_{\omega max}}{I_\omega} \omega_{max} = \frac{6Fel}{hb^2 t_f}$$

and the maximum bending stress is

$$\sigma_{b max} = \frac{Fl}{W_x} = \frac{3Fl}{2\beta h^3}$$

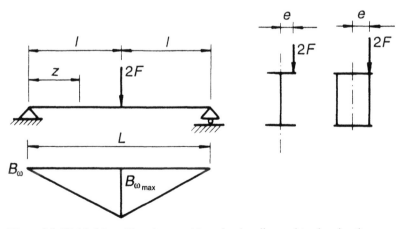

Figure 8.3. Welded I- and box-beam subjected to bending and torsional actions.

The ratio of these two stresses is

$$\frac{\sigma_{\omega max}}{\sigma_{b max}} = \frac{8e}{b}$$

which shows that, if $e > b/80$, $\sigma_{\omega max} / \sigma_{b max} > 0.1$, i.e. that *the I-beam is very sensitive to warping torsion*.

In the case of a box-beam the following formulae can be used (see Chapter 2):

$$\sigma_{\omega max} = \frac{B_{\omega max}}{I_\omega} \omega_{T max}$$

$$B_{\omega max} = \mu Fe \frac{\tanh \alpha_1 l}{\alpha_1}, \quad \alpha_1^2 = \frac{\mu G I_{tc}}{E I_\omega}, \quad \mu = 1 - \frac{I_{tc}}{I_p}$$

$$I_p = \int_A r_T^2 dA = h_1 t_{w1}(b_1/2)^2 + 2b_1 t_{f1}(h_1/2)^2 = 5.398 * 10^{-3} h_1^4$$

$$I_{tc} = 4.08 * 10^{-3} h_1^4, \quad \mu = 0.2442$$

For the corner point of a box-section we get

$$\omega_{T max} = \frac{b_1 h_1}{4} - \frac{b_1 h_1}{b_1/t_{f1} + 2h_1/t_{w1}}\left(\frac{h_1}{t_{w1}}\right) = -7.187 * 10^{-2} h_1^2$$

$$I_\omega = \omega_{T max}^2 \left(\frac{2}{3}b_1 t_{f1} + \frac{1}{3}h t_{w1}\right) = 5.5527 * 10^{-5} h_1^6, \quad \alpha_1 = 2.6145/h_1$$

If $\alpha_1 l > 1.5$, i.e. if $h_1 < 1.74l$, the following approximations can be used: $\tanh \alpha_1 l \approx 1$, and

$$|B_{\omega max}| = \mu Fe / \alpha_1 = 0.0934 Feh_1$$

The maximum warping and bending stresses are

$$\sigma_{\omega max} = 120.9 Fe / h_1^3 \quad \text{and} \quad \sigma_{b max} = 372 Fl / h_1^3$$

and their ratio is $\sigma_{\omega max} / \sigma_{b max} = 0.325e / l$. This ratio is larger than 10% if $e > 3.38l$, thus it can be concluded that *box beams are insensitive to warping torsion*.

8.1.4 *Effect of painting costs*

In the objective function, the cost of surface preparation and painting can be considered in addition to material cost. In the case of an I-beam the cost function can be written in the form

$$K = K_m + K_p = k_m \rho A + k_p(2h + 4b) \tag{8.39}$$

If the stress constraint is active, Equation (8.11) takes the form

$$A = 2W_0 / h + 2\beta h^2 / 3 \tag{8.40}$$

With the approximation $b = 0.3h$ Equation (8.39) can be written as

$$\frac{K}{k_m \rho} = \frac{2W_0}{h} + \frac{2\beta h^2}{3} + \frac{3.2 k_p h}{k_m \rho} \tag{8.41}$$

We take the data of the numerical example treated in Section 8.1.2 as follows: $W_0 = 14.0426 * 10^7$ mm^3, $\beta = 1/124$. $k_m = 0.3$ \$/kg, $\rho = 7850$ kg/m^3. According to published data, e.g. Bo et al. (1974) $k_p = 4.4$ \$/m^2. With the above data Equation (8.41) reduces to

$$\frac{K}{k_m \rho} = \frac{2.8085 * 10^7}{h} + \frac{h^2}{186} + 6.0h \tag{8.42}$$

Condition $dK/dh = 0$ yields an equation the numerical solution of which is $h_{opt} = 1215$ mm, as against 1377 mm for the case when painting cost is not considered (see numerical example in Section 8.1.2).

8.2 OPTIMUM DESIGN CONSIDERING SHEAR

We determine the optimum cross-sections for the different points or regions of the bending moment-shear force interaction diagram according to EC3 shown in Figure 8.4 considering increasing values of the shear force acting in addition to a constant given value of the bending moment (or required section modulus) (Farkas & Jármai 1996). Note that the notation 0.5 V'_{ba} is used since the actual numerical value of this shear force is not the half of V_{ba} (see the numerical example).The corresponding A_{min}-values are given in Figure 8.5 in function of the shear force.

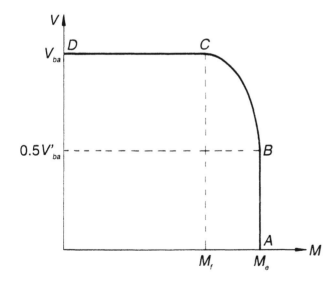

Figure 8.4. Bending moment - shear force interaction diagram according to EC3.

In the interaction diagram

$$V_{ba} = ht_w \tau_{ba} / \gamma_{M1} = ht_w \tau_Y, \quad \tau_Y = \tau_{ba} / \gamma_{M1} \tag{8.43}$$

The reduced web slenderness is

$$\overline{\lambda}_w = \frac{h/t_w}{37.4\varepsilon\sqrt{k_\tau}} = \frac{1}{86.4256\varepsilon\beta} \tag{8.44}$$

since for unstiffened webs $k_\tau = 5.34$ (see Section 3.6.1). The web buckling diagram has three regions as follows:

$$\tau_Y = \tau_{Y1} = f_{y1} / \sqrt{3} \qquad \text{for} \quad \overline{\lambda}_w \leq 0.8 \tag{8.45a}$$

$$\tau_Y = \tau_{Y1}(1.5 - 0.625\overline{\lambda}_w) \quad \text{for} \quad 0.8 < \overline{\lambda}_w < 1.2 \tag{8.45b}$$

$$\tau_Y = 0.9\tau_{Y1} / \overline{\lambda}_w \qquad \text{for} \quad \overline{\lambda}_w \geq 1.2 \tag{8.45c}$$

For elastic design with Equation (8.11), according to Equation (8.44) we get

$$\overline{\lambda}_w = 1.4348 > 1.2$$

thus

$$\tau_Y = 0.6273\tau_{Y1}$$

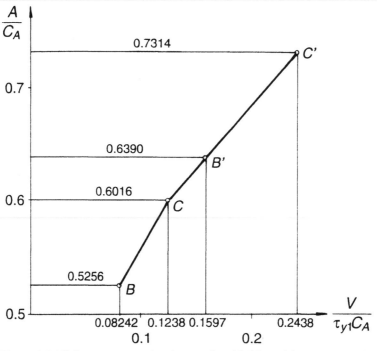

Figure 8.5. Minimum cross-sectional area of welded I- and box-beams in function of the shear force.

The region A-B is defined by

$$V \le 0.5V_{ba} = 0.5\beta h^2 * 0.6273\tau_{Y1}$$ (8.46)

Using Equations (8.10) and (8.11) and introducing the notation

$$C_A = \sqrt[3]{W_0^2 / \varepsilon}$$ (8.47)

Equation (8.46) can be written in the form

$$V / (\tau_{Y1} C_A) \le 0.08242$$ (8.48)

The corresponding A/C_A-value can be obtained from Equation (8.13)

$$A/C_A = 0.5256$$ (8.49)

For the point C, where $V = V_{ba}$, using Equations (8.46) and (8.10) with the value of $\varsigma_f = 1$ one obtains

$$V / (\tau_{Y1} C_A) = 0.1258$$ (8.50)

and the corresponding A/C_A-value from Equation (8.13) is

$$A/C_A = 0.6016$$ (8.51)

Between points B and C a linear interpolation can be used according to Figure 8.5.

When $V / (\tau_{Y1} C_A) > 0.1258$ the value of $1/\beta = 69\varepsilon$ can be used instead of 124ε. In this case Equation (8.44) gives the value of $\lambda_w = 0.8$, thus, according to Equation (8.45a), $\tau_Y = \tau_{Y1}$ and we calculate in Equation (8.46) with τ_{Y1} instead of $0.6273\tau_{Y1}$. We get the points B' and C'. Between points C and B' a linear interpolation can also be used. The value of $1/(\beta\varepsilon)$ can also be linearly interpolated between 124 and 69. Since Equation (8.13) can be written in the form

$$A / C_A = \sqrt[3]{27\varsigma\beta\varepsilon}$$ (8.52)

therefore, knowing A/C_A and $\beta\varepsilon$, the corresponding ς-value can be calculated from Equation (8.52).

When

$$V / (\tau_{Y1} C_A) > 0.2438$$

then the shear will be governing, thus from Equation (8.43) we obtain

$$h_V = \sqrt{V / (\beta\tau_{Y1})}$$ (8.53)

and, with $1/(\beta\varepsilon) = 69$ and $\varsigma = 1$, using Equation (8.13) we get

$$A_{min} = 3\varsigma\beta h_V^2 = 3V / \tau_{Y1} \quad \text{or} \quad A / C_A = 3V / (\tau_{Y1} C_A)$$ (8.54)

which means a straight line in the diagram. The corresponding flange width can be obtained from Equation (8.16):

$$b_f = h_V \sqrt{\beta / \delta}$$ (8.55)

The above derived formulae are valid also for a doubly symmetric welded box section with two webs of equal thickness $t_w/2$ with two exceptions as follows:

1. Instead of β we should calculate in all formulae with 2β since the local web buckling constraint Equation (8.9) relates to one web $t_w/2 \geq \beta h$ or $t_w \geq 2\beta h$,

2. In the local buckling constraint of the compression flange the limiting flange slenderness in Equation (8.16) is 42ε instead of 28ε. In Figure 8.5 all values should be multiplied by 1.26.

A numerical example

Optimize a welded box beam with the following data: The bending moment is $M = 630$ kNm, $f_y = 355$ MPa, $\gamma_{M1} = 1.1$. $W_0 = 1.9505*10^6$ mm^3, $\varepsilon = \sqrt{235/355} = 0.8136$, $\tau_{Y1} = 186$ MPa.

– *Point B* in Figures 8.4 and 8.5: $1/\beta = 124\varepsilon, \varsigma = 2/3$. In this case, using Equations (8.44) and (8.45) $\lambda_w = 1.4348 > 1.2, \tau_Y = 0.9\tau_{Y1}/\lambda_w = 116.7$ MPa. $h = \sqrt[3]{0.75W_0/\beta} = 528$, rounded 530 mm, $t_w/2 = 5.2$, rounded 6 mm. Using Equation (8.16) $b = h\sqrt{\beta}/\delta\sqrt{3\varsigma} - 1 = 307$, rounded 310 mm, $t_f = 9$ mm. The maximum shear force is $V_{max} = 0.5ht_w\tau_Y = 371$ kN.

– *Point C*: The only change is that $\varsigma = 1$, $h = \sqrt[3]{W_0/(2\beta)} = 462 \rightarrow 470$, $t_w/2 = 5$, $b = h\sqrt{2\beta}/\delta = 380$, $t_f = 12$ mm. The maximum shear force is $V_{max} = ht_w\tau_Y = 548$ kN.

– *Point B'*: $1/\beta = 69\varepsilon = 56.14, \lambda_w = 0.8, \tau_Y = \tau_{Y1} = 186$ MPa, $\varsigma = 2/3$, $h = 435$, $t_w/2 = 8$, $b = 340$, $t_f = 10$ mm, $V_{max} = 647$ kN.

– *Point C'*: $\varsigma = 1$, $h = 380$, $t_w/2 = 7$, $b = 420$, $t_f = 13$ mm. $V_{max} = 990$ kN.

If the shear force is larger than 990 kN, e.g. $V = 1200$ kN, then we use Equation (8.53) modified for box section: $h_V = \sqrt{V/(2\beta\tau_{Y1})} = 426 \rightarrow 430$, $t_w/2 = 8$, $b = h\sqrt{2\beta}/\delta = 470$, $t_f = 14$ mm, $V_{max} = 1280$ kN.

8.3 MULTICRITERIA OPTIMIZATION OF A WELDED BOX BEAM

The cost, mass and maximum deflection are selected as objective functions (Farkas & Jármai 1995). In the cost function the material and fabrication costs are included (see Chapter 5). The variables are the four plate dimensions of a symmetric welded box beam. The design constraints relate to the bending stress and local buckling of plate elements. The shear stress constraint and size limitations are also considered. The optimum beam dimensions are computed using several single- and multiobjective optimization methods treated in Chapter 6. The results of an illustrative numerical example show the effect of yield stress of steel and that of the weighting coefficients.

In order to stiffen the box beam against the torsional deformation of the cross-sectional shape, some transversal diaphragms should be used. Note that the minimum cost design of a welded box beam with diaphragms, considering fatigue constraints has been treated in (Farkas 1991). As shown in Figure 8.6, in our example we use 7 diaphragms, so the number of elements to be assembled is $\kappa = 11$. Note that the mass of these diaphragms is neglected. For the four longitudinal fillet welds we consider the constants $C_2 = 0.4*10^{-3}$ and $C_3 = 0.12*10^{-3}$ (min/mm$^{2.5}$), for manual-arc-welded transversal fillet welds connecting the diaphragms to the box beam we use $C_2 = 0.8*10^{-3}$ and $C_3 = 0.24*10^{-3}$ (min/mm$^{2.5}$).

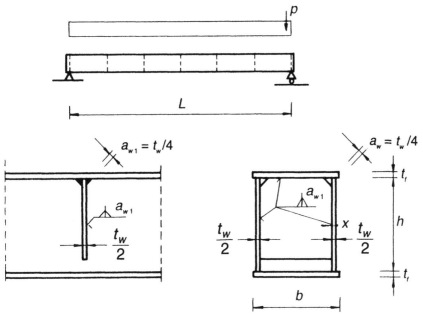

Figure 8.6. Welded box beam with transverse diaphragms.

For the difficulty factor we take $\Theta = 2$, so the final formula of the cost function as the first objective function is

$$f_1 = K/k_m \text{ (kg)} = \rho AL + k_f/k_m *$$
$$\left[2\sqrt{\rho AL}\sqrt{11} + 0.52*10^{-3}*4L\left(\frac{t_w}{4}\right)^{1.5} + 1.04*10^{-3}*7*2(2h+b)\left(\frac{t_w}{4}\right)^{1.5} \right] \quad (8.56)$$

Disregarding the fabrication costs, i.e. taking $k_f = 0$, we obtain the mass function as the second objective function

$$f_2 = \rho AL \quad (8.57)$$

The third objective function to be minimized is the maximum deflection of the beam due to the uniformly distributed normal static load p_0 neglecting the self mass

$$f_3 = \frac{5p_0L^4}{384EI_x} \quad (8.58)$$

where $E = 2.1*10^5$ MPa is the modulus of elasticity for steels, I_x is the moment of inertia

$$I_x = \frac{h^3 t_w}{12} + \frac{2bt_f^3}{12} + 2bt_f\left(\frac{h+t_f}{2}\right)^2 \quad (8.59)$$

The constraint on bending stress, according to EC3, can be expressed as

$$\sigma_{max} = \frac{\gamma_1 M_{max}}{W_x} \leq f_{y1} \tag{8.60}$$

where the safety factor is $\gamma_1 = 1.5$.

The bending moment is

$$M_{max} = \frac{pL^2}{8} \tag{8.61}$$

Considering also the self mass

$$p = p_0 + \rho A g \tag{8.62}$$

where $g = 9.81$ m/s^2 is the gravitational acceleration. Furthermore, f_y is the yield stress, for steel Fe 360 $f_y = 235$, for steel Fe 510 $f_y = 355$ MPa.

The section modulus is

$$W_x = \frac{I_x}{(h/2) + t_f} \tag{8.63}$$

Note that we consider the cross section of class 3.

Local buckling constraint for compressed upper flange

$$t_f \geq \delta b, \quad 1/\delta = 42\,\varepsilon, \quad \varepsilon = \sqrt{\frac{235}{\sigma_{max}}} \quad (\sigma_{max} \text{ in MPa}) \tag{8.64}$$

and for bent webs

$$t_w/2 \geq \beta h, \quad 1/\beta = 124\,\varepsilon \tag{8.65}$$

The shear constraint can be expressed, according to EC3 for $1/\beta = 124\,\varepsilon$, as

$$V_{max} = \frac{\gamma_1 pL}{2} \leq 0.627 * 0.5\, V_b \tag{8.66}$$

where

$$V_b = \frac{h t_w}{\gamma_{M1}} \frac{f_y}{\sqrt{3}}$$

With $\gamma_1 = 1.1$ Equation (8.66) takes the form

$$\frac{\gamma_1 pL}{2} \leq 0.1645\, h t_w f_y \tag{8.67}$$

Since the deflection minimization leads to maximization of the beam dimensions, size constraints should be defined as follows

$$h \leq h_{max}, \quad t_w \leq t_{w\,max}, \quad b \leq b_{max}, \quad t_f \leq t_{f\,max} \tag{8.68}$$

In the optimization procedure the optimum values of variables h, t_w, b and t_f should be determined which minimize the objective functions and fulfill the design constraints.

We have used the min-max, the weighting min-max, two types of global criterion, weighting global criterion, pure weighting and normalized weighting techniques. They are described in details in Chapter 6.

A numerical example

Data: $p_0 = 80$ kN/m, L = 15 m, $\rho = 7850$ kg/m^3, $h_{max} = 1800$, $b_{max} = 1000$, $t_{wmax} = 40$, $t_{fmax} = 40$ mm.

Table 8.3 shows the results of the single-objective optimization using three differ-

Table 8.3. Characteristics of beams optimized using different single-objective techniques.

Technique		h	t_w	b	t_f	f_1	f_3
			(mm)			(kg)	(mm)
Flexible		1450	22	700	19	9332	19.7
Tolerance	f_{3min}	1800	32	1000	40	20801	5.1
Direct random	f_{1min}	1400	22	650	22	9402	20.5
Search	f_{3min}	1800	32	1000	40	20801	5.1
Hillclimb	f_{1min}	1300	20	550	32	9343	21.9
	f_{3min}	1800	32	1000	40	20801	5.1

Table 8.4. Characteristics of beams optimized using different multiobjective optimization methods and various weighting coefficients.

Technique	h	t_w	b	t_f	f_1	f_3
		(mm)			(kg)	(mm)
Min-max	1800	20	750	39	13762	7.3
Global 1, P = 3	1800	20	750	40	13947	7.2
Global 2, P = 5	1800	20	900	33	13910	7.2
Weighting min-max w_1/w_3						
0.9/0.1	1750	24	700	18	10834	12.2
0.75/0.25	1800	22	950	20	11974	9.5
0.5/0.5	1800	20	750	39	13762	7.3
0.25/0.75	1800	18	1000	38	15294	6.1
0.1/0.9	1800	24	1000	40	17869	5.5
Weighting global						
0.9/0.1	1800	22	950	20	11974	9.5
0.75/0.25	1800	20	900	28	12799	8.2
0.5/0.5	1800	20	950	35	14329	7.0
0.25/0.75	1800	18	950	40	15284	6.1
0.10/0.9	1800	18	1000	40	15786	5.9
Normalized weighting						
0.9/0.1	1800	26	500	13	10136	14.0
0.75/0.25	1800	22	950	20	11974	9.5
0.5/0.5	1800	18	950	40	15284	6.1
0.25/0.75	1800	22	950	40	16658	5.9
0.1/0.9	1800	32	1000	40	20801	5.1

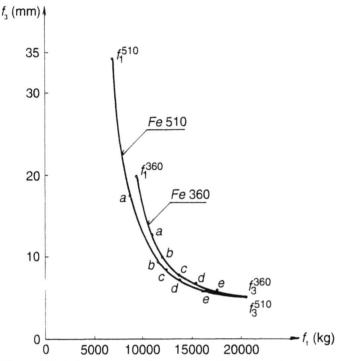

Figure 8.7. Optima in a coordinate-system $f_1 - f_3$ for various weighting coefficients, for steels Fe 360 and Fe 510, according to the weighting min-max technique. The points relate to the following weighting coefficients: a) $w_1 = 0.90$, $w_2 = 0.10$, b) $w_1 = 0.75$, $w_2 = 0.25$, c) 0.50/0.50, d) 0.25/0.75, and e) 0.10/0.90.

Table 8.5. Characteristics of optimized beams made of steel Fe 360 and Fe 510 .

Steel	Technique		h	t_w (mm)	b	t_f	f_1 (kg)	f_3 (mm)
Fe 360	Single-objective	f_{1min}	450	22	700	19	9332	19.7
	Optimization	f_{3min}	1800	32	1000	40	20801	5.1
	Min-max method		1550	30	750	25	16343	8.8
Fe 510	Single-objective	f_{1min}	1300	20	500	17	7051	34.2
	Optimization	f_{3min}	1800	32	1000	40	20301	5.1
	Min-max method		1500	24	750	40	14253	10.2

ent techniques. The differences between results are very small. All the techniques treat the variables as continuous ones and give unrounded optima. To give plate sizes available in the market, a secondary search is used for finding the discrete optima. The discrete steps for h and b are 50 mm, for thicknesses $t_w/2$ and $t_f 1$ mm.

In Table 8.4 the multiobjective – Pareto – optima are given, obtained using five different techniques, for steel Fe 360 and for the ratio $k_f/k_m = 1.5$. Figure 8.7 shows the results in the coordinate-system $f_1 - f_3$, for steels Fe 360 and Fe 510. The notation f_1^{510} means the optimum of f_1 for steel Fe 510.

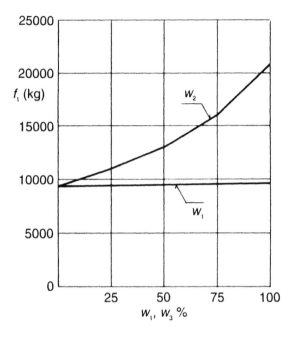

Figure 8.8. Effect of weighting coefficients w_2 and w_3 on f_1.

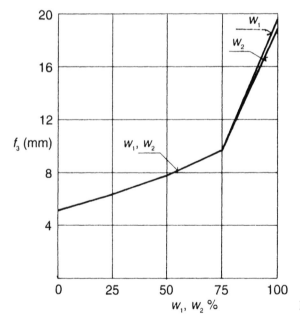

Figure 8.9. Effect of w_1 and w_2 on f_3.

The Pareto-optima for various weighting coefficients of the weighting min-max technique can be seen between the limiting points of the single-objective optima. Note that the points c/ are the same also for min-max technique. It can be seen that the single-optimum of the deflection f_3 does not depend on the steel type, i.e. $f_3^{360} = f_3^{510}$.

Table 8.6. Characteristics of beams optimized using single-objective optimization technique.

Objective function	h (mm)	t_w	b	t_f	f_1 (kg)	f_2 (kg)	f_3 (mm)
f_1	1450	22	700	19	9332	6888	19.7
f_2	1500	24	700	16	9577	6876	19.0
f_3	1800	32	1000	40	20801	16202	5.1

It can be seen from Table 8.5 that cost savings of 24% may be achieved using Fe 510 instead of Fe 360, but the deflection will be nearly doubled.

Table 8.6 shows the results of the single-objective optimization for Fe 360 and $k_f/k_m = 1.5$ for the three objective functions.

Figures 8.8 and 8.9 show the effect of the relative importance of an objective function on the value of the other objective function.

The investigated numerical example illustrates the possibilities given for designers to select the most suitable structural version considering the cost, mass and maximum deflection of a structure.

It can be seen from Table 8.6 that the fabrication cost is about 26% of the total cost and therefore does not affect significantly the optima. In other words, the mass and the cost function are only slightly conflicting. Therefore the mass f_2 is not shown in the figures. The effect of fabrication cost is much more significant in the case of a stiffened plate as it is shown in Chapter 12.

The deflection minimization leads to maximum prescribed sizes and to significant increase of cost and mass. The use of steel Fe 510 instead of Fe 360 results in 24% cost savings without deflection minimization and no savings considering the deflection minimization.

The multiobjective optimization gives structural versions for selected weighting coefficients according to Table 8.4 and Figure 8.7.

CHAPTER 9

Tubular members

9.1 COMPRESSION

9.1.1 *Centrally compressed steel struts*

In this section the advantages of circular or square hollow sections (CHS or SHS) over double angle section (Fig. 9.1) are shown.

The main advantages of CHS or SHS over angle sections are as follows: 1. The overall buckling strength is higher. This is expressed e.g. in EC3 prescribing the buckling curve (b) for CHS and SHS and curve (c) for angles, 2. The radius of gyration (r) in function of the cross-sectional area (A) is much higher than that for double angle profiles, and 3. The effective buckling length of chords and braces in trusses is smaller than that of angle profiles.

Trusses constructed from angles have several other disadvantages such as difficulties in manufacturing of nodes with gusset plates or the need of connecting the pairs of angles in prescribed distances. It should be noted that our investigations relate to trusses of larger roofs or smaller bridges and not to columns or towers in which single angle profiles can be advantageously used without gusset plates.

The radius of gyration (r) can be expressed by the cross-sectional area (A) as follows. For CHS and SHS this relationship can be exactly expressed using the local slenderness

$$\delta_C = D / t \quad \text{and} \quad \delta_S = b / t \tag{9.1}$$

where D and b are the mean diameter and mean width, t is the thickness (Fig. 9.1). Limiting values for δ_C and δ_S are given for instance by EC3 for CHS

$$\delta_{CL} = 70 * 235 / f_y \tag{9.2}$$

where f_y is the yield stress, so for $f_y = 235$ and 355 MPa $\delta_{CL} = 70$ and 50, respectively. Note that, according to CIDECT (Wardenier et al. 1991), for both steel grades $\delta_{CL} = 50$ is given. For SHS, according to CIDECT (Packer et al. 1992),

$$\delta_{SL} = 1.25 \sqrt{E / f_y} \tag{9.3}$$

thus, for steels of $f_y = 235$ and 355 MPa, $\delta_{SL} = 35$ and 30, resp. The formulae and values of a_C and a_S are summarized in Table 9.1.

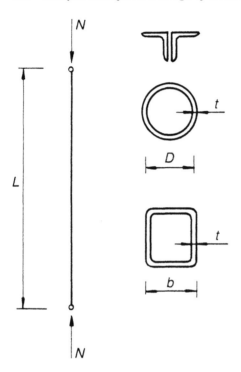

Figure 9.1. Compressed strut of CHS, SHS and double-angle section.

Table 9.1. The radius of gyration expressed by cross-sectional area for CHS and SHS.

	CHS	SHS
A	$\pi D^2 / \delta_C$	$4b^2 / \delta_S$
I_x	$\pi D^4 / (8\delta_C)$	$b^4 / (24\delta_S)$
$r = \sqrt{I_x / A} = a\sqrt{A}$	$a_C = \sqrt{\delta_C / (8\pi)}$	$a_S = \sqrt{\delta_S / 24}$
δ_L ($f_y = 235$ MPa)	$\delta_{CL} = 70$	$\delta_{SL} = 35$
δ_L ($f_y = 355$ MPa)	$\delta_{CL} = 50$	$\delta_{SL} = 30$
a ($f_y = 235$ MPa)	$a_{CL} = 1.6689$	$a_{SL} = 1.2076$
a ($f_y = 355$ MPa)	$a_{CL} = 1.4105$	$a_{SL} = 1.1180$

For hot rolled equal leg double-angle sections an approximate a_d value can be obtained using r and A values of standard sections according to ISO 657-1 (1989) or DIN 1028 (1976). The calculations result in 0.6938 for single angles and

$$a_d = 0.6938 / \sqrt{2} = 0.49, \quad r = 0.49\sqrt{A} \tag{9.4}$$

for double-angle profiles. It can be seen that a_d is much smaller than a_{CL} and a_{SL}.

In the optimum design of a centrally compressed strut the cross-sectional area should be minimized considering the overall and local buckling constraints:

$$A \rightarrow \min$$

The overall buckling constraint is defined according to EC3 as follows (Chapter 3)

$$N / A \leq \chi f_y, \quad 1 / \chi = \phi + \sqrt{\phi^2 - \bar{\lambda}^2}$$

$$\bar{\lambda} = \lambda / \lambda_E = KL / (r\lambda_E), \quad \lambda_E = \pi \sqrt{E / f_y} \tag{9.5}$$

$$\phi = 0.5 \left[1 + \alpha(\bar{\lambda} - 0.2) + \bar{\lambda}^2 \right]$$

where $\alpha = 0.34$ for CHS and SHS, $\alpha = 0.49$ for double angle sections. K is the end restraint factor, for double angles $K = 1$, for CHS and SHS in trusses, according to CIDECT (Rondal et al. 1992), for chords $K = 0.9$, for braces $K = 0.75$. For $E = 2.1 * 10^5$ MPa and $f_y = 235$ MPa $\lambda_E = 93.91$.
The local buckling constraint is

$$\delta \leq \delta_L \tag{9.6}$$

Treating Equation (9.6) as equality (active constraint) the two unknown sizes of CHS or SHS (D, t) or (b, t) can be calculated.

In the optimum design of tubular trusses these sizes should be treated separately, since the strength and geometric limitations for truss nodes rely on these unknowns (Farkas 1990). Instead of these sizes we use here the relationship $r = a\sqrt{A}$, since this method is suitable for angle sections.

Using this relationship, A will be the sole unknown in the overall buckling constraint.

Introducing the symbols

$$c_0 = 100K / \lambda_E, \quad x = 10^4 N / L^2, \quad y = 10^4 A / L^2 \tag{9.7}$$

where L is the strut length in mm, A in mm^2, N is the compressive force in [N], Equation (9.5) can be written as

$$\frac{x}{f_y} \leq \frac{y}{\phi + \sqrt{\phi^2 - \left(c_0^2 / a^2 y\right)}} \qquad \phi = 0.5 \left[1 + \alpha \left(\frac{c_0}{a\sqrt{y}} - 0.2 \right) + \frac{c_0^2}{a^2 y} \right] \tag{9.8}$$

Unfortunately, it is impossible to solve Equation (9.8) for y in closed form, therefore a computer method is used to calculate y for given x.

Table 9.2 shows the relationships $10^4 N / L^2 - 10^4 A / L^2$ for CHS, SHS and double-angle sections In the case of CHS and SHS the end restraint factors $K = 0.9$ and

Table 9.2. Required $10^4 A / L^2$ values and slendernesses $\lambda = KL / r$ in function of $10^4 N / L^2$ for CHS and SHS in the case of $K = 0.9$ and 0.75, respectively, and for double-angle section for $K = 1$. Slendernesses are given only for CHS with $K = 0.9$. All values are calculated for yield stress $f_y = 235$ MPa.

$10^4 N/L^2$	10	100	1000	10,000
CHS-0.9	0.1327	0.559	4.36	40.93
CHS-0.75	0.106	0.469	4.24	40.67
SHS-0.9	0.1792	0.672	4.51	41.4
SHS-0.75	0.1457	0.52	4.32	41.01
Angles 1.0	0.484	1.66	6.79	45.6
Slenderness for CHS-0.9	148	72.1	25.8	8.4

0.75, for double angle sections $K = 1$ is considered, so these values can be used for the design of various compression members in trusses.

It can be seen that the cross-sectional area of double-angle profiles is much greater than those of CHS and SHS. The difference or the savings in mass depends on N / L^2-value or on the strut slenderness λ, which is also given for CHS in the case of $K = 0.9$. The higher the slenderness the larger the mass savings.

The ratio between the A / L^2-values for double-angle sections and CHS $K = 0.9$ varies from 3.5 to 1.1 in the range of

$$10^4 N / L^2 = 25 - 10000, \quad \lambda = 114 - 8$$

For SHS $K = 0.9$ this ratio varies from 2.7 to 1.1. These numbers illustrate the significant mass savings.

It should be noted that, for design purposes of CHS and SHS struts, the Japanese Road Association (JRA) overall buckling curve (Hasegawa et al. 1985) can be used instead of EC3 curve (b). In this case closed formulae can be given for cross-sectional sizes.

$$N / A \leq \chi f_y$$

$$\chi = 1 \qquad \qquad \text{for} \quad 0 \leq \bar{\lambda} \leq 0.2$$

$$\chi = 1.109 - 0.545\, \bar{\lambda} \quad \text{for} \quad 0.2 \leq \bar{\lambda} \leq 1 \qquad\qquad (9.9)$$

$$\chi = \frac{1}{0.773 + \bar{\lambda}^2} \qquad \text{for} \quad \bar{\lambda} \geq 1$$

Introducing the symbols

$$\vartheta_C = 100 D / L \quad \text{and} \quad \vartheta_S = 100 b / L$$

and using $\bar{\lambda} = c / \vartheta$ the closed formulae are as follows.

For $0.2\vartheta \leq c \leq \vartheta$

$$\vartheta = 0.24572\, c \left[1 + \sqrt{1 + \frac{14.93475 v}{c^2}} \right] \qquad\qquad (9.10a)$$

and for $\vartheta \leq c$

$$\vartheta = \left[0.3865 v \left(1 + \sqrt{1 + \frac{6.69424 c^2}{v}} \right) \right]^{1/2} \qquad\qquad (9.10b)$$

for CHS

$$c_c = \frac{100 K \sqrt{8}}{\lambda_E}, \quad v_{CL} = \frac{10^4 N}{L^2} * \frac{\delta_{CL}}{\pi f_y}$$

for SHS

$$c_S = \frac{100 K \sqrt{6}}{\lambda_E}, \quad v_{SL} = \frac{10^4 N}{L^2} * \frac{\delta_{SL}}{4 f_y}$$

Using the AISC column curve given by Equations (3.15a and b) one obtains for CHS struts

for $\vartheta \geq c_c / 1.41$

$$\vartheta = 0.0455 c_c \left[1 + \sqrt{107.2674 + \frac{153.7543 \delta_{CL}}{c_c^2 f_y} \frac{10^4 N}{L^2}} \right] \tag{9.11a}$$

and for $\vartheta \leq c_c / 1.41$

$$\vartheta = 0.094831 c_c \left[-1 + \sqrt{1 + \frac{4719.4746 \delta_{CL}}{c_c^2 f_y} \frac{10^4 N}{L^2}} \right]^{1/2} \tag{9.11b}$$

On the basis of calculated optimum dimensions designers should select the actual CHS or SHS using the tables of standard available profiles. It is advisable to consider not only the optimum dimensions but also the corresponding required cross-sectional area and radius of gyration. In the numerical examples of this book the following standards have been used:

– ISO/DIS 4019/2 (1979) Draft International Standard for cold-formed steel hollow sections.

– DIN 2448 unwelded CHS, DIN 2458 (1981) welded CHS,

– DIN 59410 (1974) hot-finished SHS, DIN 59411 (1978) cold-formed welded SHS.

Now the following drafts of EN standards exist:

– Draft prEN 10210-2:1996 hot-finished steel hollow sections, tolerances. dimensions and sectional properties,

– Draft prEN 10219-2:1996 cold-formed welded steel hollow sections, tolerances, dimensions and sectional properties.

9.1.2 *Unsafe design using the Euler buckling curve*

Authors dealing with the optimum design of tubular trusses have neglected the overall buckling of compression members prescribing constant permissible stresses for tension and compression rods (e.g. Khot & Berke 1984), or the overall buckling is considered by the Euler buckling formula (e.g. Vanderplaats & Moses 1972, Saka 1980, Amir & Hasegawa 1994)

$$\sigma_E = \pi^2 E / \lambda^2, \quad \lambda = KL/r, \quad r = \sqrt{I_x / A} \tag{9.12}$$

where E is the elastic modulus, λ is the slenderness, K is the end restraint factor (for pinned ends $K = 1$), I_x is the moment of inertia, A is the cross-sectional area, r is the radius of gyration.

For CHS, using the notation $\delta = D/t = (d-t)/t$, where D is the mean diameter and d is the outside diameter, t is the thickness, the following formulae are valid

$$I_x = \frac{\pi D^3 t}{8} = \frac{\pi D^4}{8\delta}, \quad A = \frac{\pi D^2}{\delta}, \quad r = \frac{D}{\sqrt{8}} = a\sqrt{A}, \quad a = \sqrt{\frac{\delta}{8\pi}} \tag{9.13}$$

Thus,

$$\sigma_E = \frac{\pi EA}{8K^2 L^2} \delta \qquad (9.14)$$

It can be seen that the local slenderness δ plays an important role in the buckling strength, therefore the selection of the limiting value δ_L influences the optimum design significantly. The first author has verified (Farkas 1992) that the local buckling constraint is active in the optimum design of a centrally compressed CHS strut. E.g. Vanderplaats & Moses (1972) have selected for steel tubes the value of $\delta_L = 10$, and this value has been used also by Saka (1980) and Amir & Hasegawa (1994) (note that in Amir & Hasegawa (1994) in Eq. 3 the erroneous value of 3 is printed instead of 8). Since in the EC3 $\delta_L = 70*235/f_y$ is given for Class 2 sections to be used in tubular trusses, i.e. 70 for a steel of yield stress $f_y = 235$ MPa and 50 for $f_y = 355$ MPa, the value of 10 is incorrect and leads to uneconomic solutions.

In the contrary, the use of the Euler formula leads to unsafe solutions, since it does not take into account the initial crookedness and residual stresses. In (Saka 1990) the AISC buckling curve has been used. Farkas & Jármai (1994) have applied the EC3 buckling formulae and have shown that the optimum slope angle of a roof truss depends on the cross-section type of compression members and the use of CHS is much more economic than that of double angle profile (Section 11.3).

In the following we compare the cross-sectional areas of a CHS compressed strut calculated from the Euler curve and from the EC3 buckling formula. In the calculations the values of $f_y = 355$ MPa, $a_L = \sqrt{50/(8\pi)} = 1.4105$ and $K = 1$ are used. Using Equation (9.13) the slenderness can be expressed by A as follows:

$$\lambda^2 = \frac{L^2}{r^2} = \frac{L^2}{a^2 A} = \frac{10^4}{a^2} * \frac{1}{10^4 A / L^2} = \frac{5027}{10^4 A / L^2} \qquad (9.15)$$

The overall buckling constraint, using the Euler formula, is

$$\frac{N}{A} \leq \chi f_y, \quad \chi = \frac{1}{\bar{\lambda}^2} \quad \text{for} \quad \bar{\lambda} \geq 1 \qquad (9.16)$$

$$\chi = 1 \quad \text{for} \quad \bar{\lambda} \leq 1$$

where

$$\bar{\lambda} = \lambda / \lambda_E, \quad \lambda_E = \pi \sqrt{E / f_y} = 76.4091 \qquad (9.17)$$

From

$$\frac{10^4 N / L^2}{10^4 A / L^2} \leq \frac{f_y}{\bar{\lambda}^2} = \frac{f_y \lambda^2_E}{\lambda^2} \qquad (9.18)$$

using Equation (9.15) one obtains

$$\frac{10^4 A}{L^2} = \frac{1}{76.4091} \sqrt{\frac{5027}{355}} \sqrt{\frac{10^4 N}{L^2}} = 0.049247 \sqrt{\frac{10^4 N}{L^2}} \qquad (9.19)$$

valid for $\lambda \geq \lambda_E$. For $\lambda \leq \lambda_E$ taking $\chi = 1$ in Equation (9.16) we get

Table 9.3. Required $10^4 A/L^2$-values for some $10^4 N/L^2$-values in the case of a compressed CHS strut, $f_y = 355$ MPa, $K=1$.

		10	100	305.7	1000	10,000
	$10^4 A/L^2$ (N/mm^2)	10	100	305.7	1000	10,000
$10^4 A/L^2$	Euler	0.1557	0.4925	0.8610	2.8169	28.17
	EC3	0.1766	0.6273	1.3171	3.4975	30.60
	Difference %	12	21	35	19	8
λ	EC3	168	89	66	38	13

$$\frac{10^4 A}{L^2} \geq \frac{10^4 N}{L^2 f_y} \tag{9.20}$$

According to EC3 the overall buckling constraint is

$$\frac{N}{A} \leq \frac{\chi f_y}{\gamma_{M1}}, \quad \gamma_{M1} = 1.1, \quad \frac{1}{\chi} = \phi + \sqrt{\phi^2 - \bar{\lambda}^2}$$

$$\phi = 0.5\left[1 + 0.34(\bar{\lambda} - 0.2) + \bar{\lambda}^2\right] \tag{9.21}$$

Introducing the symbols $c_0 = 100K/\lambda_E$, $x = 10^4 N/L^2$ and $y = 10^4 A/L^2$, where L (mm) is the strut length, A (mm^2) is the required cross-sectional area, N is the factored compressive force in (N), Equation (9.9) can be written as

$$\frac{\gamma_{M1} x}{f_y} \leq \frac{y}{\phi + \sqrt{\phi^2 - \left(c_0^2/a^2 y\right)}}$$

$$\phi = 0.5\left[1 + 0.34\left(\frac{c_0}{a\sqrt{y}} - 0.2\right) + \frac{c_0^2}{a^2 y}\right], \quad \lambda = \frac{100K}{a\sqrt{y}} \tag{9.22}$$

A computer method is used to calculate y for a given x. Results are summarized in Table 9.3. It can be seen that the results obtained by the Euler formula are unsafe by 19-35% in the range of $\lambda = 38$-89, so the Euler formula gives incorrect solutions.

9.1.3 *Absorbed energy of CHS and SHS braces cyclically loaded in tension-compression*

Braces play an important role in the earthquake-resistant design of frames. The efficiency of bracing is characterized by the absorbed energy which can be obtained as the area of the hysteresis loop.

Studies have shown that the first critical overall buckling strength decreases during the second and third cycle, but after a few cycles the hysteresis loop becomes stable. This degradation is caused by the Bauschinger-effect and by the effect of residual camber as explained by Popov & Black (1981). Unfortunately, these effects cannot be considered by analytical derivations, thus, the characteristics of the stable hysteresis loop will be taken from the experimental data published in the literature.

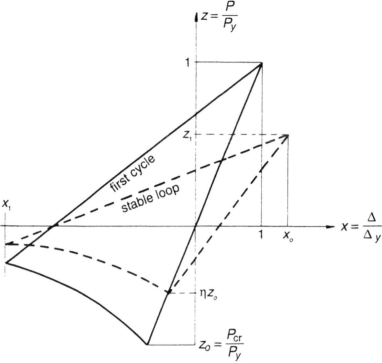

Figure 9.2. Characteristics of a stable hysteresis loop.

Table 9.4. Characteristics of the stable hysteresis loop according to Figure 9.2.

Reference	x_0	x_1	η	z_1	Cross-section
Jain (1980)	2	−12	0.5	1	SHS
Liu (1988)	2	−10	0.5	0.5	RHS
Matsumoto (1986)	2	−10	0.5	0.5	CHS
Nonaka (1977)	4	−4	0.5	0.8	Solid square
Ochi (1990)	1	−10	0.5	0.5	CHS
Papadrakakis (1987)	1	−4	0.5	0.5	CHS
Prathuangsit (1978)	1	−12	0.5	0.5	I
Shibata (1982)	5	−5	0.5	0.5	I

Our aim is to derive simple closed formulae for the calculation of the area of the stable hysteresis loop. The derived formulae enable designers to analyze the effect of some important parameters such as the yield stress of steel, end restraint and cross-sectional shape, and to work out aspects of optimization, i.e. the increasing of the energy-absorbing capacity of braces.

The stable hysteresis loop is shown schematically in Figure 9.2. The characteristics obtained by experiments are summarized in Table 9.4. It can be seen that, for η and z_1 the approximate value of 0.5 is predominantly obtained. The sum of relative axial shortenings $x + |x_1|$ varies in range 5-14. On the basis of these data we consider the values $\eta = z_1 = 0.5$ and $x_0 = 1$, $x_1 = -10$.

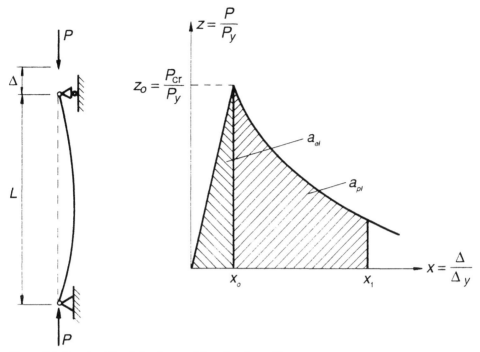

Figure 9.3. Post-buckling behaviour and the related specific areas.

Another important problem is the local buckling. According to many authors, e.g. Lee & Goel (1987), it is recommended to avoid local buckling. Unfortunately, one can find very few proposed values for the limiting D/t or b/t ratios, in the case of cyclic plastic stress. For CHS Zayas et al. (1982) proposed $(D/t)_L = 6820/f_y$ where f_y is the yield stress in MPa, thus, for yield stress of 235 and 355 MPa one obtains 29 and 20, respectively.

For SHS or RHS Liu & Goel (1988) proposed $(b/t)_L = 14$ for $f_y = 371$ MPa, thus, we take for 355 MPa the value of 15 and for 235 MPa $15(355/235)^{0.5} = 19$.

The limitation of the strut slenderness plays also an important role. API (1989) proposed $KL/r < 80$.

The relationship axial force – axial shortening $(P - \Delta)$ (Fig. 9.3) has been derived for CHS struts by Supple & Collins (1980) using the simple plastic hinge method:

The bending moment at the middle of the rod is $M = a_0 P$, thus $a_0 = M/P$. The plastic axial shortening is caused by curvature, so (Fig. 9.3)

$$\Delta_{pl} = \int_0^L (ds - dx) \cong \frac{1}{2} \int_0^L \left(\frac{dy}{dx}\right)^2 dx \quad \text{since} \quad \frac{ds}{dx} = \sqrt{\left(\frac{dy}{dx}\right)^2 + 1} \approx 1 + \frac{1}{2}\left(\frac{dy}{dx}\right)^2$$

Taking $y = a_0 \sin(\pi x / L)$ we obtain

$$\Delta_{pl} = \frac{1}{2} \int_0^L \frac{\pi^2 a_0^2}{L^2} \cos^2 \frac{\pi x}{L} dx = \frac{\pi^2 a_0^2}{4L} = \frac{\pi^2}{4L}\left(\frac{M}{P}\right)^2$$

The squash load is $P_y = 2R\pi t f_y$. The plastic stress distribution shown in Figure 9.4 can be divided into two parts, one of them is caused by the compressive force, the second is caused by the bending moment. The plastic compressive force is

$$P = 2f_y R\pi t - 2f_y R\Theta = P_y\left(1 - \frac{\Theta}{\pi}\right) \quad \text{from which} \quad \frac{\Theta}{2} = \frac{\pi}{2} - \frac{\pi P}{2P_y}$$

The bending moment of the plastic zone is

$$M = 2\int_0^{\Theta/2} R\cos\varphi\,(2f_y tRd\varphi) = 4f_y R^2 t \sin\frac{\Theta}{2} \quad \text{where} \quad \sin\frac{\Theta}{2} = \cos\left(\frac{\pi P}{2P_y}\right)$$

Thus,

$$\Delta_{pl} = \frac{D^2}{4L}\left[\frac{P_y}{P}\cos\left(\frac{\pi P}{2P}\right)\right]^2$$

and

$$\Delta = \Delta_{el} + \Delta_{pl} = \frac{PL}{AE} + \frac{\alpha D^2}{4L}\left[\frac{P_y}{P}\cos\left(\frac{\pi P}{2P_y}\right)\right]^2 \tag{9.23}$$

where α is the end restraint factor, for pinned ends $\alpha = 1$, for fixed ends $\alpha = 4$.

A numerical example shows that the plastic hinge method gives a good approximation in the post-buckling range. Figure 9.4 shows the calculation results of three methods for the following data: $D = 114.3$, $t = 2.38$ mm, $f_y = 248$ MPa, $E = 2.06 * 10^5$ MPa, $L = 3175$ mm, pinned ends. The investigated strut has the following characteristics: $A = 854.62$ mm^2, $\delta_C = D/t = 48$, $\lambda = L/r = 78.57$, $P_y = 211.95$ kN. According to Equation (9.11) $P_{cr} = 160.10$ kN. $\Delta_{el} = P_{cr}L/(AE) = 2.89$, $\Delta_y = P_y L/(AE) = 3.82$ mm. In Figure 9.4 the curve (a) represents the plastic hinge method calculated by Equation (9.23), The curve (b) was determined by Chen & Sugimoto (1987) who used the finite segment method. The curve (c) was calculated by FSM combined with the influence coefficient method by Han & Chen (1983). It can be seen that the difference between curves (a)-(b)-(c) is not significant, thus, the simple plastic hinge method can be used for optimization purposes.

Using notations $x = \Delta/\Delta_y$, $z = P/P_y$, $z_0 = P_{cr}/P_y$, $x_0 = \Delta_{el}/\Delta_y = P_{cr}/P_y = z_0$ Equation (9.23) can be written in the form

$$x - x_0 = C_1\left[\frac{\cos^2(\pi z/2)}{z^2} - C_2\right], \quad C_1 = \frac{\alpha D^2 E}{4L^2 f_y}, \quad C_2 = \frac{\cos^2(\pi z_0/2)}{z_0^2} \tag{9.24}$$

For $z < 0.4$ the following approximation is acceptable

$$\cos(\pi z/2) \approx 1 - \pi^2 z^2/8 \quad \text{and} \quad \cos^2(\pi z/2) \approx 1 - \pi^2 z^2/4 \tag{9.25}$$

and Equation (9.24) takes the form

$$x - x_0 = C_1\left(\frac{1}{z^2} - \frac{\pi^2}{4} - C_2\right) \tag{9.26}$$

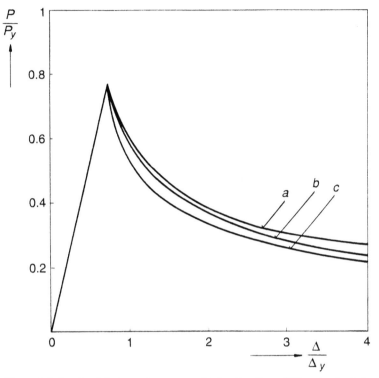

Figure 9.4. Load-axial shortening curves obtained by three different calculation methods: 1. Plastic hinge method, 2. The method of Chen & Sugimoto (1987), and 3. The method of Han & Chen (1983).

from which one obtains

$$z = C_1^{1/2}\left(x - x_0 + \frac{C_1\pi^2}{4} + C_1C_2\right)^{-1/2} \tag{9.27}$$

The areas shown in Figure 9.4 can be calculated as follows:

$$a_{el} = z_0^2 / 2 \tag{9.28}$$

and

$$a_{pl} = \int_{x_0}^{x_1} z\,dx = 2C_1^{1/2}\left[\left(x_1 - x_0 + \frac{C_1\pi^2}{4} + C_1C_2\right)^{1/2} - \left(\frac{C_1\pi^2}{4} + C_1C_2\right)^{1/2}\right] \tag{9.29}$$

It is possible to derive similar formulae for SHS struts. The results are as follows:

$$\Delta_{pl} = \frac{\alpha\pi^2}{4L}\left(\frac{M}{P}\right)^2 = \frac{\alpha\pi^2 b^2}{16L}\left(\frac{3}{4z} - z\right)^2 \tag{9.30}$$

and

$$x - x_0 = \frac{\Delta_{pl}}{\Delta_y} = C_3\left(\frac{3}{4z} - z\right)^2 + C_4, \quad C_3 = \frac{\alpha\pi^2 b^2}{16L^2 f_y}, \quad C_4 = C_3\left(\frac{3}{4z_0} - z_0\right)^2 \tag{9.31}$$

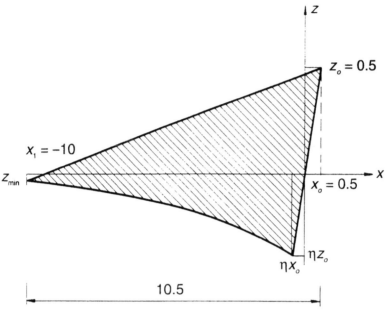

Figure 9.5. Area of the stable hysteresis loop.

Expressing z from Equation (9.31) we obtain

$$z = \frac{1}{2C_3^{1/2}}\left[\left(x - x_0 + 3C_3 + C_4\right)^{1/2} - \left(x - x_0 + C_4\right)^{1/2}\right] \tag{9.32}$$

and the area in the post-buckling range is

$$a_{pl} = \int_{x_0}^{x_1} z\,dx = \frac{1}{3C_3^{1/2}}\left[\left(x_1 - x_0 + 3C_3 + C_4\right)^{3/2} - \left(x_1 - x_0 + C_4\right)^{3/2} - \right.$$
$$\left. - \left(3C_3 + C_4\right)^{3/2} + C_4^{3/2}\right] \tag{9.33}$$

We consider the stable hysteresis loop according to Figure 9.5. The whole specific absorbed energy as the area shown in Figure 9.5 is

$$\sum_i a_i = a_{el} + a_{pl} + 10*0.5/2 - 10.5 z_{\min}/2 \tag{9.34}$$

For a_{el} and a_{pl} we use Equations (9.28), (9.29) or (9.33), but instead of $x_0 = z_0$ we calculate with $\eta x_0 = \eta z_0, \eta = 0.5$. $z_{\min}$ is calculated using Equations (9.27) or (9.32) taking $x = 10$ and instead of x_0 taking $0.5x_0$.

Numerical examples
If the design load in braces is not known, the dimensions of cross-sections can be calculated on the basis of slenderness limitations. For CHS braces the mean diameter may be obtained from the constraint on strut slenderness

$$\lambda = KL/r = \sqrt{8}KL/D \le \lambda_0 = 80, \quad D \ge \sqrt{8}KL/\lambda_0 \tag{9.35}$$

The thickness may be calculated using the local buckling constraint

$$\delta_C = D / t \leq \delta_{CL} = 6820 / f_y \qquad (9.36)$$

Similarly, for SHS braces

$$b \geq \sqrt{6} KL / \lambda_0 \qquad (9.37)$$

and

$$\delta_S = b / t \leq \delta_{SL} = 15\sqrt{355 / f_y} \qquad (9.38)$$

In Table 9.5 the calculation of the whole area of the stable hysteresis loop is summarized in the case of pinned ends, for CHS and SHS braces, made of steels with yield stress of 235 and 355 MPa, respectively. The cross-sections are taken according to ISO/DIS 4019.2 (1979) (see Section 9.1.1). The first overall buckling force P_{cr} is calculated according to EC3 curve 'b'. The total absorbed energy is calculated as

$$W_{absorb} = \left(\sum_i a_i \right) f_y^2 AL / E \qquad (9.39)$$

To complement the numerical examples treated in Table 9.5, another end restraint case has also been calculated. In the case of SHS and yield stress of 355 MPa the end restraint is changed from pinned ends to partially restrained ones with $K = 0.8$ and $\alpha = 2$. The ISO cross-section is then 160*10 mm, $A = 5660$ mm^2, $z_0 = 0.5675$, $C_3 = 0.4561$, $C_4 = 2.5379$, $a_{pl} = 1.8291$, $\Sigma a_i = 3.6299$ and $W_{absorb} = 73.978$ kNm.

The numerical examples show that, if we design the braces considering the strut slenderness limitation and the limitations on local slendernesses, there will be no significant difference between the specific absorbed energies. In the total absorbed energies are large differences. W is 38% larger for SHS than for CHS, and it is 160% larger for yield stress of 355 MPa than for 235 MPa.

If the end restraint is changed from pinned ends to partially restrained ones, in the case of SHS and 355 MPa, W decreases by 35%.

Table 9.5. Numerical examples. $L = 6$ m, pinned ends, $K = 1$, $\alpha = 1$, $\lambda_0 = 80$, dimensions in mm.

Cross-section	CHS	CHS	SHS	SHS
f_y (MPa)	235	355	235	355
Required D or b	212	212	183.7	183.7
Limiting slenderness	29	20	19	15
ISO cross-section	219.1/8	219.1/10	200*10	200*12.5
D or b	211.1	209.1	190	187.5
A (mm^2)	5310	6570	7260	8840
r	74.7	74.0	76.5	75.2
a_{el}	0.0595	0.0390	0.0617	0.0405
a_{pl}	2.0241	1.5808	2.1145	1.7148
$10.5 z_{min}/2$	0.7954	0.6288	0.8373	0.6725
Σa_i	3.7882	3.4910	3.8389	3.5828
W_{absorb} (kNm)	31.739	82.586	43.975	114.042

The main conclusion is that, to increase the total absorbed energy, the cross-sectional area and the yield stress should be increased. This conclusion contradicts to the general optimum design problems in which the minimum cross-sectional area-design is sought.

9.1.4   *Optimum design and imperfection-sensitivity of centrally compressed SHS and CHS aluminium struts*

9.1.4.1   *Introduction*

In the past several authors (e.g. Shanley 1960, Gerard 1962) have solved optimization problems with two active stability constraints using buckling formulae which do not consider the effect of initial imperfections. Some authors (e.g. Thompson & Hunt 1973, Tvergaard 1973, Van der Neut 1973) have shown that this 'naive' method leads to imperfection-sensitive solutions. Thompson (1972) has named this optimization 'as a generator of structural instability'.

Our first aim is to show that the optimization which considers the initial imperfections gives solutions of much smaller imperfection-sensitivity than that without imperfections. Rondal & Maquoi (1981) have already shown that the method considering the imperfections does not lead to severe erosion of compressive strength.

A detailed survey of the coupled instability phenomena has been worked out by Gioncu (1994).

Our second aim is to extend the optimization worked out for steel SHS and CHS struts (Section 9.1.1) to aluminium struts, since it has been mentioned in Section 3.3.1 that the overall buckling formulae given by BS 8118 (1991) are the same as those given by EC 3. Some optimum design results are given to enable designers the use of optimization in these cases.

Note that the design of open-section aluminium compressed struts has been treated by Lai & Nethercot (1992).

9.1.4.2   *Optimum design of compressed aluminium SHS struts without the consideration of initial imperfections*

In order to compare the optimal solutions obtained without and with initial imperfections we derive first the optimal formulae using buckling formulae which do not consider the initial imperfections. The objective function to be minimized is the cross-sectional area. Neglecting the effect of rounding at the corners, a simple approximate formula can be used as follows

$$A = 4bt \tag{9.40}$$

The overall buckling constraint, using the Euler formula for pinned ends, is expressed by

$$\frac{N}{4bt} \leq \frac{\pi^2 E}{\lambda^2} = \frac{\pi^2 E b^2}{6L^2} \tag{9.41}$$

The local buckling constraint, without the effect of initial imperfections, can be written as

$$\frac{N}{4bt} \le \frac{4\pi^2 E}{12(1-\upsilon^2)}\left(\frac{t}{b}\right)^2 \tag{9.42}$$

The constraint on yielding is

$$N/A \le p_0/\gamma_m \tag{9.43}$$

where p_0 is the limiting stress for yielding, γ_m is the material factor.

It can be shown (Farkas 1992) that, using a suitable coordinate-system of two unknowns, the optimum point is determined as the intersection point of the limiting curves of constraints (Eqs 9.41 and 9.42 or Eqs 9.43 and 9.42, respectively). Thus, these constraints are active and can be treated as equalities. Thus, from Equation (9.41) one obtains

$$t = 3L^2 N / \left(2\pi^2 E b^3\right) \tag{9.44}$$

and from Equation (9.42)

$$t^3 = 3(1-\upsilon^2)Nb / \left(4\pi^2 E\right) \tag{9.45}$$

Substituting Equation (9.44) into Equation (9.45) we get

$$b = \left[\frac{9L^6 N^2}{2\pi^4 E^2 (1-\upsilon^2)}\right]^{1/10} \tag{9.46}$$

and

$$t = \frac{3L^2 N}{2\pi^2 E}\left[\frac{2\pi^4 E^2 (1-\upsilon^2)}{9L^6 N^2}\right]^{3/10} \tag{9.47}$$

Our aim is to calculate the A/L^2-values in the function of N/L^2. Thus, using Equations (9.46) and (9.47) we get

$$\frac{A}{L^2} = \frac{4bt}{L^2} = 6\left(\frac{2}{9\pi^6}\right)^{1/5}\left[\frac{N^3(1-\upsilon^2)}{E^3 L^6}\right]^{1/5} \tag{9.48}$$

To scale the dimensions we multiply with 10^4

$$\frac{10^4 A}{L^2} = 6\left(\frac{2}{9\pi^6}\right)^{1/5}\left(10^4\right)^{2/5}\left(1-\upsilon^2\right)^{1/5}E^{-3/5}\left(\frac{10^4 N}{L^2}\right)^{3/5} \tag{9.49}$$

Note that N/L^2 should have a dimension of N/mm^2. For aluminium alloys it is $E = 7*10^4$ MPa and $\upsilon = 0.3$, thus

$$\frac{10^4 A}{L^2} = 0.05441\left(\frac{10^4 N}{L^2}\right)^{3/5} \tag{9.50}$$

Using data of $p_0 = 240$ MPa and $\gamma_m = 1.2$, in the plastic region Equation (9.43) can be written as

Table 9.6. Limiting stresses for heat-treatable alloys.

Alloy	Condition	Product	Thickness over (mm)	Thickness up to and including (mm)	p_0 (MPa)	Nearest equivalent to ISO 209-1
6061	T6	Extrusion	–	150	240	AlMg 1 SiCu
		Drawn tube	–	6	240	
		Drawn tube	6	10	225	
6063	T6	Extrusion	–	150	160	AlMg 0.7 Si
		Drawn tube	–	10	180	
6082	T6	Extrusion	–	20	225	AlSi 1 MgMn
		Extrusion	20	150	270	
		Drawn tube	–	6	255	
		Drawn tube	6	10	240	
7020	T6	Extrusion	–	25	280	AlZn 4.5 Mg 1

$$\frac{10^4 A}{L^2} \geq \frac{10^4 N / L^2}{p_0 / \gamma_m} = \frac{10^4 N / L^2}{200} \tag{9.51}$$

Data of some aluminium alloys, according to BS 8118, are given in Table 9.6.

9.1.4.3 Optimum design of compressed aluminium SHS and CHS struts considering the initial imperfections

On the basis of the Ayrton-Perry formulation

$$\frac{N}{A} + \frac{N(a_0 + y_0)}{W_x} \leq \frac{p_0}{\gamma_m} \tag{9.52}$$

where a_0 is the maximal initial imperfection, y_0 is the elastic deformation due to the compressive force, W_x is the elastic section modulus, the overall buckling constraint can be derived which is used in EC3 and BS 8118 as follows:

$$\frac{10^4 N / L^2}{p_0 / \gamma_m} \leq \frac{10^4 A / L^2}{\phi + \sqrt{\phi^2 - \bar{\lambda}^2}}, \quad \bar{\lambda}^2 = \frac{c_0^2}{a^2 y}, \quad c_0 = \frac{100K}{\lambda_E},$$

$$\lambda_E = \pi \sqrt{\frac{E}{p_0}}, \quad y = \frac{10^4 A}{L^2}, \quad \phi = 0.5\left[1 + \alpha(\bar{\lambda} - 0.2) + \bar{\lambda}^2\right], \tag{9.53}$$

$$x = 10^4 N / L^2. \quad \text{For} \quad \bar{\lambda} \leq 0.2 \quad \phi + \sqrt{\phi^2 - \bar{\lambda}^2} = 1$$

where K is the effective length factor, e.g. for pinned ends $K = 1$, for struts effectively held in position and restrained in direction at both ends $K = 0.7$. According to BS 8118 the imperfection factor for unwelded and symmetric profiles is $\alpha = 0.2$.

In the derivation the member proportional to the initial imperfection is expressed by the reduced slenderness

$$\frac{a_0 A}{W_x} = \alpha (\bar{\lambda} - 0.2) \tag{9.54}$$

thus, this expression should be changed in the investigation of the imperfection-sensitivity.

The local buckling constraint may be expressed by the limiting local slenderness

$$\delta \leq \delta_L \qquad (9.55)$$

for SHS $\delta_S = b/t$; for CHS $\delta_C = D/t$.

Neglecting the effect of initial imperfections, for SHS, it can be written that

$$\frac{4\pi^2 E}{12(1-\upsilon^2)}\left(\frac{t}{b}\right)^2 \geq \frac{p_0}{\gamma_m} \qquad (9.56)$$

From Equation (9.56) we get

$$\left(\frac{b}{t}\right)_L = \sqrt{\frac{\pi^2 E \gamma_m}{3(1-\upsilon^2)p_0}} \qquad (9.57)$$

With values of $E = 7*10^4$ MPa, $p_0 = 240$ MPa, $\gamma_m = 1.2$, $\upsilon = 0.3$ we obtain $(b/t)_L = 35.6$.
According to BS 8118, the limiting plate slenderness is

$$\delta_{SL} = 22\sqrt{250/p_0} = 22.45 \qquad (9.58)$$

which shows that the value of 35.6 is decreased considering the initial imperfections.
The radius of gyration is given in Table 9.1.
According to BS 8118, the limiting local slenderness for CHS is

$$\delta_{CL} = \left(\frac{22}{3}\right)^2 \frac{250}{p_0} \qquad (9.59)$$

9.1.4.4 *Comparison of the optimum designs without and with initial imperfections*
The solutions are summarized in Table 9.7. The imperfection-sensitivity is investigated for the following numerical data: $10^4 N/L^2 = 100$ N/mm^2 and $L = 6$ m ($N = 360$ kN).

The solution without imperfections is (Table 9.7) $10^4 A/L^2 = 0.8623$, thus $A = 3104$ mm^2. Using Equations (9.46) and (9.47) we get

$$\frac{b}{t} = \sqrt{\frac{A\pi^2 E}{3(1-\upsilon^2)N}} = 46.71 \quad \text{and} \quad b = \sqrt{\frac{Ab}{4t}} = 190.4 \text{ mm}$$

Based on the value of $A = 3104$ mm^2 we calculate the buckling strength of this section taking a limiting plate slenderness according to BS 8118 i.e. $b/t = 22.45$ (Eq. 9.58). Then $b = 132$ mm, with Equation (9.53) $\lambda_E = 53.65$, $\bar{\lambda} = 6000\sqrt{6}/(132*53.65) = 2.0753$.

The maximum initial imperfection using Equation (9.54) with $W_x/A = b/3$ is $a_0 = 16.5$ mm. The overall buckling force according to Equation (9.53) is $N_1 = 130.0$ kN.

Table 9.7. Optimum $10^4 A/L^2$-values in the function of $10^4 N/L^2$ for aluminium SHS struts.

$10^4 N/L^2$ (N/mm^2)	10	100	390.64	1000	10,000
Without imperfections elastic (Eq. 9.50)	0.2166	0.8623	1.9532	3.4330	13.6670
Without imperfections plastic (Eq. 9.51)	0.0500	0.5000	1.9532	5.0000	50.0000
With imperfections (Eq. 9.53)	0.4463	1.4719	–	6.1390	50.7595

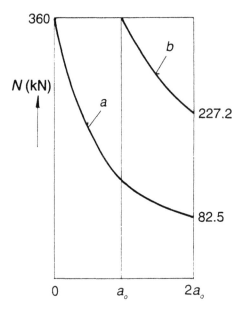

Figure 9.6. The imperfection sensitivity of compressed struts. a) Without initial imperfections, and b) With initial imperfections.

Changing the previous value of $b/t = 22.45$ to 15 and taking $2a_0$ instead of a_0 we get $N_2 = 82.5$ kN.

The solution with imperfections is (Table 9.7) $10^4 A/L^2 = 1.4719$. $A = 5299$ mm^2, with $b/t = 22.45$ we obtain $b = 172.45$ and $t = 7.68$ mm. Investigating the imperfection-sensitivity of this solution we take $b/t = 15$ and $0.4(\lambda - 0.2)$ instead of $0.2(\lambda - 0.2)$ and get $N_3 = 227.2$ kN.

Summarizing the results in a diagram (Fig. 9.6) it can be seen that the sensitivity of the solution initial imperfections is much smaller than that without imperfections. Thus, the concept of two simultaneously active buckling constraints can be applied using buckling formulae which consider the initial imperfections.

9.1.4.5 *Table for the optimum design of compressed aluminium SHS and CHS struts*

Since a closed solution of Equation (9.53) cannot be given, it is solved numerically using a computer program. In the design practice the following data can be varied: N, L, p_0, K, SHS or CHS, α (for unwelded symmetric profiles 0.2, for welded symmetric profiles 0.45).

In Table 9.8 results are given for $p_0 = 240$ and 160 MPa (see Table 9.6), for $K = 1$ and 0.7 and for SHS and CHS. If the $10^4 A/L^2$-values are plotted in the function of $10^4 N/L^2$ in a doubly logarithmic coordinate system, it can be seen that a linear interpolation can be used between the values given in Table 9.8.

It has been shown that the structural optimization with two simultaneously active overall and local buckling constraints may lead to imperfection-sensitive solutions when it is based on buckling formulae which do not consider the initial imperfections. On the contrary, the optimum design results in practically safe and non-imperfection-sensitive solutions when it is based on formulae considering the initial imperfections. The sensitivity of rods designed with initial imperfections is less since they are sensitive only against additional imperfections.

Table 9.8. Optimum 10^4 A/L^2-values in the function of 10^4 N/L^2.

p_0 (MPa)	K	Section	10^4 N/L^2 (N/mm^2)			
			10	100	1000	10,000
240	1	CHS	0.2921	0.9948	5.4534	50.0000
240	1	SHS	0.4463	1.4719	6.1390	50.7595
240	0.7	CHS	0.2070	0.7507	5.2127	50.0000
240	0.7	SHS	0.3150	1.0640	5.5316	50.0000
160	1	CHS	0.2965	1.0894	7.7846	75.0000
160	1	SHS	0.4973	1.6719	8.3891	75.0163
160	0.7	CHS	0.2117	0.9047	7.5947	75.0000
160	0.7	SHS	0.3520	1.2392	7.9248	75.0000

9.1.5 *Tubular columns prestressed by tension ties*

9.1.5.1 *Introduction*

In the design of prestressed columns the effect of initial imperfections has been predominantly neglected in the past. The design of tubular columns prestressed by wire ropes is treated now on the basis of the European column buckling formulae which consider the initial imperfections. Columns of one and three intermediate supports are designed. The necessary prestressing is calculated considering the effect of temperature change and creep of cables. The optimum angle between the core tube and cables is sought in the case of one support to minimize the material cost of the core tube, cables and supporting bars.

Prestressed columns are used as booms for erection purposes or as parts of higher towers. Their advantage is the high compressive strength and lightweight. A detailed study has been published dealing with the design of such columns by Chu & Berge (1963). On the basis of this analysis Mauch & Felton (1967) have worked out an optimum design procedure for these columns. This study has shown that the weight of the core tube can be significantly decreased by prestressing it with one or three intermediate supports.

Moreover Mauch & Felton (1967) have used a local buckling calculation for aluminium tubes which leads to unrealistic CHS dimensions. E.g. in their numerical example a tube of $D*t = 176.5*0.447$ mm with a local slenderness of 395 has been calculated. On the contrary, the British Standard BS 8118 (1991) prescribes, for an aluminium alloy strut of yield stress 240 MPa, a local slenderness limit of 56 (see Section 9.1.4.5).

The calculation method used in the book of Belenya (1975) also neglects the initial imperfection and in the numerical example only one support is considered for a 30 m long column loaded by a compressive force of 2000 kN. Round steel bars have been applied as prestressing elements of diameter 30 mm with a limiting tensile stress of 170 MPa instead of high-strength steel wire ropes.

A calculation method based on the Euler solution is treated in Petersen's book (1980) without the effect of initial imperfections. Schock (1976) has considered the initial crookedness and used wire ropes as ties. Several structural versions have been calculated using different complicated systems of ties. For the comparison the material mass has been used since the cost of assembly was not known.

In the case of prestressing with high-strength cables it is necessary to take into account the decrease of prestress during the service time due to creep and temperature changes. These effects have not been treated in the publications mentioned above.

A relatively simple design method is proposed here for steel CHS columns prestressed by high-strength cables with one or three intermediate supports. In the calculation of the necessary prestressing the effect of creep and temperature change is also considered. In the optimization the optimum angle between the core tube and cables is sought which minimizes the total cost. Since the cost of assembly of such prestressed structures is not known, in the cost function only the material cost of core tube, cables and supporting bars is considered. The shortening of the core tube and that of supporting bars is neglected. The design method is illustrated by a numerical example.

9.1.5.2 *Design of columns with one support (Fig. 9.7)*
Assume that the core tube is fixed by ties at its midpoint so that it buckles in the second mode (Fig. 9.7). Thus, the core tube can be designed for a given factored external axial compressive force F with an effective buckling length of $L/2$ and for a prescribed limiting local slenderness according to EC3

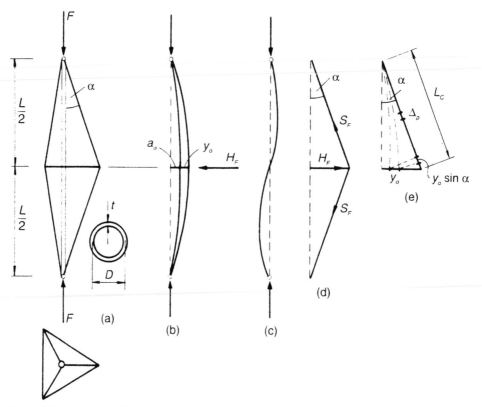

Figure 9.7. a) Prestressed column with one support, b) Initial crookedness and the elastic deformation of the column without prestressing, c) Second mode of column buckling, d) Forces in ties, and e) Measure of prestressing.

$$\delta_L = (D/t)_L = 90*235/f_y \tag{9.60}$$

where f_y is the yield stress. The design inequality is expressed by

$$\frac{F_T}{\pi D^2/\delta_L} \le \frac{\chi f_y}{\gamma_{M1}} \tag{9.61}$$

where $\gamma_{M1} = 1.1$ is the partial safety factor,

$$F_T = F + 3(F_p + S_F)\cos\alpha$$
$$1/\chi = \phi + \sqrt{\phi^2 - \bar\lambda^2}$$
$$\phi = 0.5\,[1 + 0.34\,(\bar\lambda - 0.2) + \bar\lambda^2]$$
$$\bar\lambda = K_e L\sqrt{8}/(2D\lambda_E), \quad \lambda_E = \pi\sqrt{E/f_y}$$

E is the elastic modulus, K_e is the effective length factor, here for pinned ends $K_e = 1$. Note that F_p and S_F are cable forces calculated below (Eqs 9.66 and 9.72) but not known in advance, so we should guess these values and then correct them using an iteration process.

Equation (9.61) can be solved for D numerically using a computer program. Knowing D and t the Euler overall buckling force for pinned ends can be calculated

$$F_E = \pi^2 E I_x/(L/2)^2, \quad I_x = \pi D^3 t/8 \tag{9.62}$$

where I_x is the moment of inertia, for which the table given by the standard DIN 2448 or 2458 can be used (Dutta & Würker 1988) (Section 9.1.1). The EC3 formula is based on the Ayrton-Perry formulation (Maquoi & Rondal 1978). In this calculation the initial imperfection a_0 and the elastic deformation y_0 at the midpoint (Fig. 9.7), which could occur without prestressing, is (Section 3.3.1)

$$a_0 = 0.34\left(\frac{L}{r\lambda_E} - 0.2\right)\frac{W_x}{A}, \quad \frac{W_x}{A} = \frac{D}{4} \tag{9.63}$$

$$y_0 = F_T a_0/(F_E - F_T) \tag{9.64}$$

where W_x is the elastic section modulus.

The horizontal force H_F which eliminates the bending moment $F_T(a_0 + y_0)$ (Fig. 9.7) is obtained from the equation $F_T(a_0 + y_0) = H_F L/4$

$$H_F = \frac{4F_T a_0 F_E}{L(F_E - F_T)} \tag{9.65}$$

The force H_F causes forces in cables (Fig. 9.7d) of

$$S_F = H_F/(2\sin\alpha) \tag{9.66}$$

where α is the angle between the core tube and cables.

The prestressing of wire ropes is realized by turnbuckles in each cable. The measure of prestressing to avoid y_0 at the midpoint should compensate also the elastic elongation of cables due to the prestressing force F_p, the effect of creep and temperature change

Table 9.9. Constants for the calculation of creep in cables (Kmet' 1989).

% of 1370 MPa	c_1	c_2
25	0.0018746	0.424056
30	0.0026588	0.373725
35	0.0054372	0.295726
40	0.0110466	0.248691
45	0.0201812	0.227701
50	0.0335550	0.218934
55	0.0470003	0.196371
60	0.0580456	0.180206
65	0.0662387	0.171571
70	0.0705592	0.168776
75	0.0717165	0.169366
80	0.0.737764	0.176920

$$\Delta_p \geq F_p L_C / (E_C A_C) + \Delta_C + \Delta_T + y_0 \sin \alpha \tag{9.67}$$

where L_C, E_C and A_C are the length, elastic modulus and cross-sectional area of a cable, respectively. The temperature change can be taken as $\Delta T = \pm 15\,°C$, thus

$$\Delta_T = \alpha_T \Delta T L_C, \quad \alpha_T = 12 * 10^{-6} \tag{9.68}$$

The creep effect can be calculated using the empirical formula

$$\Delta_C = L_C \varepsilon_C(t_C) = L_C c_1 \exp(c_2 \log t_C) \tag{9.69}$$

where c_1 and c_2 are constants for given stress levels in percent of the cable strength 1370 MPa (Table 9.9). If in Equation (9.69) the time t_C is given in log of minutes, then ε_C is obtained in percent. E.g. for $t_C = 10^6$ minutes (approx. 2 years) and 50% of 1370 MPa it is

$$\varepsilon_C = 0.033555 \exp(0.218934 * 6) = 0.1248\% \tag{9.70}$$

Assuming that

$$\Delta_p = \eta(\Delta_C + \Delta_T + y_0 \sin \alpha), \quad (\eta > 1) \tag{9.71}$$

(η can be optionally chosen between 1 and 2), from Equation (9.67) we obtain

$$F_p = \frac{E_C A_C}{L_C} (\eta - 1)(\Delta_C + \Delta_T + y_0 \sin \alpha) \tag{9.72}$$

The required cross-sectional area of a cable can be calculated from the condition

$$(F_p + S_F) / A_C \leq f_C / \gamma \tag{9.73}$$

where f_C is the cable ultimate tensile strength, and γ is a safety factor. Using Equations (9.72) and (9.73) one obtains

$$A_C = \frac{S_F}{(f_C / \gamma) - (E_C / L_C)(\eta - 1)(\Delta_C + \Delta_T + y_0 \sin \alpha)} \tag{9.74}$$

The supporting bars are designed as compressed CHS struts of an effective length factor $K_e = 2$ and subject to an axial compression force

$$F_S = 2(F_p + S_F)\sin\alpha \tag{9.75}$$

Note that approximate closed-form equations for the design of compressed CHS struts can be found in Section 9.1.1.

The material cost of the prestressed column is

$$K = k_{CHS}\rho_{CHS}\left(LA_{core} + 3L_S A_S\right) + k_C\rho_C A_C L_C \tag{9.76}$$

where k_{CHS} and k_C are material cost factors (\$/kg) for CHS column, supporting bars and cables, respectively, ρ are material densities (kg/m^3), $L_S = (L\tan\alpha)/2$ is the length and A_S is the cross-sectional area of tubular supporting bars.

9.1.5.3 *Design of a column with three supports (Fig. 9.8)*

The design method, detailed for one support in the previous section, can be generalized for more symmetrically arranged and equally spaced supports. The method is treated here for three supports.

First the core tube is designed for a compressive force

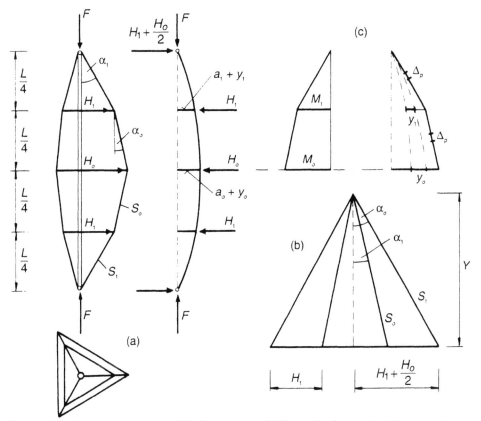

Figure 9.8. a) Prestressed column with three suports, b) Forces in ties, and c) Measures of prestressing.

$$F_T = F + 3(F_{p1} + S_1) \cos \alpha_1 \tag{9.77}$$

and buckling length $L/4$. The second term in Equation (9.77) should be guessed in advance and then corrected. Then calculate the Euler buckling force

$$F_E = \pi^2 EI_x / (L/4)^2 \tag{9.78}$$

and a_0 and y_0 with Equations (9.63) and (9.64). For the forces H_1 and H_0 which eliminate the bending moments M_1 and M_0 (Fig. 9.8) a system of two equations can be written as follows

$$M_1 = \left(H_1 + \frac{H_0}{2}\right)\frac{L}{4} = F_T \left(a_0 + y_0\right) \sin \frac{\pi}{4} \tag{9.79a}$$

$$M_0 = \left(H_1 + \frac{H_0}{2}\right)\frac{L}{2} - \frac{H_1 L}{4} = F_T \left(a_0 + y_0\right) \tag{9.79b}$$

Solving Equation (9.79) we obtain

$$H_0 = \frac{8F_T}{L}\left(a_0 + y_0\right)\left(1 - \sin\frac{\pi}{4}\right) \tag{9.80a}$$

$$H_1 = \frac{8F_T}{L}\left(a_0 + y_0\right)\left(\sin\frac{\pi}{4} - \frac{1}{2}\right) \tag{9.80b}$$

The tension forces in cables due to H_1 and H_0 are as follows (Fig. 9.8)

$$S_1 = \left(H_1 + \frac{H_0}{2}\right)/\sin\alpha_1 \tag{9.81a}$$

$$S_0 = H_0 / \left(2\sin\alpha_0\right) \tag{9.81b}$$

For a given angle α_1 the angle α_0 can be calculated (Fig. 9.8)

$$\tan\alpha_0 = H_0 / (2Y), \quad Y = \left(H_1 + \frac{H_0}{2}\right)/\tan\alpha_1 \tag{9.82}$$

The measure of prestressing in cables, necessary to avoid the elastic deformations y_1 and y_0 can be calculated similarly to Equation (9.67)

$$\Delta_{p1} \geq \frac{F_{p1}L_1}{A_{C1}E_C} + \Delta_{C1} + \Delta_{T1} + y_1 \sin\alpha_1 \tag{9.83a}$$

$$\Delta_{p0} \geq \frac{F_{p0}L_0}{A_{C0}E_C} + \Delta_{C0} + \Delta_{T0} + \left(y_0 - y_1\right)\sin\alpha_0 \tag{9.83b}$$

Assuming that

$$\Delta_{p1} = \eta_1 \left(\Delta_{C1} + \Delta_{T1} + y_1 \sin\alpha_1\right) \tag{9.84a}$$

$$\Delta_{p0} = \eta_0 \left[\Delta_{C0} + \Delta_{T0} + \left(y_0 - y_1\right)\sin\alpha_0\right] \tag{9.84b}$$

$\eta_1 > 1$, $\eta_0 > 1$ the prestressing forces can be calculated from Equations (9.83) and (9.84) similarly to Equation (9.72)

$$F_{p1} = \frac{A_{C1}E_C}{L_1}(\eta_1 - 1)(\Delta_{c1} + \Delta_{T1} + y_1 \sin \alpha_1) \qquad (9.85a)$$

$$F_{p0} = \frac{A_{C0}E_C}{L_0}(\eta_0 - 1)[\Delta_{C0} + \Delta_{T0} + (y_0 - y_1)\sin \alpha_0] \qquad (9.85b)$$

Using the condition Equation (9.73) we get

$$A_{C1} = \frac{S_1}{(f_C/\gamma) - (E_C/L_1)(\eta_1 - 1)(\Delta_{C1} + \Delta_{T1} + y_1 \sin \alpha_1)} \qquad (9.86a)$$

$$A_{C0} = \frac{S_0}{(f_C/\gamma) - (E_C/L_0)(\eta_0 - 1)[\Delta_{C0} + \Delta_{T0} + (y_0 - y_1)\sin \alpha_0]} \qquad (9.86b)$$

The compressive forces acting on supporting bars are as follows: At midpoint

$$F_{S0} = 2(S_0 + F_{p0})\sin \alpha_0 \qquad (9.87a)$$

others

$$F_{S1} = (S_1 + F_{p1})\sin \alpha_1 - (S_0 + F_{p0})\sin \alpha_0 \qquad (9.87b)$$

9.1.5.4 *Numerical example*
Data: $F = 440$ kN, $L = 10$ m, $f_y = 355$ MPa, $E_C = 1.5*10^5$ MPa, $f_C = 1500$ MPa, $\gamma = 1.5, \eta = 2.$

The case of one support
We guess that the total compressive force will be $F_T = 590$ kN. We calculate first with an angle of $\alpha = 15°$, but this angle will be varied to find an optimum value corresponding to the minimum total material cost of the structure.

According to Equation (9.60) $\delta_L = 60$. From Equation (9.61) we obtain a CHS of 215.1*4 with an outside diameter of 219.1 mm. The radius of gyration is $r = 76.1$ mm, the cross-sectional area is $A = 2700$ mm^2. Check of Equation (9.61): $218.5 < 221.7$ MPa, OK.

Using Equation (9.62) we get $F_E = 1296.6$ kN. Equations (9.63) and (9.64) yield $a_0 = 27.79$ and $y_0 = 23.20$ mm. With Equations (9.65) and (9.66) one obtains $H_F = 12.0$ and $S_F = 23.2$ kN. According to Equations (9.68) and (9.69), with $L_C = 5176.4$ mm, it is $\Delta_T = 0.93$, $y_0 \sin \alpha = 6.00$ mm.

The creep effect is calculated for a year i.e. for 525,600 minutes for a stress level $f_C/\gamma = 1000$ MPa, it is approx. 75% of 1370 MPa, thus from Table 9.9 $c_1 = 0.0717165$ and $c_2 = 0.169366$, $\varepsilon_C = c_1 \exp(5.7206c_2) = 0.1890\%$, $\Delta_C = 9.78$ mm. The required prestressing is (Eq. 9.71) $\Delta_p = 2*16.71 = 33.42$ mm. Equation (9.74) gives $A_C = 45$ mm^2.

Using an approximate formula valid for steel wire ropes $A_C = 0.79 \pi d_c^2/4$ one obtains $d_C = 8.5$ mm. According to the data of the German standard DIN 3052 (for

Table 9.10. Cost calculation as a function of the cable slope angle.

$\alpha°$	10	15	20	25	30
A_C (mm^2) Equation (9.74)	60.73	45.03	37.76	33.65	30.99
Rounded d_C (mm)	10	9	8	8	7
A_C for rounded d_C	62.05	50.26	39.71	39.71	30.40
F_p (kN) Equation (9.88)	26.65	24.35	21.21	22.92	18.60
F_T (kN) Equation (9.61)	588	578	565	565	549
K_C ($) Equation (9.89)	18.6	15.3	12.4	12.9	10.3
F_S (kN) Equation (9.75)	17.36	24.67	30.32	38.88	41.67
CHS supporting bars	51*1.2	63.5*1.4	82.5*1.6	101.6*2	108*2
K_S ($) Equation (9.90)	3.9	8.6	17.4	34.3	45.3
K ($) Equation (9.76)	234.5	235.9	241.8	259.2	267.6

helical ropes of one stranding) the mass of a cable of diameter 9 mm is $m = 0.407$ kg/m and the cost of this cable (with zinc covering of wires) is approx. 1.25 \$/kg.

With $A_C = 50.3$ mm^2 Equation (9.72) yields

$$F_p = 30A_C(10.35 + 23.2 \sin \alpha \cos \alpha) = 24.35\text{kN} \qquad (9.88)$$

With Equation (9.61) $F_T = 578$ kN, thus the guess of 590 kN was good and an iteration is not needed.

The material cost of cables is (Eq. 9.76)

$$K_C = 6\,k_C \rho_C A_C L_C \qquad (9.89)$$

and the material cost of tubular supporting bars is (Eq. 9.76)

$$K_S = 3\,k_{CHS} \rho_{CHS} A_S L_S \qquad (9.90)$$

To find the optimal angle α, the total cost is calculated for several angles between 10 and 30° and the calculated values are summarized in Table 9.10. It can be seen that the cost of cables decreases and the cost of supporting bars increases with the increase of the angle. Furthermore the total cost slightly increases when the angle increases, so that the optimum angle may be selected in the range of 10-20°.

The case of three supports

We calculate with the angle $\alpha_1 = 20°$. For an approx. compressive force $F_T = 520$ kN and an effective buckling length $L/4 = 2.5$ m one obtains for the core tube a CHS of 193.7*3.2 mm with $A = 1920$ mm^2, $r = 67.4$ mm, $I_x = 8.69*10^6$ mm^4. With Equation (9.78) we get $F_E = 2.8817*10^3$ kN. Equation (9.63) gives $a_0 = 28.20$ mm, $a_0 + y_0 = 34.41$ mm. Equation (9.80) give $H_0 = 4.19$ and $H_1 = 2.96$ kN. Equation (9.82) yields $\alpha_0 = 8.6°$. Equation (9.81) give $S_1 = 14.78$ and $S_0 = 14.01$ kN. Furthermore $L_1 = 2660$ and $L_0 = 2528$ mm, $\Delta_{T1} = 0.48$, $\Delta_{T0} = 0.46$ mm, $y_1 = y_0 \sin(\pi/4) = 4.39$, $y_0 - y_1 = 182$ mm. As in the case of one support $\varepsilon_C = 0.189\%$, $\Delta_{C1} = 5.03$, $\Delta_{C0} = 4.78$ mm. From Equation (9.86) we get $A_{C1} = 24.44$ and $A_{C0} = 20.82$ mm^2.

For all ties we select wire ropes of $d_C = 7$ mm, $A_C = 30.4$ mm^2, $m = 0.246$ kg/m, $k_C = 1.25$ \$/kg. Equation (9.84) give $\Delta_{p1} = 14.0$, $\Delta_{p0} = 11.0$ mm. With Equation (9.85) one obtains $F_{p1} = 12.02$ and $F_{p0} = 9.94$ kN, with Equation (9.87) $F_{S0} = 7.16$ and $F_{S1} = 5.59$ kN.

For a supporting bar at the midpoint we use a CHS profile designed for a compressive force $F_{S0} = 7.16$ kN and an effective buckling length 2*1288 = 2576 mm. The selected CHS profile for all supporting bars is 42.4*1.2 mm with $A_S = 155$ mm^2. The total length of supporting bars is 3*1288+6*910 = 9324 mm.

The total material cost of the prestressed structure, with $k_{CHS} = 1.0$ $/kg, $L_C = 31.128$ m, is

$$K = K_{core} + K_C + K_S = 150.7+9.6+11.4=171.7\$$$

For the column without prestressing the CHS profile of 273*4.5 mm is selected with the material cost of 298.3 $, thus, it can be concluded that, in this numerical example, the material cost savings achieved by using one support with $\alpha = 10°$ is 21% and with three supports is 42%.

9.1.5.5 *Conclusions*
The mass of a compressed tubular column can be significantly decreased by prestressing it with ties. The larger the number of supports the smaller the mass with the same compressive strength.

The necessary prestressing can be calculated by relatively simple closed-form equations based on the overall buckling relations considering the initial crookedness.

In the calculation of prestress the temperature change and the creep of cable ties is taken into account. The supporting CHS bars are designed for overall buckling.

The mass of these bars and the cable ties is small compared with the core tube and the change of their mass with the varied angle between the core tube and ties does not significantly affect the total mass or total material cost. The material cost calculations in the case of one support show that the optimum value of this angle is about 10-20°.

9.2 BENDING OF CHS BEAMS

9.2.1 *Elastic range*

Consider a simply supported CHS beam subject to a uniformly distributed normal load. The objective function is the cross-secional area

$$A = \pi Dt = \pi D^2 / \delta_C, \quad \delta_C = D / t \tag{9.91}$$

Using the graphoanalytical optimization method (see Chapters 6 and 8) we select the coordinate-system of $\delta_C - D^2$ (Fig. 9.9) in which the contours of the objective function are straight lines starting from the origo.

The local buckling constraint can be defined according to EC3

$$\delta_C \le \delta_{CL} = 70\varepsilon^2, \quad \varepsilon^2 = 235 / f_y \tag{9.92}$$

The stress constraint is expressed by

$$\sigma_{max} = M_{max} / W_x \le f_y / \gamma_{M1}, \quad W_x = \pi D^3 / (4\delta_C), \quad \gamma_{M1} = 1.1 \tag{9.93}$$

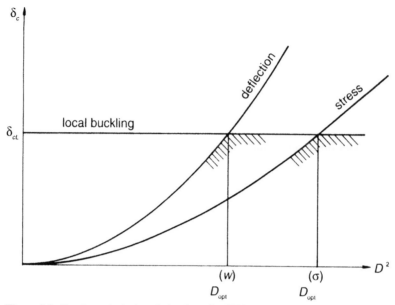

Figure 9.9. Graphoanalytical optimization of a CHS beam subject to bending.

or in another form

$$\delta_C^{(\sigma)} \leq \frac{\pi}{4W_0}\left(D^2\right)^{3/2} \qquad W_0 = M_{max} / (f_y / \gamma_{M1}) \tag{9.94}$$

The deflection constraint, according to Equations (8.23)-(8.25) and using the formula

$I_x = \pi D^4 / (8\delta_C)$ can be written as

$$\delta_C^{(w)} \leq \frac{\pi}{8I_0}\left(D^2\right)^2 \tag{9.95}$$

The limiting lines of the constraints Equations (9.92, 9.94 and 9.95) are shown in Figure 9.9. It can be seen that the optima are determined by the intersection points of the relevant lines. Thus,

$$D_{opt}^{(\sigma)} = \sqrt[3]{4W_0\delta_{CL} / \pi} \tag{9.96}$$

and

$$D_{opt}^{(w)} = \sqrt[4]{8I_0\delta_{CL}} \tag{9.97}$$

and the governing solution is the larger D_{opt}.

9.2.2 *Plastic range*

According to Chen & Sohal (1988), the minimum cross-sectional areas required for a given bending moment capacity can be calculated in function of the required rota-

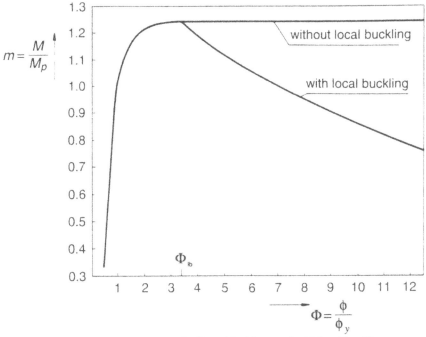

Figure 9.10. Moment-curvature curve for $\delta_C = 48$ without and with local buckling.

tional capacity without and with local buckling. Figure 9.10 shows the curves for bending moment normalized with respect to yield moment

$$m = M / M_p, \quad M_p = f_y D^2 t = f_y D^3 / \delta_C \tag{9.98}$$

in function of curvature normalized by the yield curvature of the cross-section

$$\Phi = \phi / \phi_y, \quad \phi = 2\varepsilon / D, \quad \phi_y = 2\varepsilon_y / D \tag{9.99}$$

where ε_y is the yield strain. For the case without local buckling

$$m = 1.25 - \frac{0.043729}{(\phi - 0.76729)^2} \tag{9.100}$$

and with local buckling

$$m = \left(m_{max} - m_{min}\right) \exp\left[-\frac{(\Phi - \Phi_{lb})(EI)_{max}}{m_{max} - m_{min}} \right] + m_{min} \tag{9.101}$$

where, for $36 \le \delta_C \le 72$

$$m_{min} = 0.75 - 0.00514\delta_C \tag{9.102}$$

$$(EI)_{max} = -0.094 + 0.00367\delta_C \tag{9.103}$$

$$1 / \Phi_{lb} = -0.292 + 0.0106\delta_C + 0.336 * 10^{-4}\delta_C^2 \tag{9.104}$$

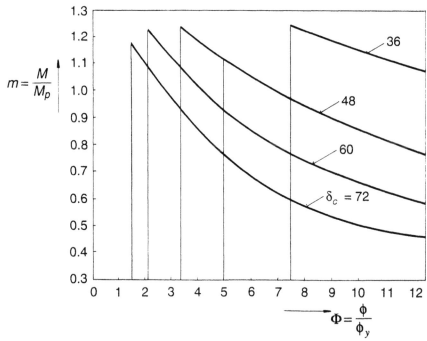

Figure 9.11. Moment-curvature curves with local buckling in plastic range for different δ_C values.

$$m_{max} = 1.25 - \frac{0.043729}{\left(\Phi_{lb} - 0.76729\right)^2} \qquad (9.105)$$

Figure 9.11 shows the curves with local buckling in plastic range for different δ_C values according to Equations (9.101-9.105).

In the optimum design, for the case of tubes without local buckling, we have to minimize the cross-sectional area (Eq. 9.91) subject to the constraint on bending moment capacity expressed from Equation (9.98)

$$M = mf_y D^3 / \delta_C \ge M_0 \qquad (9.106)$$

where M_0 is the given bending moment. The constraint on local buckling is

$$\delta_C \le \delta_{crit} \qquad (9.107)$$

According to Chen & Sohal (1988)

$$\delta_{crit} = \frac{1}{2}\left[-\frac{500}{\Phi - 4.1} \pm \sqrt{\left(\frac{500}{\Phi - 4.1}\right)^2 + \frac{90000}{\Phi - 4.1}} \right] \qquad (9.108)$$

In the coordinate-system of $\delta_C - D^2$ the feasible region is determined by the curve of the constraint (9.106)

$$\delta_C \le \frac{mf_y}{M_0}\left(D^2\right)^{3/2} \qquad (9.109)$$

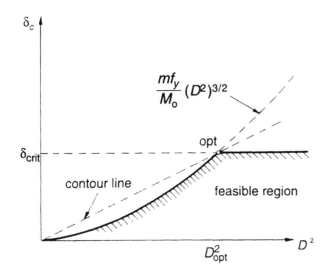

Figure 9.12. Feasible region and optimum point for the case of tubes without local buckling.

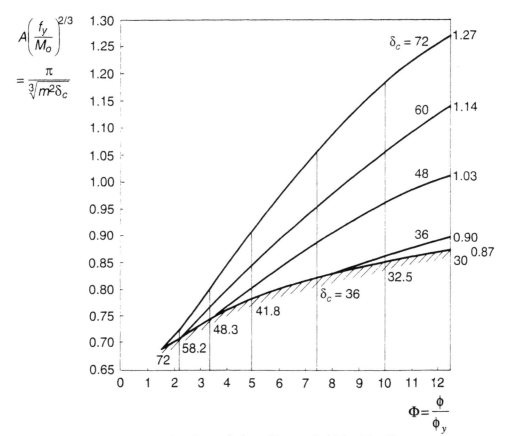

Figure 9.13. Minimum cross-sectional area of tubes without and with local buckling.

and by the value of δ_{crit} in the constraint (Eq. 9.107). Figure 9.12 shows the feasible region and the optimum point. Thus,

$$D_{opt} = \sqrt[3]{\frac{M_0 \delta_C}{m f_y}} \qquad (9.110)$$

and the corresponding minimum cross-sectional area is

$$A_{min} = \left(\frac{M_0}{f_y}\right)^{2/3} \frac{\pi}{m^{2/3} \delta_{crit}} \qquad (9.111)$$

Figure 9.13 shows the $A_{min} - \Phi$-curve for the case without local buckling according to Equation (9.111). The curves for cases with local buckling may be obtained for different δ_C-values also from (Eq. 9.111) taking δ_C instead of δ_{crit}.

It can be concluded that, for the minimum mass design, we have to choose δ_C-ratios without local buckling. When we apply larger δ_C-values with local buckling, the cross-sectional area increases depending on the required rotational capacity and the applied δ_C-ratio.

It is worth noting that this conclusion is not self-evident, because there are cases of other types of structures, for which it is economic to increase the slenderness allowing local buckling.

Note that the optimum design of CHS beam-columns has been treated in (Farkas 1992).

CHAPTER 10

Steel and aluminium structural components with fatigue constraints for welded joints

10.1 FACTORS INFLUENCING THE FATIGUE OF WELDED JOINTS

As a tragic case of a structural fatigue failure, for instance, the catastrophe of the semi-submersible Kielland offshore platform may be mentioned (Almar-Ness et al. 1984). The platform capsized suddenly on March 27, 1980 with loss of 123 lives while 89 persons were rescued. Hydrophones were attached to the underside of tubular bracings. Installing the hydrophones has been performed by cutting a hole in the brace and by poor quality fillet welds, from which fatigue cracks propagated in bracings, the fracture of bracings caused a complete separation of a main leg and the disaster of the platform.

The factors are as follows:

Parent material: Steel, aluminium alloy. The fatigue strength data given by EC3 are the same for steels Fe 360 and Fe 510. Fatigue strengths of aluminium alloy welded joints are given in BS 8118. Design rules are given also in Recommendations (1995).

Welding technology: For some new technologies fatigue strength data do not exist yet.

Residual welding stresses: The fatigue is influenced by the high tensile residual stresses around the welds since the fatigue cracks propagate faster in this zone than in parent material.

Type of joint: A detailed categorization of joints with the corresponding fatigue strength is given by EC3, BS 8118 and Recommendations.

Weld geometry: The geometry of toes of butt and fillet welds can be improved by several methods such as grinding or TIG remelting, but an additional fabrication cost arises (Farkas 1991).

Weld defects: At the ends of welds, at points of electrode changing, lack of penetration in cruciform joints. The weld quality and inspection (by non-destructive testing) play an important role, since, if we have a reliable information on the weld quality produced by a technology, significant inspection cost savings may be achieved.

Load spectra: The introduction of an equivalent constant wave load using the Palmgren-Miner rule enables designers to consider the load in a more realistic mode for the investigated structural application.

Stress range: The most important factor for the fatigue crack propagation. The recent studies showed that the fatigue life depends on the stress range $\Delta\sigma =$

$\sigma_{max} - \sigma_{min}$ rather than on the σ_{max}, thus, the fatigue life is characterized by the $\Delta\sigma - N$ curves, where N is the number of load cycles.

Number of cycles: Recent experiments showed that the fatigue of some welded structural parts should be characterized by curves different from the classical $\Delta\sigma - N$-curve having an asymptote at $N = 2*10^6$ cycles. For instance, for welded joints of hollow sections, EC3 gives a diagram which has a cut-off point at $N = 10^8$.

Plate thickness: Recent experiments with specimens of realistic dimensions showed that the older experimental results obtained with small size specimens should be corrected. EC3 prescribes a correction for thicknesses over 25 mm.

Stress state: In most cases not only a normal stress, but also a shear stress component arises. For these cases EC3 gives an interaction formula.

Post-welding treatments: Used for the decreasing the residual welding stresses or for improving the weld toe geometry (Gregor 1989) (see also Chapter 1).

Special phenomena: The interaction of fatigue crack propagation and local buckling due to the breathing of a thin web plate in a welded plate girder (Maeda & Okura 1983); fatigue cracks arising in crane runway girders due to local pulsating torsional loading caused by the wheel load (Farkas 1983, Ingraffea et al. 1987).

Environmental effects: The fatigue strength data given by EC3 are valid only for temperatures less than 150°C and for structures with suitable corrosive protection. The corrosion fatigue is a very dangerous failure mode, it should be considered using data from special publications.

Joints of hollow sections in tubular trusses are treated in EC3, but improved international rules are in preparation considering the recent research results (van Wingerde et al.) 1995.

In the case of *aluminium* welded structures some differences emerge as compared to the design of welded steel structures as follows:

– There are *several aluminium alloys*, thus, the material selection is an important task,

– There are *different types of profiles* depending on the fabrication procedure: Extrusions, drawn tubes, profiles welded from plates, open profiles with bulbs, etc.,

– The *welded profiles have different strength* as compared to unwelded ones,

– In static design the *softening effect of the heat affected zone* (HAZ) should be considered,

– Some *characteristic data are different* as compared to steels: The elastic modulus and the material density is the third of the data for steels, the thermal expansion coefficient is double of steels. The material and welding costs are also different.

10.2 FATIGUE DESIGN RULES OF EC3

In Part 1.1 of EC3 general rules and rules for steel buildings are given. The welded joints are grouped into detail categories. The number of the category $\Delta\sigma_C, \Delta\tau_C$ means the fatigue stress range in MPa at number of cycles $N = 2*10^6$. The fatigue stress ranges corresponding to another given number of cycles ($\Delta\sigma_N, \Delta\tau_N$) are given graphically (straight lines in a log-log coordinate system). (Fig. 10.1) On this basis the $\Delta\sigma_N, \Delta\tau_N$ values can be calculated by linear interpolation as follows:

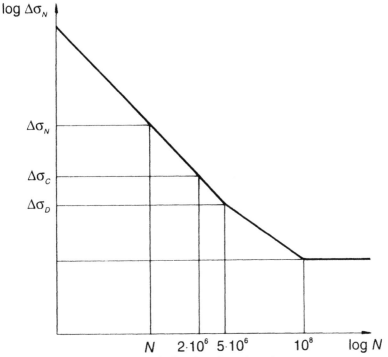

Figure 10.1. Fatigue stress range in function of number of cycles.

$$\text{for } N \leq 5*10^6 \qquad \log \Delta\sigma_N = \frac{1}{m} \log \frac{2*10^6}{N} + \log \Delta\sigma_C \qquad (10.1)$$

where m is the slope constant, $m = 3$
and

$$\text{for } 5*10^6 \leq N \leq 10^8 \quad \log \Delta\sigma_N = \frac{1}{m} \log \frac{5*10^6}{N} + \log \Delta\sigma_D \quad m = 5 \quad (10.2)$$

$\Delta\sigma_D$ is the fatigue stress range corresponding to $N = 5*10^6$.
For shear stress fatigue ranges

$$\text{for } N \leq 10^8 \qquad \log \Delta\tau_N = \frac{1}{m} \log \frac{2*10^6}{N} + \log \Delta\tau_C \quad m = 5 \quad (10.3)$$

and for modified strength curves (categories of 36*, 45* and 50*)

$$\text{for } N \leq 10^7 \qquad \log \Delta\sigma_N = \frac{1}{m} \log \frac{2*10^6}{N} + \log \Delta\sigma_C \quad m = 3 \quad (10.4)$$

and

$$\text{for } 10^7 \leq N \leq 10^8 \qquad \log \Delta\sigma_N = \frac{1}{m} \log \frac{10^7}{N} + \log \Delta\sigma_S \quad m = 5 \quad (10.5)$$

$\Delta\sigma_S$ corresponds to $N = 10^7$.
The safety factors for fatigue γ_{Mf} are given in Table 10.1.

Table 10.1. Safety factors for fatigue according to EC3.

	Fail-safe components	Non fail-safe components
Accessible joint	1.00	1.25
Poor accessibility	1.15	1.35

'Fail-safe' structural components are such that their local failure does not result in failure of the whole structure. The failure of a 'non fail-safe' component leads to the failure of the whole structure.

The problem is that it cannot be known in advance, which constraints are active, therefore all constraints should be formulated and investigated for an actual case. Special problem is the interaction of fatigue and local stability constraints. Local stability constraints are formulated by means of the limiting plate slenderness which depends on the maximum static stress. In static design, when the deflection constraint is passive, this maximum stress is calculated from limiting stress for steel or an aluminium alloy. When the fatigue constraint is active, the design stress level may be much lower than this limiting stress, thus we can calculate with this actual stress level, to achieve more economic structures.

10.3 COMPRESSION SQUARE HOLLOW SECTION (SHS) STRUT CONNECTED TO GUSSET PLATES (FIG. 10.2)

10.3.1 *Steel structure*

The strut is connected to a gusset plate by means of a groove and fillet welds on both sides. The strut of length $L = 7.5$ m is cyclically loaded in tension- compression by a pulsating force $F = 190$ kN. The number of cycles is $N = 5*10^5$. The cross-section should be a square hollow section designed for overall buckling. For maximal stress in overlapping plate elements is $\Delta\sigma_C = 45*$ MPa. For given number of cycles, according to Equation (10.4), $\Delta\sigma_N = 71.4$ MPa.

In general case a permanent load F_G and a variable load F act for which the safety factors γ_G, γ_Q should be considered. According to EC3 $\gamma_G = 1.35$, $\gamma_Q = 1.50$. In our example $F_G = 0$ and we suppose that the given F force already contains the safety factor. Depending on the ratio F_G/F the fatigue or the overall buckling constraint can be active. Since it is not known in advance which constraint will be active, in a hand calculation we suppose that the fatigue constraint is active, solve the optimization problem for the active constraint and check the solution for overall buckling. When the solution does not satisfy the overall buckling constraint, the optimization should be solved considering the overall buckling constraint. In a computerized calculation this procedure is not needed.

The fatigue constraint is

$$\Delta\sigma = \frac{2F}{\gamma_Q A} \leq \frac{\Delta\sigma_N}{\gamma_{Mf}}, \quad A = 4\,bt \tag{10.6}$$

It should be noted that F is a factored load, so we should divide it with the safety

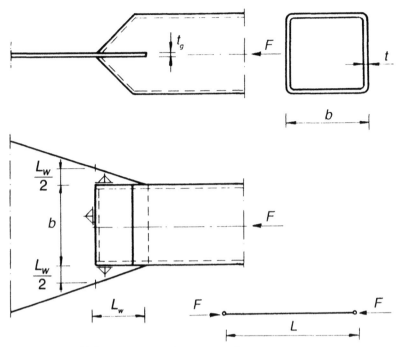

Figure 10.2. A SHS strut connected to a gusset plate.

factor, since in the fatigue constraint the pulsating load should not be multiplied by the safety factor. $\gamma_{Mf} = 1.25$ is the safety factor for fatigue.

The overall buckling constraint is

$$F / A \leq \chi f_y / \gamma_{M1} \tag{10.7}$$

χ is the overall buckling factor. For the calculation we use, as an approximation to the EC3 buckling curve 'b', the simpler Japanese JRA buckling curve as follows (see Chapter 3)

$$\chi = 1.109 - 0.545\overline{\lambda} \quad \text{for} \quad 0.2 < \overline{\lambda} < 1$$

$$\chi = 1/(0.773 + \overline{\lambda}^2) \quad \text{for} \quad \overline{\lambda} > 1$$

where

$$\overline{\lambda} = KL / (r\lambda_E), \quad r = \sqrt{I / A}, \quad \lambda_E = \pi\sqrt{E / (f_y / \gamma_{M1})}$$

From the fatigue constraint one obtains the required cross-sectional area $A_{req} = (2F\gamma_{Mf} / \gamma_Q \Delta_N) = 44.35 \text{ mm}^2$.

σ_{max} is the maximal compressive stress, the smaller value from $\Delta\sigma / 2$ or F / A. Since this value is not known in advance, we suppose that the fatigue constraint is active and $\sigma_{max} = \Delta\sigma_N / (2\gamma\sigma_{Mf}) = 71.4 / (2*1.25) = 28.6 \text{ MPa}$.

The limiting plate slenderness is $(b / t)_L = 42\sqrt{235 / \sigma_{max}} = 42\sqrt{235 / 28.6} = 120$.

This value is too large so we take a profile from ISO/DIS 4109.2 for cold-formed

SHS with maximal b/t and with A near the A_{req}. We select the profile 285*285*4, $A = 4440$ mm^2 and the radius of gyration is $r = 114$ mm.

Check for overall buckling: $\sigma_{max} = 190,000 / 4440 = 42.8$ MPa, thus

$$\lambda_E = \pi\sqrt{2.1*10^5 / 213.6} = 98.50, \bar{\lambda} = L/(r\lambda_E) = 7500/(114*98.5) = 0.6679,$$

$$\chi = 0.8015 \qquad 42.8 < 0.8015*235/1.1 = 171 \text{ MPa, OK}$$

Calculation of the overlapping length L_w for the fillet welded connection according to the Figure 10.2, with the total weld length $4L_w$. (We neglect the transverse fillet welds.) For shear it is $\Delta\tau_C = 80$ MPa and for given number of cycles $\Delta\tau_N = 105$ MPa. We take $a_w = 4$ mm, thus, from

$$\frac{2F/\gamma_Q}{4L_w a_w} \leq \frac{\Delta\tau_N}{\gamma_{Mf}} = \frac{105}{1.25} = 84 \text{ MPa} \tag{10.8}$$

$L_w = 189$ rounded 190 mm.

Calculation of the required gusset plate thickness t_g according to EC3 with the category $\Delta\sigma_C = 63$ MPa, and for given number of cycles $\Delta\sigma_N = 100$ MPa. From

$$\frac{2F/\gamma_Q}{(L_w+b)t_g} \leq \frac{\Delta\sigma_N}{\gamma_{Mf}} = \frac{100}{1.25} = 80 \text{ MPa} \tag{10.9}$$

one obtains $t_g = 5.5$, rounded 6 mm.

10.3.2 *Aluminium structure*

The strut of length $L = 5$ m is cyclically loaded in tension-compression by a pulsating force $F = 50$ kN, the number of cycles is $N = 5*10^5$.

In general case a permanent load F_G and a variable load F_Q act, for which the safety factors, according to BS, are $\gamma_G = 1.20$, $\gamma_Q = 1.33$.

The objective function is the cross-sectional area, considering the effect of corner roundings approximately by a factor of 0.9

$$A = 0.9*4bt \tag{10.10}$$

The fatigue constraint is expressed by

$$2F/A \leq \Delta\sigma_N/\gamma_{mf} \tag{10.11}$$

where $\Delta\sigma_N$ is the fatigue stress range corresponding to the number of cycles N, γ_{mf} is the fatigue material factor. Since this factor is not given in BS, we take it according to the EC3, which prescribes for important (non fail-safe) and accessible joints a factor of 1.25. The fatigue stress range $\Delta\sigma_C$ corresponding to $N = 2*10^6$ is given in BS or in Recommendations in tables for various welded connections. When $N \leq 5*10^6$ (Eq. 10.1) can be used. In our example, according to BS, for the edge point of the gusset plate it is $\Delta\sigma_C = 17$ MPa and $m = 3$, thus, one obtains $\Delta\sigma_N = 27$ MPa. It should be noted that, in Recommendations the fatigue stress range for such aluminium joint is not given.

The overall buckling constraint is, according to BS (the permanent load is taken as $F_G = 0$)

$$\gamma_Q F / A \le \chi p_{01}, \quad p_{01} = p_0 / \gamma_m \tag{10.12}$$

Note that the overall buckling formulae given in BS are the same as those in EC3, thus, we use the EC3 formulae as follows:

$$1/\chi = \phi + \sqrt{\phi^2 - \overline{\lambda}^2}, \quad \phi = 0.5\left[1 + \alpha\left(\overline{\lambda} - 0.2\right) + \overline{\lambda}^2\right] \tag{10.13}$$

where for unwelded profiles $\alpha = 0.2$, for welded ones $\alpha = 0.4$

$$\overline{\lambda} = \lambda / \lambda_E, \quad \lambda = KL / r, \quad \lambda_E = \pi\sqrt{E / p_{01}} \tag{10.14}$$

where K is the effective length factor, for pinned ends $K = 1$, $r = \sqrt{I_x / A}$ is the radius of gyration, $E = 7*10^4$ MPa is elastic modulus for aluminium alloys, p_0 is the limiting stress, $\gamma_m = 1.2$ is the material factor. We select the aluminium alloy 6061-T6 and extruded SHS profiles for which $p_0 = 240$ MPa and $\lambda_E = 58.77$.

The local buckling constraint for unwelded profiles is defined by

$$\delta_S = b / t \le \delta_{SL} = 18\sqrt{250 / p_{01}} \ \ (p_{01} \text{ in MPa}) \tag{10.15}$$

Since the actual maximum static stress can be lower than p_{01} we can calculate with $\sigma_{max} = \gamma_Q F/A$ instead of p_{01}.

Depending on the ratio of F_G/F_Q the fatigue or the overall buckling constraint may be active. Since the active constraint is not known in advance, we assume that in our example the fatigue constraint is active, thus, we perform the design for fatigue and then check the result for overall buckling.

The required cross-sectional area from Equation (10.11) is $A_{req} = 2F\gamma_{mf}/\Delta\sigma_N = 4630$ mm². The actual static stress is $\sigma_{max} = \gamma_Q F/A_{req} = 14.4$ MPa and the limiting b/t is $\delta_{SL} = 75$. We select the SHS profile 260*5 mm with $A = 4680$ mm² and $b/t = 52$, thus, this profile fulfils the fatigue and local buckling constraint. Check for overall buckling: $r = b/\sqrt{6} = 106$, $\lambda = 5000 / (58.77*106) = 0.8026$, $\chi = 0.8007$ and Equation (10.12) is in our case $14.2 < 160$ MPa, OK.

Calculation of the overlapping length for the fillet welded connection according to Figure 10.2, neglecting the transverse fillet welds. For shear it is $\Delta\tau_C = 14$ MPa and for the given N using Equation (10.3) $\Delta\tau_C = 22.2$ MPa. The fatigue constraint for the welds is

$$2F / \left(4L_w a_w\right) \le \Delta\tau_N / \gamma_{mf} = 22.2 / 1.25 = 17.76 \text{ MPa} \tag{10.16}$$

Taking the size of the fillet welds $a_w = 5$ mm, from Equation (10.16) one obtains $L_w = 281$, rounded 290 mm.

Calculation of the required gusset plate thickness. For the gusset plate welded on both sides it is $\Delta\sigma_C = 20$, $\Delta\sigma_N = 31.7$ MPa. Assuming a gusset plate width of $L_w + b$ (Fig. 10.2), from

$$\frac{2F}{\left(L_w + b\right)t_g} \le \frac{\Delta\sigma_N}{\gamma_{mf}} = \frac{31.7}{1.25} = 25.36 \text{ MPa} \tag{10.17}$$

we get $t_g = 7.3$ rounded 8 mm.

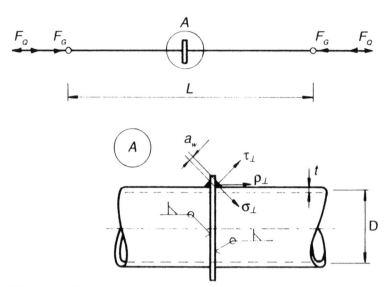

Figure 10.3. Compressed CHS strut with a welded splice.

10.4　COMPRESSION ROD OF CIRCULAR HOLLOW SECTION WITH A WELDED SPLICE (FIG. 10.3)

10.4.1　*Steel structure*

A CHS strut is centrally compressed by a permanent force $-F_G$ (minus denotes compression) and a variable load F_Q pulsating between $+F_Q$ and $-F_Q$. F_Q contains also a dynamic factor. The strut is constructed with a splice fillet welded end-to-end with an intermediate transverse plate. In the optimization of the strut section the unknown dimensions D and t are sought to minimize the strut cross-sectional area.

In the case of two unknowns the graphoanalytical optimization method can be advantageously applied (Chapters 6 and 8).

In the optimum design of a compression strut we use the unknowns $\vartheta = 100D/L$ and $\delta_C = D/t$ instead of D and t since the objective function can be expressed by

$$A = \pi Dt = \frac{\pi D^2}{\delta_C} = \frac{\pi L^2 \vartheta^2}{10^4 \delta_C} \tag{10.18}$$

and the contours of the objective function are defined by

$$\delta_C = \text{const} * \vartheta^2 \tag{10.19}$$

which are straight lines in the coordinate-system $\delta_C - \vartheta^2$, so it is easy to find the touching point.

The fatigue constraint is expressed by

$$2F_Q / A \le \Delta\sigma_N / \gamma_{Mf} \quad \text{or} \quad \delta_C \le \frac{L^2 \pi \Delta\sigma_N}{2 * 10^4 F_Q \gamma_{Mf}} \vartheta^2 \tag{10.20}$$

According to the Recommendations (1995) the fatigue stress range at $N = 2*10^6$ is $\Delta\sigma_C = 50$ MPa for splice of CHS with intermediate plate, toe crack, wall thickness smaller than 8 mm.

The local buckling constraint is defined by using the limiting local slenderness (according to EC3 for section class 1)

$$\delta_C \leq \delta_{CL} = 50*235/f_{yl}, \quad f_{yl} = f_y/\gamma_{M1}, \quad \gamma_{M1} = 1.1 \tag{10.21}$$

The static overall buckling constraint is given by

$$\left(\gamma_G F_G + \gamma_Q F_Q\right)/A \leq \chi f_{yl} \tag{10.22}$$

where γ_G, γ_Q are the safety factors for permanent and variable load, respectively.
According to EC3

$$\chi = \frac{1}{\phi + \sqrt{\phi^2 - \bar{\lambda}^2}}, \quad \phi = 0.5\left[1 + 0.34\left(\bar{\lambda} - 0.2\right) + \bar{\lambda}^2\right],$$

$$\bar{\lambda} = \frac{KL\sqrt{8}}{D\lambda_E} = \frac{100K\sqrt{8}}{\lambda_E \vartheta}, \quad \lambda_E = \pi\sqrt{\frac{E}{f_{yl}}} \tag{10.23}$$

$K = 1$ for pinned ends. Thus, Equation (10.22) can be written as

$$\delta_C \leq \frac{\pi\chi f_{yl}}{10^4\left(\gamma_G F_G + \gamma_Q F_Q\right)L^2}\vartheta^2 \tag{10.24}$$

In a *numerical example* the given data are as follows: $F_G = 300$ kN, $2F_Q = 145$ kN, $\Delta\sigma_C = 50$ MPa, with Equation (10.1) $\Delta\sigma_N = 43.7$ MPa, $L = 5$ m, $f_y = 235$ MPa, $\gamma_{M1} = 1.1$, $\gamma_{Mf} = 1.25$, $\gamma_G = 1.35$, $\gamma_Q = 1.50$. $N = 3*10^6$.
The fatigue constraint Equation (10.20) is

$$\delta_C \leq 1.89580\vartheta^2 \tag{10.25}$$

The local buckling constraint Equation (10.21) is

$$\delta_C \leq 55 \tag{10.26}$$

In the overall buckling constraint the buckling factor can be calculated with simpler formulae of the Japanese Road Association which give values near to the EC3 column curve 'b' (Chapter 3)

$$\chi = 1 \qquad \text{for} \quad 0 \leq \bar{\lambda} \leq 0.2 \tag{10.27a}$$
$$\chi = 1.109 - 0.545\bar{\lambda} \quad \text{for} \quad 0.2 \leq \bar{\lambda} \leq 1 \tag{10.27b}$$
$$\chi = 1/\left(0.773 + \bar{\lambda}^2\right) \quad \text{for} \quad \bar{\lambda} \geq 1 \tag{10.27c}$$

In our example from Equation (10.24) we get

$$\delta_C \leq 3.26540\vartheta^2(1.109 - 1.5649/\vartheta) \qquad \text{for} \quad \vartheta^2 \geq 8.2447 \tag{10.28}$$

$$\delta_C \leq 3.26540\vartheta^2/\left(0.773 + 8.2447/\vartheta^2\right) \text{ for} \quad \vartheta^2 \leq 8.2447 \tag{10.29}$$

Figure 10.4 shows the limiting lines of constraints in the coordinate-system of the

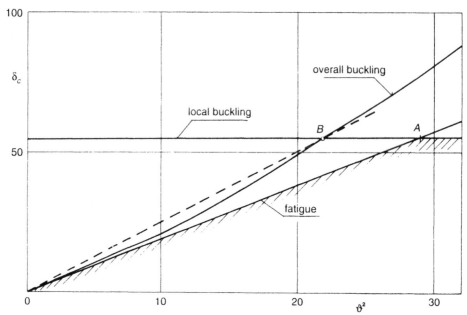

Figure 10.4. Graphoanalytical optimization of the steel strut shown in Figure 10.3.

two unknowns. These lines define the feasible region and the optimum point at the intersection of the constraint on fatigue and local buckling (point A), since the contour touches the feasible region in this point. This means that the overall buckling constraint is in this case passive. The result is

$$\vartheta^2_{opt} = 2.9.01, \quad \vartheta_{opt} = 5.3862, \quad D_{opt} = 269.3, \quad t = D/55 = 4.9 \text{ mm}$$

we use the available section 273*5 mm. It should be noted that, in this special case, all points lying on the straight line O-A between the points O and A give optima, since the feasible region coincides with the contour.

The point B gives optimum when the fatigue constraint is not considered. The solution of Equation (10.28) with $\delta_C = 55$ is $\vartheta = 4.6660$, $\vartheta^2 = 21.77$, $D = 233.3$, $t = 4.24$ mm.

If the value of F_Q is kept constant, it is easy to find F_G corresponding to point A, this means, when all three constraints are active ($\vartheta = 5.3862$). From Equation (10.28) $F_G = 455.9$ kN.

Finally, *the required fillet weld size* can be calculated from the fatigue constraint for root cracking. The pulsating load causes a stress range in the section

$$\Delta\sigma = 2F_Q / A = 145000 / 4210 = 34.4 \text{ MPa}$$

and stress components in the fillet weld (Fig. 10.3)

$$\Delta\rho_\perp = \Delta\sigma t / a_w = 172.2 / a_w, \quad \Delta\sigma_\perp = \Delta\tau_\perp = \Delta\rho_\perp / \sqrt{2} = 121.8 / a_w$$

According to the Recommendations (1995) $\Delta\sigma_C = 40, \Delta\tau_C = 80$ MPa. For the given number of cycles $N = 3*10^6$ one obtains $\Delta\sigma_N = 34.9, \Delta\tau_N = 73.8$ MPa. Using the EC3 interaction formula

$$\left(\frac{\gamma_{Ff}\Delta\sigma}{\Delta\sigma_N/\gamma_{Mf}}\right)^3 + \left(\frac{\gamma_{Ff}\Delta\tau}{\Delta\tau_N/\gamma_{Mf}}\right)^5 \le 1 \tag{10.30}$$

we get

$$83.02/a_w^3 + 37.37/a_w^5 \le 1$$

from which $a_w = 5$ mm.

Note that the static stress constraint for fillet welds is in this case also passive.

10.4.2 *Aluminium structure*

The unknowns and the objective function are the same as for steel structure (Section 10.4.1).

The fatigue constraint is expressed by

$$2F_Q/A \le \Delta\sigma_N/\gamma_{mf} \quad \text{or} \quad \delta_C \le \frac{L^2\pi\Delta\sigma_N}{2*10^4 F_Q\gamma_{mf}}\vartheta^2 \tag{10.31}$$

$F_Q = 12$ kN, $L = 5$ m. According to BS the fatigue stress range at $N = 2*10^6$ is $\Delta\sigma_C = 20$ MPa for splice fillet weld toe crack (Recommendations give 22 MPa). $\gamma_{mf} = 1.25$. For a given $N = 3*10^6$, using Equation (10.1), one obtains $\Delta\sigma_N = 17.5$ MPa, thus Equation (10.31) takes the form

$$\delta_C \le 4.5815\vartheta^2 \tag{10.32}$$

This gives a straight line in our coordinate-system (Fig. 10.5).

The local buckling constraint, according to BS for compact sections, is

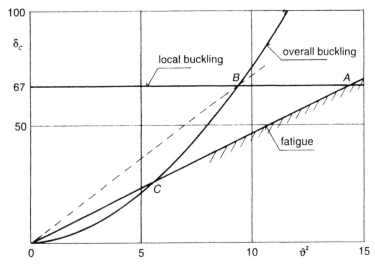

Figure 10.5. Graphoanalytical optimization of the aluminium strut.

$$D/t \leq \delta_{CL} = \left(\frac{22}{3}\right)^2 \varepsilon^2 = 53.78 * 250 / 200 = 67 \tag{10.33}$$

Note that we could calculate with a maximum static stress instead of p_{01}, but a value larger than 67 is not realistic.

The static overall buckling constraint is expressed by

$$(\gamma_G F_G + \gamma_Q F_Q) / A \leq \chi p_{01} \tag{10.34}$$

χ is given by Equation (10.27), $F_G = 50$ kN. Equation (10.34) takes the form

$$\delta_C \leq \frac{20.68}{\phi + \sqrt{\phi^2 - \overline{\lambda}^2}} \vartheta^2 \tag{10.35}$$

where with $r = D / \sqrt{8}, \lambda_E = 58.77$, with $K = 1$ for pinned ends $c_0 = 100K / \lambda_E = 1.7014$ and

$$\overline{\lambda} = c_0 \sqrt{8} / \vartheta = 4.8124 / \vartheta \tag{10.36}$$

Figure 10.5 shows the limiting lines of constraints in the coordinate-system of the two unknowns. These lines define the feasible region and the optimum point at the intersection of the constraint on fatigue and local buckling (point A), since the contour touches the feasible region in this point. This means that the overall buckling constraint is in this case passive. The result is $\vartheta_{opt}^2 = 14.62$, $\vartheta_{opt} = 3.824$, $D_{opt} = 191.2$, $t = 2.85$ mm.

We take a CHS profile 200*3 mm ($A = 1885$ mm^2) (see remark in Section 10.4.1). In this case all points on the straight line C-A between C and A give optima).

The point B gives optimum when the fatigue constraint is not considered. The solution of Equation (10.35) with $\delta_{CL} = 67$ is $\vartheta^2 = 9.3$, $\vartheta = 3.05$, $D = 152$, $t = 2.3$ mm. If the value of F_Q is kept constant, it is easy to find F_G corresponding to point A, this means, when all three constraints are active ($\vartheta = 3.824$). From Equation (10.35) $F_G = 129$ kN.

Finally, *the required fillet weld size* can be calculated from the fatigue constraint for root cracking. According to BS $\Delta\rho_{\perp C} = 14$ MPa (Recommendations give 16 MPa). With Equation (10.1), for the given N one obtains $\Delta\rho_{\perp N} = 12.2$ MPa. The fatigue constraint is

$$\Delta\rho_{\perp} = 2F_Q / (\pi D a_w) \leq \Delta\rho_{\perp N} / \gamma_{mf} = 12.2 / 1.25 = 9.8 \text{ MPa} \tag{10.37}$$

From Equation (10.37) $a_w = 3.89$ rounded 4 mm.

The constraint on the static fillet weld resistance according to BS is

$$\sqrt{\sigma_{\perp}^2 + 3(\tau_{\perp}^2 + \tau_{//}^2)} \leq 0.85 p_w / \gamma_m \tag{10.38}$$

where p_w is the limiting stress for weld metal, for alloy 6061-T6 it is 190 MPa.

Since $\sigma_{\perp} = \tau_{\perp} = \rho_{\perp} / \sqrt{2}, \tau_{//} = 0$, Equation (10.38) can be written as

$$\sqrt{2}\rho_{\perp} = \sqrt{2} \frac{\gamma_G F_G + \gamma_Q F_Q}{\pi D a_w} = \frac{\sqrt{2} * 75.96 * 10^3}{200 \pi a_w} \leq \frac{0.85 * 190}{1.2} = 134.6 \text{ MPa} \tag{10.39}$$

from which $a_w = 1.3$ mm, this constraint is not active.

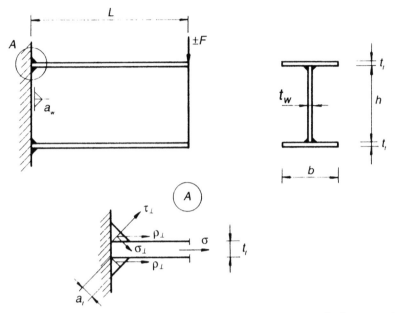

Figure 10.6. Welded I-section cantilever and the stress components in the connecting fillet welds.

10.5 WELDED I-SECTION CANTILEVER CONNECTED TO A COLUMN BY FILLET WELDS (FIG. 10.6)

10.5.1 *Steel structure*

For the calculation of the optimum dimensions of the welded I-section cantilever the constraint on fatigue is used. The objective function is the cross-sectional area

$$A = ht_w + 2bt_f \tag{10.40}$$

The fatigue constraint for the parent material at the toes of fillet welds connecting the flanges to a column, in the case of a force F fluctuating between $+F$ and $-F$, is defined by

$$\Delta\sigma = 2FL / W_x \le \Delta\sigma_N / \gamma_{Mf} \tag{10.41}$$

The moment of inertia is expressed by

$$I_x \cong h^3 t_w / 12 + 2bt_f (h/2)^2$$

Equation (10.41) can be written in the form

$$W_x \cong I_x / (h/2) = h^2 t_w / 6 + bt_f h \ge W_0 = 2FL / \left(\Delta\sigma_N / \gamma_f\right) \tag{10.42}$$

where $\Delta\sigma_N$ is the fatigue stress range, $\gamma_{Mf} = 1.25$ is the safety factor, W_0 is the required section modulus. From Equation (10.40) one obtains

$$bt_f = A/2 - ht_w / 2 \tag{10.43}$$

Substituting Equation (10.43) into Equation (10.42) we get

$$A \geq 2W_0 / h + 2ht_w / 3 \tag{10.44}$$

The local buckling constraint can be expressed by means of the limiting plate slenderness (Chapter 3).

For a bent web plate of an I-beam the theoretically calculated value of $k = 23.9$, with the elastic modulus of $E = 2.1*10^5$ MPa, $\upsilon = 0.3$, $\sigma_{max} = f_y = 235$ MPa one obtains

$$h / t_w \leq 138.9 \tag{10.45}$$

This value should be decreased according to EC3 to 124 considering the effect of initial imperfections and residual welding stresses. For another f_y values and different design stresses Equation (10.45) can be generalized as

$$\frac{h}{t_w} \leq \frac{1}{\beta} = 124 \sqrt{\frac{235}{f_y}} \sqrt{\frac{f_y}{\sigma_{max}}} \tag{10.46}$$

or using the notation of EC3

$$1/\beta = 124\varepsilon, \quad \varepsilon = \sqrt{235/\sigma_{max}} \tag{10.47}$$

EC3 gives a smaller value for the case when the designer wants to neglect the shear buckling check of the web

$$1/\beta = 69\varepsilon \tag{10.48}$$

Since in the present example the maximum compressive stress is caused by $-F$, so

$$\sigma_{max} = \Delta\sigma_N / \left(2\gamma_{Mf}\right) \tag{10.49}$$

and *the local buckling constraint for the web* can be written as

$$t_w \geq \beta h, \quad 1/\beta = 69\varepsilon, \quad \varepsilon = \sqrt{\frac{235}{\Delta\sigma_N / (2\gamma_{Mf})}} \tag{10.50}$$

The calculations show that this constraint is always active thus it can be treated as equality. Then the objective function (Eq. 10.44) is

$$A = 2W_0 / h + 2\beta h^2 \tag{10.51}$$

and the condition $dA/dh = 0$ gives the optimum web height

$$h_{opt} = \sqrt[3]{3W_0 / (2\beta)} \tag{10.52}$$

The local buckling constraint for the compression flange is given by

$$t_f \geq \delta b \tag{10.53}$$

EC3 gives the value of

$$1/\delta = 28\varepsilon \tag{10.54}$$

and from Equation (10.43) one obtains

$$b_{\text{opt}} = h_{\text{opt}} \sqrt{\frac{\beta}{2\delta}}$$
(10.55)

Finally, using Equation (10.51) we get

$$A_{\min} = \sqrt[3]{18\beta W_0^2}$$
(10.56)

which shows that the cross-sectional area (mass) can be decreased using higher values of $1/\beta$.

Numerical data: $F = 150$ kN fluctuating between $+F$ and $-F$, $L = 2$ m, the number of cycles $N = 3*10^5$. The fatigue detail category for toe cracking is $\Delta\sigma_C = 71$ MPa, for the given number of cycles, using the interpolation formula (Eq. 10.1) we get $\Delta\sigma_N = 133.6$ MPa, thus, with Equation (10.42) $W_0 = 5.61*10^6$ mm^3. Using Equation (10.50) one obtains $1/\beta = 144.7$ and Equation (10.52) gives $h_{\text{opt}} = 1068$ rounded 1070 mm. Furthermore $t_w = 8$ mm, $1/\delta = 28\varepsilon = 58.7$, $b_{\text{opt}} = 485$, $t_f = 9$ mm.

The stress range in the upper flange is (with $\beta = 8/1070$)

$$\Delta\sigma = \frac{M_{\max}}{W_x} = \frac{300*2*10^6}{2\beta h^3/3} = 98.3 \text{ MPa}$$

To calculate the required fillet weld throat size a_f we reduce the normal stress in parent metal into the fillet weld based on the following equality (see Fig. 10.6)

$$\Delta\sigma t_f = 2\rho_\perp a_f$$

from which

$$\rho_\perp = \Delta\sigma t_f / \left(2a_f\right)$$

and the two stress components are

$$\Delta\sigma_\perp = \Delta\tau_\perp = \frac{\rho_\perp}{\sqrt{2}} = \frac{312.8}{a_f}$$

For root cracking of partial penetration fillet welds according to EC3 $\Delta\sigma_C = 36*$ and for $N = 3*10^5$ cycles we use $\Delta\sigma_N = 67.7$ and $\Delta\tau_N = 116.9$ MPa. According to the EC3, the fillet weld should be checked using the interaction formula (Eq. 10.30).

With

$$\Delta\sigma_\perp = \Delta\tau_\perp = 312.8/a_f, \quad \gamma_{Ff} = 1.0 \quad \text{and} \quad \gamma_{Mf} = 1.25$$

one obtains

$$193/a_f^3 + 418/a_f^5 \leq 1$$

from which $a_f = 6$ mm (rounded value).

In the fillet welds connecting the web plate, in addition to the perpendicular stress components

$$\sigma_\perp = \tau_\perp = \frac{\Delta\sigma t_w}{2\sqrt{2}a_w} = \frac{278}{a_w}$$

a parallel stress component should be calculated as well

$$\tau_{//} = \frac{2F}{2a_w h} = \frac{140.2}{a_w}$$

In this case we use, according to the EC3, the following stress components in the interaction formula

$$\Delta\sigma = \sqrt{\sigma_\perp^2 + \tau_\perp^2} = \rho_\perp = \sqrt{2}\sigma_\perp = 393/a_w$$

and

$$\Delta\tau = \tau_{//} = 140.2/a_w$$

From

$$\left(\frac{393}{67.7a_w / 1.25}\right)^3 + \left(\frac{140.2}{116.9a_w / 1.25}\right)^5 \leq 1$$

we obtain $a_w = 7.3$, rounded 8 mm.

10.5.2 *Aluminium structure*

Equations (10.40)-(10.44) treated for the steel structure are valid also for the aluminium one.

The local buckling constraint can be expressed by means of the limiting plate slenderness:

$$\left(\frac{b}{t}\right)_L = \sqrt{\frac{k\pi^2 E}{12(1-\upsilon^2)\sigma_{max}}} \tag{10.57}$$

For a bent web plate of an I-beam the theoretically calculated value of $k = 23.9$, with the elastic modulus of $E = 7*10^4$ MPa, $\upsilon = 0.3$, $\sigma_{max} = p_{01} = 250$ MPa one obtains

$$h/t_w \leq 77.8 \tag{10.58}$$

This value should be decreased according to BS to 51.4 considering the effect of initial imperfections and residual welding stresses. For another p_{01} values and different design stresses Equation (10.58) can be generalized as

$$\frac{h}{t_w} \leq \frac{1}{\beta} = 51.4\varepsilon, \quad \varepsilon = \sqrt{250/\sigma_{max}} \tag{10.59}$$

Since in the present example the maximum compressive stress is caused by $-F$, so $\sigma_{max} = FL/W_x$, or, assuming that the fatigue constraint (Eq. 10.41) is active

$$\sigma_{max} = \Delta\sigma_N / (2\gamma_{mf}) \tag{10.60}$$

and *the local buckling constraint for the web* can be written as

$$t_w \geq \beta h, \quad 1/\beta = 51.4\sqrt{\frac{250}{\Delta\sigma_N / (2\gamma_{mf})}} \tag{10.61}$$

The calculations show that this constraint is always active thus it can be treated as equality. Similar to the calculation described for the steel structure one obtains

$$h_{opt} = \sqrt[3]{3W_0 / (2\beta)} \tag{10.62}$$

The local buckling constraint for the compression flange as an outstand with $k = 0.4$ is given by

$$b / (2t_f) \leq 10.1\varepsilon \tag{10.63}$$

this value is decreased to 6ε (for semi-compact cross-section), thus

$$t_f \geq \delta b; 1 / \delta_L = 12\varepsilon \tag{10.64}$$

Equations (10.55) and (10.56) are also valid.

Numerical data: $F = 50$ kN fluctuating between $+F$ and $-F$, $L = 2$ m, the number of cycles $N = 3*10^5$. The fatigue detail category for toe cracking of fillet welds connecting the flanges to a rigid column, according to BS is $\Delta\sigma_C = 20$ MPa (Recommendations give 22 MPa). For the given number of cycles we get $\Delta\sigma_N = 37.6$ MPa, thus, with Equation (10.42) $W_0 = 6.6489*10^6$ mm³. Using Equation (10.61) one obtains $1 / \beta = 210$, this is unrealistic, so we use the value of 100 and Equation (10.62) gives $h_{opt} = 1000$ mm. Furthermore $t_w = 10$ mm, $1 / \delta = 49$, we use 20, $b_{opt} = 320$, $t_f = 16$ mm.

For the root cracking of fillet welds connecting the flanges to the column $\Delta\rho_{\perp C} = 14$, $\Delta\rho_{\perp N} = 26.3$ MPa. The stress range in a flange is approximately (with $1 / \beta = 100$).

$$\sigma_{max} = \frac{2FL}{2\beta h^3 / 3} = 30 \text{ MPa}$$

This stress is reduced to fillet welds to obtain the fatigue constraint

$$\rho_\perp = \sigma_{max} \frac{t_f}{2a_f} \leq \frac{\Delta\rho_{\perp N}}{\gamma_{mf}} \tag{10.65}$$

From Equation (10.65) we get $a_f = 11.4$ rounded 12 mm.

For fillet welds connecting the web to the column the fatigue constraint is defined by a vector resultant of the stress components

$$\Delta\rho_\perp = \sigma_{max} \frac{t_w}{2a_w} = \frac{150}{a_w} \quad \text{and} \quad \Delta\tau_{//} = \frac{2F}{2ha_w} = \frac{50}{a_w}$$

From the fatigue constraint

$$\sqrt{\left(\Delta\rho_\perp\right)^2 + \left(\Delta\tau_{//}\right)^2} \leq \Delta\rho_{\perp N} / \gamma_{mf} = 21.0 \text{ MPa} \tag{10.66}$$

we get $a_w = 7.5$ rounded 8 mm.

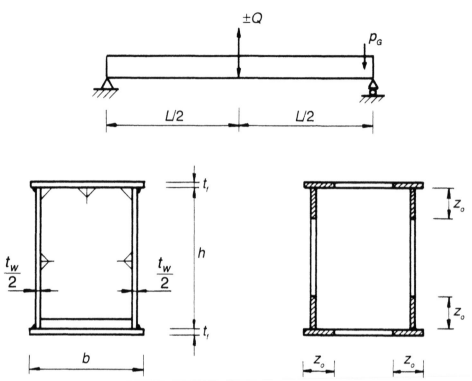

Figure 10.7. Main dimensions of a welded box beam and the reduced cross-section of an aluminium beam.

10.6 WELDED BOX BEAMS (FIG. 10.7)

10.6.1 *Steel structure*

A simply supported beam is loaded by a constant, uniformly distributed static load p_G and by a variable force Q pulsating between $+Q$ and $-Q$ (Fig. 10.7). In order to stiffen the box beam against torsional deformation of the cross-sectional shape, transversal diaphragms should be welded inside by fillet welds. The dimensions of the box beam h, $t_w/2$, b and t_f should be optimized to minimize the cross-sectional area

$$A = ht_w + 2bt_f \tag{10.67}$$

and fulfil the design constraints as follows.

   The fatigue constraint is defined by

$$\Delta\sigma = \psi_d \frac{2QL}{4W_x} \leq \frac{\Delta\sigma_N}{\gamma_{Mf}} \tag{10.68}$$

$$W_x = 2I_x/(h+t_f), \quad I_x = h^3 t_w/12 + 2bt_f\left(h+t_f\right)^2/4 \tag{10.69}$$

where ψ_d is a dynamic factor, W_x is the elastic section modulus, I_x is the moment of inertia, $\Delta\sigma_N$ is the fatigue stress range corresponding to the given number of cycles N. According to EC3, for diaphragms of box girders welded to the flange or web, for diaphragm thickness $t < 12$ mm the detail category (the stress range at $N = 2*10^6$) is $\Delta\sigma_C = 80$ MPa. For another number of cycles, for $N < 5*10^6$ the stress range can be calculated using Equation (10.1). γ_{Mf} is the safety factor for fatigue. According to EC3, for non 'fail-safe' components with poor accessibility (Table 10.1) it is $\gamma_{Mf} = 1.35$.

The static stress constraint is expressed by

$$\sigma_{max} = \left(\gamma_G p_G L^2 / 8 + \gamma_Q \psi_d QL / 4\right) / W_x \leq f_y / \gamma_{M1} \tag{10.70}$$

where, according to EC3, $\gamma_G = 1.35, \gamma_Q = 1.50$ are safety factors for permanent and variable actions, respectively, f_y is the yield stress, $\gamma_{M1} = 1.1$ is the partial safety factor.

The local buckling constraints are as follows:

for flange buckling

$$(b-40) / t_f \leq 42\sqrt{235 / \sigma_{max}} \quad (\sigma_{max} \text{ in MPa}) \tag{10.71}$$

and for web buckling

$$2h / t_w \leq 124 \sqrt{235 / \sigma_{max}} \tag{10.72}$$

It should be noted that we use in Equations (10.71) and (10.72) the maximum static stress instead of yield stress, since the static stress may be much smaller than the yield stress when the fatigue constraint is active. Note that the shear buckling of webs should also be checked, but, in our numerical example, this constraint will be always passive.

The deflection constraint is, according to EC3 for floor beams

$$\frac{5p_G L^4}{384 E_S I_x} + \frac{QL^3}{48 E_S I_x} \leq \frac{L}{300} \tag{10.73}$$

where E_S is the elastic modulus of steel.

10.6.2 *Aluminium structure*

The fatigue constraint is the same as for steel beams (Eqs 10.68 and 10.69). The difference is, that, according to the IIW Recommendations (1995), for plates with transverse stiffeners $\Delta\sigma_C = 28$ MPa. Note that, according to the BS 8118 (1991), in the fatigue constraint the section properties need not to be reduced for HAZ (heat affected zone).

The static stress constraint. In Equation (10.70), according to BS 8118 $\gamma_G = 1.20$, $\gamma_Q = 1.33$, and the section modulus should be reduced considering the HAZ softening effect. According to BS 8118

$$W_{x.red} = \frac{2I_{x.red}}{h + t_f}, \quad I_{x.red} = I_x - 4z_0 \frac{t_f}{2}\left(\frac{h + t_f}{2}\right)^2 - 4z_0 \frac{t_w}{4}\left(\frac{h}{2} - \frac{z_0}{2}\right)^2 \tag{10.74}$$

where the width of the HAZ is

$$z_0 = 3t_B^2 / t_A \tag{10.75}$$

if

$$t_w/2 \le t_f \qquad t_B = t_w/2$$

if

$$t_w/2 > t_f \qquad t_B = t_f$$

if

$$0.5\,(t_w/2 + t_f) \le 1.5 t_B \qquad t_A = 0.5\,(t_w/2 + t_f)$$

if

$$0.5\,(t_w/2 + t_f) > 1.5 t_B \qquad t_A = 1.5 t_B$$

Furthermore, in Equation (10.70), instead of f_y / γ_{M1} the value of p_0 / γ_m should be used, where p_0 is the limiting stress for bending and overall yielding, $\gamma_m = 1.2$ is the material factor.

The local buckling constraints, according to BS 8118, are as follows:
for flange

$$(b-40)/t_f \le 18\sqrt{250/\sigma_{max}} \tag{10.76}$$

and for webs

$$2h/t_w \le 18/0.35\,\sqrt{250/\sigma_{max}} \tag{10.77}$$

The deflection constraints, according to BS 8118 for beams in buildings, are as follows:

$$\frac{5p_G L^4}{384 E_a I_x} \le \frac{L}{200} \tag{10.78}$$

$$\frac{5p_G L^4}{384 E_a I_x} + \frac{Q L^3}{48 E_a I_x} \le \frac{L}{100} \tag{10.79}$$

where E_a is the elastic modulus for aluminium alloys.

10.6.3 *Numerical example*

The following data are given: $Q = 6$ kN, $L = 12$ m, $\psi_d = 2$, the values of p_G are varied to show that for low p_G/Q ratios the fatigue constraint, for large ratios the static stress or deflection constraint is active. $N = 3*10^6$, so, using Equation (10.1), we get for steel beam $\Delta\sigma_N = 69.8$ MPa, for aluminium beam $\Delta\sigma_N = 24.5$ MPa. Take $f_y = 235$ MPa for steel Fe 360 and $p_0 = 240$ MPa for 6082-T6 heat treatable aluminium alloy (ISO: AlSi1MgMn) plates with thicknesses of $t = 3 - 25$ mm. For steel $E_S = 2.1*10^5$ and for aluminium alloys $E_a = 7*10^{4.}$ MPa.

The Rosenbrock's Hillclimb mathematical programming method has been used

Table 10.2. Optimum dimensions (mm) and minimum cross-sectional areas of welded steel box beams for $Q = 6$ kN and various values of p_G.

p_G (N/mm)	3	6	9	10	12
$h*t_w/2$	595*6	535*7	510*8	540*9	550*9
$b*t_f$	230*8	245*8	230*9	200*9	215*10
A (mm^2)	7250	7665	8220	8460	9250
Active constraint	Fatigue	Fatigue	Fatigue	Fatigue and static stress	Static stress

Table 10.3. Optimum dimensions (mm) and minimum cross-sectional areas of welded aluminium box beams for $Q = 6$ kN and various values of p_G.

p_G (N/mm)	3	12	21	25	30
$h*t_w/2$	705*10	625*14	665*19	665*19	695*20
$b*t_f$	245*18	290*17	280*14	260*19	290*19
A (mm^2)	15,870	18,610	20,275	22,515	24,920
Active constraint	Fatigue	Fatigue	Fatigue and deflection	Deflection	Deflection

for computerized optimization. Rounded values are computed by a complementary special program. The results are summarized in Tables 10.2 and 10.3. It can be seen that, depending on the p_G/Q ratio, the fatigue or the static stress as well as the deflection constraint is active.

CHAPTER 11

Tubular trusses

11.1 EFFECT OF CROSS-SECTIONAL SHAPE ON THE OPTIMUM GEOMETRY OF TRUSS STRUCTURES

Several authors do not take into account the shape of cross-sections of compression members in the optimum design of trusses. Our aim is to show this effect on a simplest two-bar structure (Fig. 11.1).

In this structure one rod is compressed and should be designed against overall buckling. In the overall buckling constraint the radius of gyration depends on the cross-sectional shape. In our example two different shapes are selected, namely a circular hollow section (CHS) and a double-angle section (DA). The radius of gyration and the cross-sectional area for CHS is

$$r_C = a_C \sqrt{A}, \quad a_C = \sqrt{\delta/(8\pi)}, \quad \delta = (d-t)/t, \quad A = (d-t)\pi t \tag{11.1}$$

for tubular trusses

$$\delta_{\lim} = 50, \quad \text{thus}, \quad a_C = 1.41 \tag{11.2}$$

For DA, according to ISO 657-1 (1989) or DIN 1028 (1976)

$$r_D = a_D \sqrt{A}, \quad a_D = 0.49 \tag{11.3}$$

For given F, L and yield stress f_y the optimum $\xi = x/L$ is sought, which minimizes the total volume of the structure

$$V = A_1 L + A_2 \sqrt{x^2 + L^2} \quad \text{or} \quad V/L = A_1 + A_2 \sqrt{\xi^2 + 1} \tag{11.4}$$

The cross-sectional areas can be expressed from the design constraints as follows.

The overall buckling constraint for rod 1 is

$$S_1/A_1 \leq \chi f_y/\gamma_{M1} \quad \text{where} \quad S_1 = F/\xi \tag{11.5}$$

Since it is impossible to solve by hand the overall buckling constraint according to EC3 for A_1, it is better to use the JRA (Japan Road Association) overall buckling curve instead of EC3 curve 'b' (Chapter 3)

$$\chi = 1.109 - 0.545\overline{\lambda} \quad \text{for} \quad 0.2 \leq \overline{\lambda} \leq 1 \tag{11.6}$$

$$\chi = 1/(0.773 + \overline{\lambda}^2) \quad \text{for} \quad \overline{\lambda} \geq 1 \tag{11.7}$$

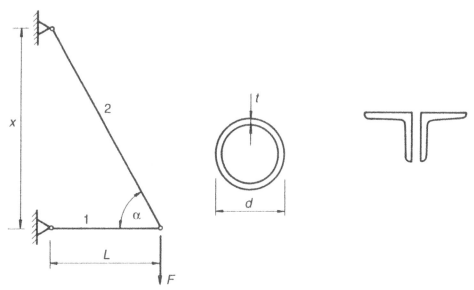

Figure 11.1. Two-bar structure.

Assuming that $\bar{\lambda} \geq 1$ Equation (11.5) can be written in the form

$$\frac{F}{\xi A_1} \leq \frac{f_y}{(0.773 + \bar{\lambda}^2)\gamma_{M1}} \qquad \bar{\lambda} = \frac{KL}{\lambda_E a \sqrt{A_1}}, \quad \lambda_E = \pi\sqrt{\frac{E}{f_y}} \tag{11.8}$$

Solving Equation (11.8) for A_1 one obtains

$$A_1 = \frac{0.773 F\gamma_{M1}}{2\xi f_y}\left[1 + \sqrt{1 + \left(\frac{2KL}{0.773\lambda_E a}\right)^2 \frac{\xi f_y}{F\gamma_{M1}}}\right] \tag{11.9}$$

The stress constraint for tension strut is

$$S_2 / A_2 \leq f_y / \gamma_{M0} \quad \text{where} \quad S_2 = F\sqrt{\xi^2 + 1}/\xi \tag{11.10}$$

Equation (11.10) may be written in the form

$$A_2 \geq \frac{F\gamma_{M0}\sqrt{\xi^2 + 1}}{\xi f_y} \tag{11.11}$$

The final form of the objective function is (for $\bar{\lambda} \geq 1$)

$$\frac{V}{L} = \frac{0.773 F\gamma_{M1}}{2\xi f_y}\left[1 + \sqrt{1 + \left(\frac{2KL}{0.773\lambda_E a}\right)^2 \frac{\xi f_y}{F\gamma_{M1}}}\right] + \frac{F\gamma_{M0}(\xi^2 + 1)}{\xi f_y} \tag{11.12}$$

The ς_{opt} for V_{min} can be calculated only numerically. We take the following numerical data: Factored load $F = 9*10^5$ (N), $f_y = 355$ MPa, $\gamma_{M1} = \gamma_{M0} = 1.1, \lambda_E = 76.4091,$

$K = 1$ (pinned ends), $L = 8000$ mm. For two cases of $a_C = 1.41$ and $a_D = 0.49$, respectively, one obtains

$$\frac{V}{L}(\text{CHS}) = \frac{1077.85}{\xi}\left(1 + \sqrt{1 + 13.2357\xi}\right) + 2788.73\left(\xi + \frac{1}{\xi}\right) \tag{11.13}$$

and

$$\frac{V}{L}(\text{DA}) = \frac{1077.85}{\xi}\left(1 + \sqrt{1 + 109.5953\xi}\right) + 2788.73\left(\xi + \frac{1}{\xi}\right) \tag{11.14}$$

Differentiating Equations (11.13) and (11.14) we get equations for ξ_{opt} as follows:
  for CHS

$$2.5873\xi^2 + \frac{6.61785\xi}{\sqrt{1 + 13.2357\xi}} = 3.5873 + \sqrt{1 + 13.2357\xi} \tag{11.15}$$

and for DA

$$2.5873\xi^2 + \frac{54.79765\xi}{\sqrt{1 + 109.5953\xi}} = 3.5873 + \sqrt{1 + 109.5953\xi} \tag{11.16}$$

Solutions are as follows:
  for CHS:

$$\xi_{opt} = 1.522, \quad \alpha = \arctan\xi = 56.7^\circ, \quad x = 12.18 \text{ m}$$
$$A_1 = 3964.6 \text{ mm}^2 (d = 250, t = 5 \text{ mm}), \quad \bar{\lambda} = 1.1793 > 1, \text{ OK}$$
$$A_2 = 3336.8 \text{ mm}^2, \quad V/L = 10041.3 \text{ mm}^2$$

and for DA:

$$\xi_{opt} = 2.079, \quad \alpha = 64.3^\circ, \quad x = 16.63 \text{ m}$$
$$A_1 = 8361.3 \text{ mm}^2 (2 \times 160 \times 160 \times 14), \quad \bar{\lambda} = 2.3367 > 1, \text{ OK}$$
$$A_2 = 3094.6 \text{ mm}^2, \quad V/L = 15500 \text{ mm}^2$$

It can be seen that there is a considerable difference between the α_{opt}-values and a large difference between the minimal volumes: $100(15500-10041)/10041 = 54\%$.

   It can be concluded that, to obtain realistic optimum geometries of trusses, the effect of the cross-sectional shape of compression members should be taken into account.

11.2 A TUBULAR TRUSS OF PARALLEL CHORDS

In order to illustrate the role of stability constraints we select a simple planar, statically determinate, K-type truss with parallel chords and gap joints, welded from CHS rods (Fig. 11.2). In the optimum design the optimal distance of chords h is sought which minimizes the total volume of the structure and the dimensions of rods fulfil the design constraints. The structural members are divided to 4 groups of equal

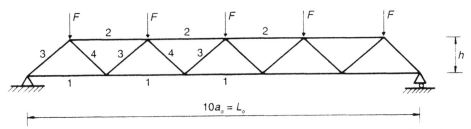

Figure 11.2. A planar truss with parallel chords. The numbering relates to groups of members of equal cross-section.

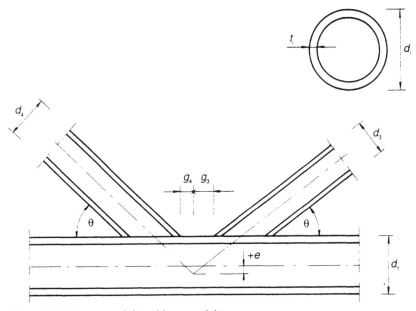

Figure 11.3. K-type gap joint with eccentricity *e*.

cross-section as follows: 1. Lower chord, 2. Upper chord, 3. Compression braces, and 4. Tension braces.

According to DIN 2448 and DIN 2458 (Dutta & Würker 1988) the available CHS have the following dimensions (discrete values)

d = 133, 139.7, 152.4, 159, 168.3, 177.8, 193.7, 219.1, 244.5, 273, 298.5, 323.9.

t = 2.9, 3.2, 3.6, 4, 4.5, 5, 5.6, 6.3, 7.1, 8, 8.8, 10.

All members are made from steel Fe 510 with ultimate strength f_u = 510 MPa and yield stress f_y = 355 MPa.

The load is shown in Figure 11.2, the factored value of the static forces is F = 200 kN. Calculate the required cross-sections for various values of $\omega = h/a_0$ to select the ω_{opt} which minimizes the total volume V. The variables are as follows: d_i and t_i (i = 1, 2, 3, 4). The objective function is expressed as

$$\frac{V}{2\pi a_0} = 5\left(d_1 - t_1\right)t_1 + 4\left(d_2 - t_2\right)t_2 + 3\sqrt{\omega^2 + 1}\left(d_3 - t_3\right)t_3 +$$
$$+ 2\sqrt{\omega^2 + 1}\left(d_4 - t_4\right)t_4 \tag{11.17}$$

The constraints are as follows.

Local buckling constraints for all sections according to Wardenier et al. (1991) are

$$d_i / t_i \le 50 \tag{11.18}$$

Stress constraint for tension members are

$$\frac{S_{1\max}}{\pi\left(d_1 - t_1\right)t_1} \le \frac{f_y}{\gamma_{M0}}, \quad S_{1\max} = \frac{6.5F}{\omega}, \quad \gamma_{M0} = 1.1 \tag{11.19}$$

$$\frac{S_{4\max}}{\pi\left(d_4 - t_4\right)t_4} \le \frac{f_y}{\gamma_{M0}}, \quad S_{4\max} = \frac{1.5F}{\omega}\sqrt{\omega^2 + 1} \tag{11.20}$$

Overall buckling constraints for compression members according to EC3 are as follows

Upper chord:

$$\frac{S_{2\max}}{\pi\left(d_2 - t_2\right)t_2} \le \frac{\chi_2 f_y}{\gamma_{M1}}, \quad S_{2\max} = \frac{6F}{\omega}, \quad \gamma_{M1} = 1.1$$
$$\chi_2 = \frac{1}{\phi_2 + \sqrt{\phi_2^2 - \overline{\lambda}_2^2}}, \quad \phi_2 = 0.5\left[1 + 0.34\left(\overline{\lambda}_2 - 0.2\right) + \overline{\lambda}_2^2\right] \tag{11.21}$$
$$\overline{\lambda}_2 = \frac{\lambda_2}{\lambda_E} = \frac{K_2 L_2}{\lambda_E r_2} = \frac{0.9 * 2a_0 \sqrt{8}}{\lambda_E\left(d_2 - t_2\right)}$$

With $E = 2.1 * 10^5$ MPa and $f_y = 355$ MPa $\lambda_E = \pi\sqrt{E / f_y} = 76.4091$.

$K_2 = 0.9$ is the end restraint factor according to Rondal et al. (1992), $r_2 = (d_2 - t_2) / \sqrt{8}$ is the radius of gyration.

Compression braces:

$$\frac{S_{3\max}}{\pi\left(d_3 - t_3\right)t_3} \le \frac{\chi_3 f_y}{\gamma_{M1}}, \quad S_{3\max} = \frac{2.5F}{\omega}\sqrt{\omega^2 + 1}$$
$$\chi_3 = \frac{1}{\phi_3 + \sqrt{\phi_3^2 - \overline{\lambda}_3^2}}, \quad \phi_3 = 0.5\left[1 + 0.34\left(\overline{\lambda}_3 - 0.2\right) + \overline{\lambda}_3^2\right] \tag{11.22}$$
$$\overline{\lambda}_3 = \frac{\lambda_3}{\lambda_E} = \frac{K_3 L_3}{\lambda_E r_3} = \frac{0.75 a_0 \sqrt{\omega^2 + 1}\sqrt{8}}{\lambda_E\left(d_3 - t_3\right)}$$

In order to ease the fabrication the diameter of braces should be smaller than those of chords:

$$d_3 = 0.92 d_1, \quad d_3 \le 0.92 d_2, \quad d_4 \le 0.92 d_1, \quad d_4 \le 0.92 d_2 \tag{11.23}$$

Prescription for the joint eccentricity to avoid too large additional bending moment in the vicinity of nodes is as follows (Fig. 11.3.):

$$e \leq 0.25 d_1, \quad e \leq 0.25 d_2 \tag{11.24}$$

The eccentricity can be expressed by d_i, angle θ and gap parts g_3 and g_4 as follows:

$$\tan \theta = \frac{e + d_1/2}{g_3 + d_3/(2 \sin \theta)} \quad \text{or} \quad \tan \theta = \frac{e + d_1/2}{g_4 + d_4/(2 \sin \theta)} \tag{11.25}$$

Assuming that

$$g_3 = g_4 = 0.05 \, d_1 \quad \text{or} \quad 0.05 \, d_2 \tag{11.26}$$

the geometry constraints can be given by:

$$\frac{d_3}{2}\sqrt{\omega^2 + 1} + d_1\left(0.05\omega - 0.75\right) \leq 0 \tag{11.27}$$

and

$$\frac{d_3}{2}\sqrt{\omega^2 + 1} + d_2\left(0.05\omega - 0.75\right) \leq 0 \tag{11.28}$$

Constraint on static strength of welded joints between chords and braces according to EC3 is

$$\sqrt{\sigma_\perp^2 + 3\left(\tau_\perp^2 + \tau_{II}^2\right)} \leq f_u / \left(\beta_w \gamma_{MW}\right) \tag{11.29}$$
$$f_u = 510 \text{ MPa}, \quad \beta_w = 0.9, \quad \gamma_{Mw} = 1.25$$

From the force S in a brace the following stress components arise in welds:

$$\sigma_\perp = \tau_\perp = \frac{S \sin \theta}{\pi d a_w} \cdot \frac{\sqrt{2}}{2}, \quad \tau_{II} = \frac{S \cos \theta}{\pi d a_w} \tag{11.30}$$

where a_w is the fillet weld dimension. Substituting Equation (11.30) into Equation (11.29) we get

$$\frac{S}{\pi d a_w}\sqrt{\frac{2\omega^2 + 3}{\omega^2 + 1}} \leq 453 \text{ MPa} \tag{11.31}$$

For the maximal value of a_w the corresponding brace thickness can be taken. This constraint should be fulfilled for S_3 and S_4.

For the node strength the following constraints should be fulfilled (Wardenier et al. 1991).

Constraints on chord plastification. In the joint of rods 1 and 3:

$$S_{3\max} \leq S_{31}^* = \frac{f_y t_1^2}{\sin \theta}\left(1.8 + 10.2\frac{d_3}{d_1}\right)f_1\left(\gamma_1, g_1'\right)$$

$$f_1\left(\gamma_1, g_1'\right) = \gamma_1^{0.2}\left[1 + \frac{0.024\gamma_1^{1.2}}{\exp\left(0.5g_1' - 1.33\right) + 1}\right], \quad \gamma_1 = \frac{d_1}{2t_1} \tag{11.32}$$

$$g_1' = g_1/t_1, \quad \text{we assume that} \quad g_1 = g_3 + g_4 = 0.1\, d_1$$

Table 11.1. Optimal discrete dimensions (mm) and $V/(2\pi a_0)$-values (mm²) for various $\omega = h/a_0$-values.

$\omega = h/a_0$	0.8	0.9	1.0	1.1	1.2	1.3	1.4
d_1/t_1	244.5/8	244.5/8	244.5/8	219.1/8	273/8	273/8	298.5/8.8
d_2/t_2	273/8	244.5/8	244.5/8	219.1/8.8	273/8	273/8	298.5/8.8
d_3/t_3	219.1/4.5	219.1/4.5	219.1/4.5	193.7/4.5	219.1/4.5	219.1/4.5	293.7/4.5
d_4/t_4	159/3.6	152.4/3.6	152.4/3.2	152.4/3.2	139.7/3.2	139.7/3.2	139.7/2.9
$V/(2\pi a_0)$	23083	22367	22475	21063	24970	25264	28704

Table 11.2. Check of the constraints for the optimal solution $\omega = 1.1$.

Constraint	Dimension	Equation 11	Rod 1	2	3	4	Remarks
Local buckling	–	(18)	27 < 50	25 < 50	43 < 50	48 < 50	Active for rods 3, 4
Tensile stress	MPa	(19), (20)	223 < 323	–	–	270 < 323	Near active for rod 4
Overall buckling	MPa	(21), (22)	–	188 < 204	240 < 261	–	Active for rods 2, 3
Fabrication	mm	(23)	–	–	194 < 202	152 < 202	Active for rod 3
Eccentricity	mm	(27), (28)	–	–	–8.32	–	Near active for rods 1, 2, 3
Weld strength	MPa	(31)	–	–	368 < 453	414 < 453	Near active for rod 4
Chord plastification	kN	(32)	–	–	642 < 713	405 < 586	Active for rods 3-1
Punching shear	kN	(33)	–	–	642 < 1744	–	Passive

Constraints on chord plastification for joints of rods 1-4, 2-3 and 2-4 can be formulated similarly to Equation (11.32), therefore these constraints are not detailed here.

Constraints on punching shear. In the joint of rods 2 and 3:

$$S_{3max} \leq \frac{f_y}{\sqrt{3}} t_2 \pi d_3 \frac{1+\sin\theta}{2\sin^2\theta} \tag{11.33}$$

Note that the constraint on punching shear was in our calculations always passive, so it is not necessary to investigate it for other joints.

For the computations the Rosenbrock's hillclimb mathematical programming method has been used treating the unknowns as continuous variables. After the determination of the optimal dimensions the discrete optima have been found by using an additional search. The results are summarized in Table 11.1.

The optimum value is $\omega = 1.1$, the difference between the best and worst solution in the range of $\omega = 0.8$-1.4 is $100/(28704-21063)/21063 = 36\%$. The checks of constraints are summarized in Table 11.2.

It can be seen that the overall buckling constraint is always active, the local buckling constraint is passive only for chord 2, since for thickness t_2 the chord plastification is governing. Thus, it can be stated that the effect of stability constraints in the optimum design of tubular trusses is significant.

The significant role of the stability constraints in the optimum design of tubular trusses is illustrated by a numerical example. In this optimum design procedure the dimensions of CHS truss members and the optimal distance of chords are determined which give the minimum volume (weight) of the structure and fulfil the design constraints. The constraints relate to the overall buckling of compression members, to the joint eccentricity and static strength of joints. For the final optimal version realistic available discrete tube dimensions are determined.

11.3 A TRUSS WITH NON-PARALLEL CHORDS

The design method described in Section 9.1 is applied for compression members of a statically determinate roof truss with non-parallel chords to illustrate the savings in weight in the case of trusses by using CHS or SHS instead of double-angle sections.

Consider the truss shown in Figure 11.4. Four different cross-sections (1-4) are designed for each case. To find the optimal truss height (h) or the optimal slope angle of the upper chord, the truss is designed for heights $h = 2.5, 3.5, 4.5, 6.0$ and 7.5 m (corresponding slope angles are 4.76°, 9.46°, 14.04°, 20.56° and 26.56°).

In the design of CHS and SHS struts, section properties of the ISO/DIS 4019.2 as well as the tables given by Dutta & Würker (1988) (DIN 2448, DIN 2458, DIN 59411) have been used. The results of the calculations are summarized in Table 11.3.

It can be seen that the optimal truss height (slope angle) giving the minimal total

Table 11.3. Total volumes of trusses of various heights.

Height h (m)	Slope angle $\beta°$	Section	Upper chord 1	Lower chord 2	Outside columns 3	Braces 4	Total volume 10^{-7} (mm³)
2.5	4.76	CHS	152.4/2.9	133/3.2	108/2.3	108/2.3	10.72
		SHS	115/3.2	110/3	70/3.2	70/3.2	11.04
		angles	$2 \times 80 \times 8$	$2 \times 50 \times 7$	$2 \times 50 \times 6$	$2 \times 70 \times 6$	18.14
3.5	9.46	CHS	152.4/2.3	139.7/2.3	101.6/2	101.6/2	9.24
		SHS	115/2.6	80/3.2	70/2.6	70/2.6	9.71
		angles	$2 \times 70 \times 7$	$2 \times 50 \times 5$	$2 \times 50 \times 6$	$2 \times 55 \times 6$	15.36
4.5	14.04	CHS	139.7/2	127/2	101.6/2	101.6/2	8.95
		SHS	90/3	80/2.6	70/2.6	70/2.6	9.80
		angles	$2 \times 65 \times 7$	$2 \times 45 \times 5$	$2 \times 50 \times 6$	$2 \times 60 \times 6$	17.18
6.0	20.56	CHS	127/2	101.6/2	101.6/2	88.9/2	8.79
		SHS	90/2.6	90/2	70/2.6	70/2.6	10.45
		angles	$2 \times 70 \times 6$	$2 \times 40 \times 4$	$2 \times 50 \times 6$	$2 \times 60 \times 6$	18.87
7.5	26.56	CHS	114.3/2	88.9/2.3	101.6/2	88.9/1.8	8.84

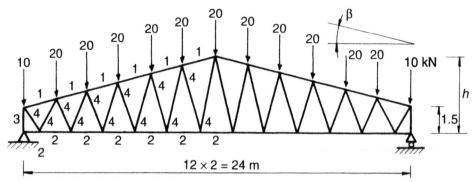

Figure 11.4. Numerical example of a roof truss. 1. Section for upper chord, 2. Section for lower chord, 3. Section for outside columns, and 4. Section for braces. The height h varies with the slope angle β of the upper chord.

volume of the structure depends on the cross-sectional shape. In the investigated numerical example the optimal slope angles are as follows: For double-angle sections $\beta \cong 10°$, for SHS $\cong 12°$ and for CHS $\cong 20°$.

The savings in weight by using CHS or SHS instead of double-angle sections, according to Table 11.3, are 41-53% or 39-45%, respectively

e.g. $100 \ (18.14 - 10.72/18.14) = 41\%$ etc.

These differences are larger than the difference between the material costs of CHS, SHS and rolled angles, thus material cost savings can also be achieved.

Note that the sensitivity of the volume functions for CHS and SHS is relatively small, but the difference between the volumes for the heights $h = 2.5$ and the optimum $h_{opt} = 6$ m (for CHS) is

$100 \ (10.72-8.79/10.72) = 18\%$

so, for economic design, it is important to choose the optimum truss slope.

CHAPTER 12

Stiffened and cellular plates

12.1 MAIN CHARACTERISTICS OF STIFFENED AND CELLULAR PLATES

Two different forms of stiffened plates can be constructed as follows (Fig. 12.1):

– Figure 12.1b: Plates stiffened by ribs on one side in one or more directions, in the following simply 'stiffened plates', and

– Figure 12.1c: Cellular plates consisting of two face plates and a grid of ribs welded between them.

These two types are similar to welded I- and box-beams: I-beams and stiffened plates (with ribs of open section) have open section with low torsional stiffness but of easier fabrication possibility, box-beams and cellular plates are closed structures with large torsional stiffness but their fabrication is more difficult.

The difference between torsional stiffnesses can be illustrated by means of a simplified comparison using the advantage of square symmetry as follows: The partial differential equation for the deflections $w(x,y)$ of an isotropic rectangular plate subject to a lateral load of intensity p can be written in the form

$$\frac{\partial^4 w}{\partial x^4} + 2\frac{\partial^4 w}{\partial x^2 \partial y^2} + \frac{\partial^4 w}{\partial y^4} = \frac{p}{B} \qquad B = \frac{Et^3}{12(1-v^2)} \tag{12.1}$$

where B is the bending stiffness of the plate of constant thickness t. The solution of Equation (12.1), for a simply supported plate is

$$w(x, y) = \sum_{m=1,2,\dots} \sum_{n=1,2,\dots} a_{mn} \sin\frac{m\pi x}{b_x} \sin\frac{n\pi y}{b_y} \tag{12.2}$$

The simplified form of the above solution for a square plate ($b_x = b_y = b$), considering only the first term of series is

$$w(x, y) = a_{11} \sin\frac{\pi x}{b} \sin\frac{\pi y}{b} \tag{12.3}$$

for which it is valid that

$$\frac{\partial^4 w}{\partial x^4} = \frac{\partial^4 w}{\partial x^2 \partial y^2} = \frac{\partial^4 w}{\partial y^4} \tag{12.4}$$

thus, Equation (12.1) can be written as

249

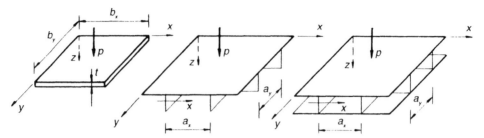

Figure 12.1. a) Isotropic, b) Stiffened, and c) Cellular plate.

$$\frac{\partial^4 w}{\partial x^4} = \frac{p}{4B} \tag{12.5}$$

For stiffened and cellular plates (Fig. 12.1b and c) Equation (12.1) is applied in generalized form valid for orthotropic (orthogonally anisotropic) plates, called Huber equation:

$$B_x \frac{\partial^4 w}{\partial x^4} + 2H \frac{\partial^4 w}{\partial x^2 \partial y^2} + B_y \frac{\partial^4 w}{\partial y^4} = p \tag{12.6}$$

where B_x, B_y and H are the bending and torsional stiffnesses, respectively.
　For stiffened plates (Fig. 12.1b)

$$B_x = EI_y / a_y, \quad B_y = EI_x / a_x$$

and

$$H = B_{xy} + B_{yx}, \quad B_{xy} = GI_{ty} / a_y, \quad B_{yx} = GI_{tx} / a_x \tag{12.7}$$

where I_x and I_y are the moments of inertia, I_{ty} and I_{tx} are the torsional constants (see Eq. 2.30) of cross-sections consisting of a rib and a part of the cover plate with an efficient width. It can be shown (see Chapter 8) that the torsional constants of open sections are very small, thus, approximately $H \approx 0$. Using the solution of Equation (12.3) for a square plate, Equation (12.6) can be expressed as

$$\frac{\partial^4 w}{\partial x^4} = \frac{p}{2B_x} \tag{12.8}$$

For cellular plates (Fig. 12.1c)

$$B_{x1} = E_1 I_{y1} / a_{y1}, \quad B_{y1} = E_1 I_{x1} / a_{x1}, \quad E_1 = E / (1 - \upsilon^2)$$

and

$$H_1 = B_{xy1} + B_{yx1} + \frac{\upsilon}{2}(B_{x1} + B_{y1}) \tag{12.9}$$

In I_{x1} and I_{y1} the effect of ribs can be neglected.
　In the calculation of B_{xy1} and B_{yx1} the moments of inertia I_{x1} and I_{y1} can be used since the shear stresses act similarly than the normal stresses due to bending as it is

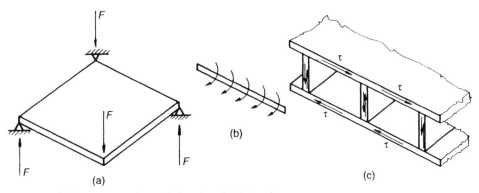

Figure 12.2. Shear stresses in a cellular plate due to torsion.

shown in Figure 12.2 for the case of a loading which causes torsion in all edges of the plate. Thus

$$B_{xy1} = GI_{y1} / a_{y1}, \quad B_{yx1} = GI_{x1} / a_{x1}, \quad G = E / [2(1+\upsilon)] \tag{12.10}$$

In the case of square symmetry one obtains

$$H_1 = 2B_{xy1} + \upsilon B_{x1} = 2GI_1 / a_1 + E_1 I_1 / a_1 = B_{x1} = B_{y1} \tag{12.11}$$

and, using the solution of Equation (12.3), Equation (12.6) can be simplified to

$$\frac{\partial^4 w}{\partial x^4} = \frac{p}{4B_{x1}} \tag{12.12}$$

The following conclusions may be drawn:
   – Comparing Equation (12.12) with Equation (12.5) it can be seen that a square cellular plate may be calculated as an isotropic one. This conclusion has been verified by some experiments carried out on glued plexiglas models for other kinds of supports (e.g. point supports at four corners) and loadings,
   – Comparing Equation (12.12) with Equation (12.8) it can be seen that, if the bending stiffnesses are the same ($B_x = B_{x1}$) the deflections of a square stiffened plate are twice as large as in the case of a square cellular plate. This is caused by the difference between the torsional stiffnesses.

12.2 OPTIMUM DESIGN OF A CELLULAR PLATE OF SQUARE SYMMETRY BY SIMPLIFIED HAND CALCULATION

Consider a square cellular plate (Fig. 12.3) consisting of an orthogonal grid of channel or I-section ribs and of upper and lower cover plate elements welded to the grid by fillet welds. The approximate volume of the plate is

$$V = 2b^2 t_f + 2bht_r \varphi \tag{12.13}$$

It is assumed that the number of ribs in one direction is $\varphi = b / a$, b is the side length

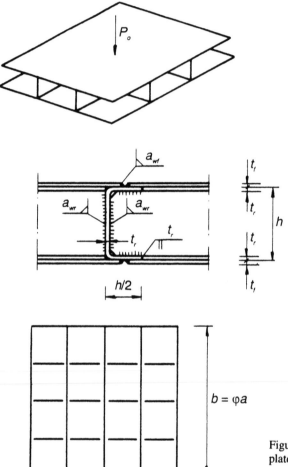

Figure 12.3. Welded square cellular plate with an orthogonal grid of channel-section ribs.

of the plate, a is the distance between ribs. The flanges of the ribs are neglected in Equation (12.13).

Introducing the dimensionless variables $\alpha = a/h, \vartheta = a/t_f, \mu = t_f/t_r$, Equation (12.13) can be written as

$$\frac{V}{2b^3} = \frac{t_f}{b} + \frac{ht_r\varphi}{b^2} = \frac{1}{\varphi\vartheta} + \frac{1}{\alpha\varphi\mu\vartheta} \tag{12.14}$$

It can be seen from Equation (12.14) that the unconstrained problem to find V_{min} does not give any optima. To decrease V, φ should be increased and t_f decreased. This leads to a dense grid of ribs and to very thin plates similar to a sandwich plate with a honeycomb core.

This situation is significantly changed by introducing stress, shear buckling and size constraints which have been active in the detailed optimization procedure. Assuming that the effective plate width factor $\psi = a_e/a$ may be considered as a constant, the stress constraint is

$$\sigma_{\max} + \sigma_{f.\max} \le \sigma_{\mathrm{adm}} \tag{12.15}$$

$\sigma_{\max}$ is caused by the bending of the whole plate, $\sigma_{f.\max}$ is the normal stress due to the local bending of the face plate elements, σ_{adm} is the admissible stress. As it is detailed in Section 12.3.3, Equation (12.14) may be written as

$$c_1 \alpha \vartheta \varphi^2 + c_2 \vartheta^2 \le c_3, \quad c_1 = 4.79 * 10^{-2} p / \psi, \quad c_2 = 0.3078 p_0, \quad c_3 = \sigma_{\mathrm{adm}} \tag{12.16}$$

p_0 is the intensity of the uniformly distributed normal load, $p = 1.1 p_0$, the self mass is considered by a factor of 1.1.

The thickness limitation is

$$t_r \ge t_0 \quad \text{or} \quad \varphi \mu \vartheta \le c_4, \quad c_4 = b / t_0 \tag{12.17}$$

The constraint on shear buckling of ribs is

$$\frac{0.42 \, pba}{h t_r} \le \frac{5.34 \pi^2 E_1}{12 \gamma_b} \left(\frac{t_r}{h} \right)^2 \quad \text{or} \quad \frac{\varphi \mu^3 \vartheta^3}{\alpha} \le c_5, \quad c_5 = \frac{5.34 \pi^2 E_1}{12 * 0.42 \gamma_b P} \tag{12.18}$$

From Equation (12.17) one obtains $\mu = c_4 / (\varphi \vartheta)$. Substituting it into Equation (12.18) we get

$$\frac{1}{\alpha} = \frac{c_5}{c_4^3} \varphi_2 \tag{12.19}$$

and Equation (12.16) takes the form

$$\vartheta^2 + c_7 \vartheta - c_8 = 0, \quad c_7 = c_6 / c_2, \quad c_6 = c_1 c_4^3 / c_5, \quad c_8 = c_3 / c_2 \tag{12.20}$$

Solving Equation (12.20) one obtains

$$\vartheta = 0.5 \left(-c_7 + \sqrt{c_7^2 + 4 c_8} \right) = c_9 \tag{12.21}$$

Now Equation (12.14) can be written as

$$\frac{V}{2 b^3} = \frac{1}{c_9 \varphi} + c_{10} \varphi^2 \qquad c_{10} = \frac{c_5}{c_4^4} \tag{12.22}$$

This shows that an optimum exists and it can be calculated using the condition $dV / d\varphi = 0$ which yields

$$\varphi_{\mathrm{opt}}^V = (2 c_9 c_{10})^{-1/3} \tag{12.23}$$

Complement now the objective function with an approximate fabrication cost which can be considered proportional to the weld length

$$L_W = 8 \varphi b \tag{12.24}$$

Thus, the new cost function is

$$\frac{K}{k_m} = \rho V + \frac{k_f}{k_m} C a_W^{1.5} L_W \tag{12.25}$$

a_W is the fillet weld size, C is a constant, k_m and k_f are the material and fabrication cost factors, respectively (see Chapter 5). Using Equations (12.21) and (12.24) we obtain

$$\frac{K}{k_m} = 2b^3\rho\left(\frac{1}{c_9\varphi} + c_{10}\varphi^2\right) + c_{11}\varphi, \quad c_{11} = \frac{8k_f Ca_W^{1.5}b}{k_m} \tag{12.26}$$

The condition $dK / d\varphi = 0$ yields a cubic equation for φ_{opt}

$$\frac{1}{c_9\varphi^2} = 2c_{10}\varphi + c_{12}, \quad c_{12} = \frac{c_{11}}{2b^3\rho} \tag{12.27}$$

which can be numerically solved. c_{12} expresses the effect of fabrication cost. An increase of c_{12} results in decrease of the optimum number of ribs.

Numerical example
Data: $p = 1.1p_0 = 5.5*10^{-3}$ N/mm^2, $t_0 = 2$ mm, $b = 8000$ mm, $\sigma_{adm} = 120$ MPa, $\rho = 7850$ kg/m^3, $E_1 = 2.2637*10^5$ MPa, $a_W = 2$ mm, $\gamma_b = 1.35$, $C = 2.5*10^{-3}$ min/mm$^{2.5}$, $k_f/k_m = 1$ kg/min. According to the previous calculations $\psi = 0.2 - 0.5$, we calculate with 0.2. All dimensions are in N, mm and min.

$c_1 = 1.31725*10^{-3}$, $c_2 = 1.539*10^{-3}$, $c_3 = 120$, $c_4 = 4000$, $c_5 = 3.1881*10^8$, $c_6 = 0.26443$, $c_7 = 0.17182*10^3$, $c_8 = 0.77973*10^5$, $c_9 = 206.24$, $c_{10} = 1.2453*10^{-6}$.

$\varphi_{opt}^v = 12$, $c_{11} = 452.8$ kg, $c_{12} = 56.3296*10^{-6}$. The solution of Equation (12.27) is $\varphi_{opt} = 8$, which clearly shows the effect of fabrication cost.

12.3 COST COMPARISONS OF STIFFENED AND CELLULAR PLATES

12.3.1 *Introduction*

Cellular plates have the following advantages over stiffened ones: 1. The torsional stiffness is much larger, 2. The height of ribs can be much smaller, 3. Because of the symmetry of the structure, initial imperfections due to the shrinkage of welds are much smaller, and 4. The planar surfaces can be more easily protected against corrosion.

Because of their advantages, cellular structures may be applied in ships, aircraft, bridges, dock gates, floating roofs of storage tanks, elements of machine structures (press tables, mounting desks, base plates), mining shields, floors of buildings and store houses, lightweight roofs, structural parts vehicles, hovercraft, excavators, bucket wheel stacker-reclaimers etc.

The disadvantage of cellular plates is that the fabrication is more complicated than in the case of stiffened plates. When the height of ribs is smaller than approximately 800 mm, difficulty arises because the ribs cannot be welded from inside. The connection between face sheets and ribs can be realized from outside by means of arc-spot welding, electron-beam welding, slot or plug welds (Hicks 1979).

12.3.2 *A brief literature survey*

A very stiff base plate for transportation of heavy structures may be built by using an orthogonal grid welded from rolled I-beams. The lower face plate has been joined to the grid by plug welds (Sahmel 1978).

In the revolving frame of surface mining equipment (dragline) a platform for boom, cab, power unit and other structural parts forms an all-welded multi-cell structure. Welded shoes raise tub and frame, moving the dragline along (Birchfield 1981). A large multi-stiffener cellular plate has been designed for a platform of a floating nuclear plant (Tsai & Orr 1977).

Laser welding technology has been used for welding of 'Norsial' metallic sandwich plates consisting of two face plates and a corrugated sheet sandwiched between them (Haroutel 1982).

Cellular plates made of glass fibre reinforced plastic have been used for lightweight roofs (Isler 1977).

A large research project was performed by Williams (1969) who used a welded cellular plate model for double bottoms of ships.

From theoretical works we mention here only some articles relating to cellular plates.

Pettersen (1979) has worked out a detailed analysis of double-bottom plates of ships.

Evans & Shanmugam (1981, 1984, 1986) have also treated the analytical problems of cellular plates relating to the ship construction.

Cellular plate made of fibre reinforced composite elements has been analysed by Rao (1988).

Optimum design of welded cellular plates has been treated by Farkas (1976, 1982, 1984).

In a discussion of the article by Evans & Shanmugam (1984), Farkas (1985) has shown that the torsional stiffness of a square cellular plate equals its bending stiffness, therefore it may be treated as an isotropic plate (see Section 12.1).

In the following, illustrative numerical examples are treated. Square plates are considered, carrying a uniformly distributed normal load, with simply supported edges. The square symmetry has two advantages: 1. The plate can be more easily treated mathematically and 2. The carrying capacity of the plate is utilized in both directions to the highest degree.

A square cellular and a stiffened plate is optimized taking into account the same design constraints and the two structural versions are compared to each other.

It should be mentioned that in Sections 12.3.3 and 12.3.4 the fabrication cost is calculated using original constants proposed by Pahl and Beelich and not on the basis of COSTCOMP software described in Chapter 5.

12.3.3 *Minimum cost design of a square cellular plate*

The cost function
It is assumed that the fabrication has the following steps. First the grid of ribs is welded from cold-formed channels or from welded I-beams. The grid nodes should be completely welded to be able to carry bending moments and shear forces. Then the square elements of the upper and lower cover plates are welded to the grid from outside with fillet welds (Fig. 12.4).

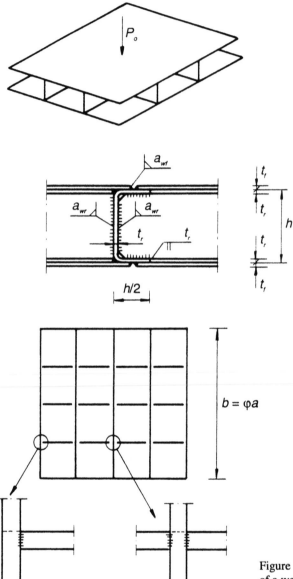

Figure 12.4. General view and node details
of a welded cellular plate.

Note that other joining methods may also be used, e.g. arc-spot welding or plug
welds. Unfortunately one cannot find sufficient data about the fabrication cost of
such welds.

The ribs are continuous in one direction, in the other direction they are intermittent. There are two types of nodes, one for peripheral and another for internal nodes.

The cross-sectional area of a rib is approximately $2ht_r$, where h is the height, t is
the constant thickness. The number of spaces between ribs in one direction is $\varphi =
b/a$, b is the side length of the plate. Thus, the number of ribs in one direction is

$\varphi+1$. Assuming that all ribs have the same cross-sectional area, the whole volume of the square cellular plate is

$$V = 2b^2 t_f + 4bht_r (\varphi+1) \tag{12.28}$$

where t_f is the thickness of face plates.

The whole objective function is written in the form as treated in Chapter 5

$$\frac{K}{k_m} (\text{kg}) = \rho V + \frac{k_f}{k_m} (T_1 + T_2 + T_3) \tag{12.29}$$

Time for preparation, assembly and tacking is

$$T_1 = C_1 \Theta \sqrt{\kappa \rho V}, \quad C_1 = 1.0 \text{ min/kg}^{0.5} \tag{12.30}$$

We take for the difficulty factor $\Theta = 3$. The number of face plate elements is $2\varphi^2$, number of internal ribs is $\varphi(\varphi-1)$ and number of peripheral ribs is $\varphi+3$, thus

$$\kappa = 3(\varphi^2 + 1) \tag{12.31}$$

$$T_1 (\text{min}) = 460.38028 \sqrt{V(\varphi^2 + 1)} \ (V \text{ in m}^3) \tag{12.32}$$

It is assumed that the grid nodes are joined by manual-arc-welded fillet welds and the face plate elements are connected to the grid by CO_2-welded fillet welds. Since the number of internal nodes is $(\varphi-1)^2$, the welding time of these nodes is

$$T_2' = 0.8 * 10{-2} (\varphi-1)2 \left(4ha_{Wr}^{1.5} + 2ht_r^{1.5}\right) \tag{12.33}$$

The number of peripheral nodes is $(\varphi+1)^2 - (\varphi-1)^2 = 4\varphi$, the corresponding welding time is

$$T_2'' = 0.8 * 10^{-3} * 4\varphi \left(2ha_{Wr}^{1.5} + ht_r^{1.5}\right) \tag{12.34}$$

The weld length of fillet welds joining the face plate elements is $L_{Wf} = 2\varphi^2 * 4b/\varphi = 8b\varphi$ thus, the corresponding welding time is

$$T_2''' = 0.5 * 10^{-3} a_{Wf}^{1.5} * 8b\varphi \tag{12.35}$$

The total welding time is

$$T_2 = 4 * 10^{-3} b\varphi a_{Wf}^{1.5} + 1.6 * 10^{-3} h \left(\varphi^2 + 1\right) \left(2a_{Wr}^{1.5} + t_r^{1.5}\right) \text{ (all sizes in mm)} \tag{12.36}$$

Time for electrode changing, weld deslagging and chipping is

$$T_3 = 1.2 * 10^{-3} \sqrt{\Theta} \sum_i a_{Wi}^{1.5} L_{wi} \tag{12.37}$$

with actual values it is

$$T_3 = 2.07846 * 10^{-3} \left[8b\varphi a_{Wf}^{1.5} + 2h \left(\varphi^2 + 1\right) \left(2a_{Wr}^{1.5} + t_r^{1.5}\right)\right] \tag{12.38}$$

Note that the cost of electrodes is neglected, and the costs of painting and corrosion protection are not included. The fabrication cost of the individual ribs (cold-formed channels or welded I-beams) are not considered here.

The design constraints
– *Constraint on compressive stress* in the central upper face plate element is

$$\sigma_{max} + \sigma_{f.max} \leq \sigma_{adm} \tag{12.39}$$

σ_{max} is caused by the bending of the whole plate, $\sigma_{f.max}$ is the normal stress due to the local bending of the face plate elements, σ_{adm} is the admissible stress.

σ_{max} can be calculated by means of the orthotropic plate theory, treating the square cellular plate as isotropic (Timoshenko & Woinowsky-Krieger 1959)

$$\sigma_{max} = \frac{4.79*10^{-2} pb^2}{B_1} E_1 e_1, \quad e_1 = \frac{h}{1+\psi}, \quad B_1 = \frac{E_1 h^2 t_f \psi}{1+\psi} \tag{12.40}$$

Thus,

$$\sigma_{max} = \frac{4.79*10^{-2} pb^2}{h t_f \psi} \tag{12.41}$$

For the effective width ratio $\psi = a_e / a$ of a compressed upper face plate element the Faulkner's formula may be used (Farkas 1984):

$$\psi = \frac{2}{\lambda_p} - \frac{1}{\lambda_p^2} - \frac{6}{\vartheta - 6} \quad \lambda_p = \vartheta \sqrt{\frac{\sigma_{max}}{E}}, \quad \vartheta = \frac{a}{t_f} \tag{12.42}$$

Substituting Equation (12.41) into Equation (12.42) we get a quadratic equation for ψ, the solution of which is

$$\psi = Y_0 + \sqrt{Y_0^2 - Z_0}, \quad Y_0 = 2 \left(\frac{Z_2}{1+Z_2^2} \right)^2 - \frac{Z_2^2 Z_3}{1+Z_2^2}, \quad Z_0 = \left(\frac{Z_2^2 Z_3}{1+Z_2^2} \right)^2$$

$$Z_1 = \frac{4.79*10^{-2} pb^2}{h t_f}, \quad Z_2 = \frac{b}{\varphi t_f} \sqrt{\frac{Z_1}{E}}, \quad Z_3 = \frac{6}{b/(\varphi t_f) - 6} \tag{12.43}$$

The local bending stress can be calculated by means of formulae valid for isotropic plates with clamped edges (Timoshenko & Woinowsky-Krieger 1959)

$$\sigma_{f.max} = \frac{5.13*10^{-2} p_0 a^2}{t_f^2 / 6} = \frac{0.3078 p_0 b^2}{\varphi^2 t_f^2} \tag{12.44}$$

p_0 is the intensity of the uniformly distributed normal load. $p = 1.1 p_0$ since the self mass is considered approximated by a factor of 1.1.
– *Constraint on shear buckling of rib webs* at the edges can be formulated as

$$\tau = \frac{0.42 pb^2}{h t_r \varphi} \leq \frac{\tau_{ub}}{\gamma_b} = \frac{5.34 \pi^2 E_1}{12 \gamma_b} \left(\frac{t_r}{h} \right)^2 \quad \text{for} \quad \frac{\tau_{ub}}{\gamma_b} \leq \tau_{adm} \tag{12.45a}$$

and

$$\tau = \frac{0.42 pb^2}{h t_r \varphi} \leq \tau_{adm} \quad \text{for} \quad \frac{\tau_{ub}}{\gamma_b} > \tau_{adm} \tag{12.45b}$$

The factor of 0.42 is considered since the distribution of edge reactions is not uniform along the edges (Timoshenko & Woinowsky-Krieger 1959).

τ_{adm} is the admissible shear stress. With a safety factor of $\gamma_b = 1.35$ and Poisson's ratio $\upsilon = 0.3$ Equation (12.45a) takes the form

$$\frac{0.11748\, pb^2 h}{E\varphi t_r} \leq 1 \qquad (12.46)$$

– *Constraint on deflection*

$$w_{max} = \frac{4.06 * 10^{-3} p_0 b^4 (1+\psi)}{E_1 h^2 t_f \psi} \leq w^* = c * b \qquad (12.47)$$

w^* is the admissible deflection.
   – *Size constraints* are the thickness limitations

$$(t_f, t_r) \geq t_0 \qquad (12.48)$$

12.3.4 *Minimum cost design of a square stiffened plate*

The cost function
In order to make a realistic comparison, we assume that the fabrication method is the same as described for a cellular plate, but without lower face plate elements (Fig. 12.5). Thus Equation (12.28) changes to

$$V = b^2 t_f + 4bht_r \,(\varphi + 1) \qquad (12.49)$$

The number of structural elements is now

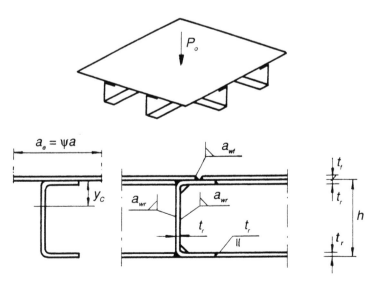

Figure 12.5. General view and a node of a stiffened plate.

$$\kappa = \varphi^2 + \varphi + 3 + \varphi \, (\varphi - 1) = 2\varphi^2 + 3 \qquad (12.50)$$

For fabrication times we get

$$T_1 = 3 \sqrt{\rho V \, (2\varphi^2 + 3)} = 265.80068 \sqrt{V \, (2\varphi^2 + 3)} \qquad (12.51)$$

$$T_2 = 2 * 10^{-3} b\varphi a_{Wf}^{1.5} + 1.6 * 10^{-3} h \, (\varphi^2 + 1) \left(2a_{Wr}^{1.5} + t_r^{1.5}\right) \qquad (12.52)$$

$$T_3 = 2.07846 * 10^{-3} \left[4b\varphi a_{Wf}^{1.5} + 2h \, (\varphi^2 + 1) \left(2a_{Wr}^{1.5} + t_r^{1.5}\right)\right] \qquad (12.53)$$

The objective function is the same as defined by Equation (12.29).

The design constraints
Constraint on compressive stress in the central face plate element can be formulated similar to Equation (12.39)

$$\sigma_{max.1} + \sigma_{f.max} \leq \sigma_{adm} \qquad (12.54)$$

where

$$\sigma_{max.1} = \frac{c_M \, pb^2}{I_x / a} \, y_C \qquad (12.55)$$

Since the torsional stiffness of the open section ribs is very small, the stiffened plate can be calculated as an orthotropic one having a torsional stiffness $H = 0$. Schade (1941) has calculated for this case $c_M = 0.1102$.

According to Figure 12.5 the distance of the centroidal axis y_C can be calculated as

$$y_C = \frac{2ht_r}{2ht_r + a_e t_f} \frac{h}{2} = \frac{h}{2 \, (1 + \alpha)}, \quad \alpha = \frac{a_e t_f}{2ht_r}, \quad a_e = \psi a \qquad (12.56)$$

Here we use for ψ, instead of Equation (12.42), a simpler formula proposed by Usami & Fukumoto (1982)

$$\psi = \frac{1.426}{\lambda_p}, \quad \lambda_p = \frac{a}{t_f} \sqrt{\frac{\sigma_{max.1}}{E}}, \quad \psi \leq 1 \qquad (12.57)$$

The moment of inertia can be expressed as

$$I_x = a_e t_f \, y_C^2 + 2ht_r \left(\frac{h}{2} - y_C\right)^2 + \frac{h^3 t_r}{3} = \frac{h^3 t_r}{6} \frac{2 + 5\alpha}{1 + \alpha} \qquad (12.58)$$

Substituting Equation (12.55) into Equation (12.57) we get again a quadratic equation, the solution of which is

$$\psi = \frac{5 * 1.426^2 \varphi^3 t_f^3 hE}{12 c_M \, pb^4} \left(1 + \sqrt{1 + \frac{96 c_M \, pb^3 t_r}{25 * 1.426^2 \varphi^3 t_f^4 E}}\right) \qquad (12.59)$$

The formula for the local bending stress is the same as Equation (12.44).

Constraint on maximum tensile stress in the central ribs is

$$\sigma_{max.2} = \sigma_{max.1} \frac{h - y_C}{y_C} = \sigma_{max.1}(1 + 2\alpha) \le \sigma_{adm} \tag{12.60}$$

Constraint on shear buckling of rib webs at the edges is the same as formulated by Equations (12.45) and (12.46).

Deflection constraint can be written as

$$w_{max} = \frac{c_W p_0 b^4}{EI_x / a} \le w^* = c * b \tag{12.61}$$

According to Schade (1941), for torsional stiffness $H = 0$, $c_W = 0.0082$.

12.3.5 *Numerical example for both structural versions*

Data: $t_0 = 2$ mm, $p_0 = 5$kN/m^2, $E = 2.06*10^5$ MPa, steel Fe 360, $\sigma_{adm} = 120$ MPa, $b = 8$ m, $\rho = 7850$ kg/m^3, $\tau_{adm} = \sigma_{adm}/\sqrt{3} = 69$ MPa, $w^* = 4$ mm.

In the optimum design procedure the optimum values of h, t_f, t_r and ϕ should be determined which minimize the cost function and fulfil the design constraints. The optimization has been performed by using the 'hillclimb' algorithm (see Chapter 6). The results are given in Tables 12.1 and 12.2. Here we give only the rounded optimum dimensions for both structural versions. We have used the weld sizes $a_{Wf} = 0.7t_f$ and $a_{Wr} = 0.7t_r$, these are rounded after the optimization.

It can be seen from Tables 12.1 and 12.2 that the optimum number of ribs depends on fabrication costs. For cellular plates ϕ_{opt} decreases from 12 to 9 when the k_f/k_m ratio increases. In the case of stiffened plates it decreases from 14 to 7.

Another important conclusion is that the fabrication cost plays an important role in the total cost. For the cellular plate with $k_f/k_m = 1$, $100 K_f/K = 100*11902/17177 = 69\%$. For the stiffened plate it is $100*7646/12720 = 60\%$.

From the comparison of Tables 12.1 and 12.2 it can be seen that the total cost of

Table 12.1. Optimum rounded dimensions in mm and K/k_m (kg) values for cellular plates.

k_f/k_m	ϕ_{opt}	h	t_f	t_r	a_{Wf}	a_{Wr}	K/k_m	w_{max}
1	9	250	4	2	3.0	1.5	17177	0.32
0.75	9	260	4	2	3.0	1.5	14287	0.41
0.50	10	280	4	2	3.0	1.5	12331	0.32
0.25	12	300	3	2	2.5	1.5	8441	0.48
0	12	300	3	2	2.5	1.5	4974	0.48

Table 12.2. Optimum rounded dimensions in mm and K/k_m values (kg) for stiffened plates.

k_f/k_m	ϕ_{opt}	h	t_f	t_r	a_{Wf}	a_{Wr}	K/k_m	w_{max}
1	7	460	5.5	2.5	4.0	2.0	12720	1.17
0.75	7	460	5.5	2.5	4.0	2.0	10812	1.17
0.50	8	430	5.0	2.5	3.5	2.0	9003	1.30
0.25	13	370	3.5	2.0	2.5	1.5	6861	1.48
0	14	355	3.5	2.0	2.5	1.5	4433	1.48

Table 12.3. Optimum rounded dimensions in mm and K/k_m values (kg) for stiffened plates with prescribed $h_{max} = h$ cellular, $a_{Wf} = 0.7t_f$, $a_{Wr} = 0.7t_r$.

k_f/k_m	h_{max}	φ_{opt}	t_f	t_r	K/k_m	w_{max}
1	250	7	6.5	8.0	19000	2.53
0.5	280	9	5.0	5.0	11560	2.35
0	300	17	3.0	2.5	4898	2.21

stiffened plates is less than that of cellular plates, but the rib height is larger and the maximum deflection is much larger.

The advantage of a cellular plate over a stiffened one can be demonstrated by formulating a size constraint on rib height for stiffened plate prescribing that $h_{stiffened}$ should be equal to $h_{opt.cellular}$. To fulfil this constraint larger thicknesses should be used which increase the costs. Table 12.3 shows the results of this optimization. It can be seen that, for $k_f/k_m = 1$ the total cost is larger than that of cellular plate.

Note that, in this optimization, the two constraints on normal stresses are active while the others are passive.

The comparison of the maximum deflections shows that the stiffness of cellular plates is much larger than that of stiffened ones.

The computations also show that the sensitivity of the objective function to changes in the main parameter is small.

12.4 EFFECT OF FABRICATION COST ON THE OPTIMUM DIMENSIONS OF A STIFFENED PLATE

The welding technology has a significant effect on optimum structural versions. The COSTCOMP software (Chapter 5) allows us to calculate the optimum structural versions and their cost for various welding technologies. The numerical example of a stiffened plate illustrates this possibility. Some articles may be mentioned in this context.

The article of Drews & Starke (1990) deals with the economy of robotization. The efficiency of automation should be increased by reducing the time of fixturing, tooling, programming and testing. Horikawa et al. (1992) proposed various modifications in structural design for efficient application of welding robots. The study of Fern & Yeo (1990) compared the effective deposition rates of various semi-automated and mechanised welding processes considering flat, horizontal, vertical and overhead welding positions. Helpful hints have been given to improve the design. Chalmers (1986) dealt with fabrication costs of ship structures analyzing the material and labour costs and giving useful comments for design.

Forde et al. (1984) have treated the design/fabrication interaction and have proposed an information system to give designers more information about costs. Sen et al. (1989) have treated the minimum weight and cost design of stiffened, corrugated and sandwich panels used in ship structures, but a detailed cost analysis has not been given. The study of Malin (1985) gives a good view on effective automation of welding operation and describes some economic aspects for automation. Pedersen &

Nielsen (1987) have treated the minimum weight and cost design of a stiffened plate used in ships considering also the cost of welding without any cost analysis. Ramirez & Touran (1991) have described the EXSYS expert system which has two main modules. The first module selects an appropriate welding method and the second one estimates the welding costs.

Stiffened panels are widely used in bridge and ship structures, so it is of interest to study the minimum cost design of such structural elements.

The design rules of API (1987) are used here for the formulation of the global buckling constraint for uniaxially compressed plate longitudinally stiffened by equally spaced uniform flat stiffeners of equal cross sections (Fig. 12.6). The cost function is defined according to Equation (12.62) in which $A = b_0 t_f + \varphi h_s t_s$, $\Theta = 3$, $\kappa = \varphi + 1$, $L_w = 2L\varphi$, φ is the number of stiffeners.

$$\frac{K}{k_m} = \rho L A + \frac{k_f}{k_m} \left(\Theta \sqrt{\kappa \rho L A} + 1.3 C_2 a_w^n L_w \right) \tag{12.62}$$

The flat stiffeners are welded by double fillet welds, the size of welds is taken as $a_w = 0.5 t_s$. The welding costs are calculated for SMAW (shielded metal-arc welding), GMAW-C (gas metal-arc welding with CO_2) and SAW (submerged arc welding) (Chapter 5).

In the optimization procedure the given data are as follows. The modulus of elasticity for steel is $E = 2.1*10^5$ MPa, the material density is $\rho = 7.85*10^{-6}$ kg/mm^3, the Poisson's ratio is $\upsilon = 0.3$, the yield stress is $f_y = 235$ MPa, the plate width is $b_0 = 4200$ mm, the length is $L = 4000$ mm. The axial compressive force is

$$N = f_y b_0 t_{fmax} = 235*4200*20 = 1.974*10^7 \text{ (N)}$$

The variables to be optimized are as follows (Fig. 12.6): The thickness of the base

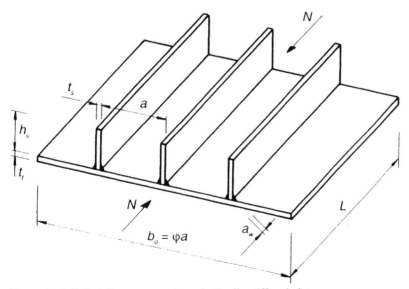

Figure 12.6. Uniaxially compressed longitudinally stiffened plate.

plate t_f, the sizes of stiffeners h_s and t_s and the number of stiffeners $\varphi = b_0/a$.
The overall buckling constraint is given by

$$N \le \chi f_y A \tag{12.63}$$

where the buckling factor χ is given in function of the reduced slenderness $\overline{\lambda}$

$$\chi = 1 \qquad \text{for} \quad \overline{\lambda} \le 0.5 \tag{12.64a}$$

$$\chi = 1.5 - \overline{\lambda} \quad \text{for} \quad 0.5 \le \overline{\lambda} \le 1 \tag{12.64b}$$

$$\chi = 0.5/\overline{\lambda} \qquad \text{for} \quad \overline{\lambda} \ge 1 \tag{12.64c}$$

where

$$\overline{\lambda} = \frac{b_0}{t_f} \sqrt{\frac{12(1-\upsilon^2)f_y}{E\pi^2 k}} \tag{12.65}$$

$$k = \min(k_R, k_F), \qquad k_R = 4\varphi^2 \tag{12.66a,b}$$

$$k_F = \frac{(1+\alpha^2)^2 + \varphi\gamma}{\alpha^2(1+\varphi\delta_P)} \quad \text{when} \quad \alpha = \frac{L}{b_0} \le \sqrt[4]{1+\varphi\gamma} \tag{12.66c}$$

$$k_F = \frac{2(1+\sqrt{1+\varphi\gamma})}{1+\varphi\gamma} \quad \text{when} \quad \alpha \ge \sqrt[4]{1+\varphi\gamma} \tag{12.66d}$$

$$\delta_P = \frac{h_s t_s}{b_0 t_f}, \quad \gamma = \frac{EI_s}{b_0 D}, \quad I_s = \frac{h_s^3 t_s}{3}, \quad D = \frac{E t_f^3}{12(1-\upsilon^2)} \tag{12.66e}$$

so

$$\gamma = 4(1-\upsilon^2)\frac{h_s^3 t_s}{b_0 t_f^3} = 3.64 \frac{h_s^3 t_s}{b_0 t_f^3} \tag{12.66f}$$

I_s is the moment of inertia of one stiffener about an axis parallel to the plate surface at the base of the stiffener, D is the flexural stiffness of the base plate.

The constraint on local buckling of a flat stiffener is defined by means of the limiting slenderness ratio according to EC3

$$\frac{h_S}{t_S} \le \frac{1}{\beta_S} = 14 \sqrt{\frac{235}{f_y}} \tag{12.67}$$

The computational results are summarized in Table 12.4.

The optimization procedure is carried out by using the software for the feasible sequential quadratic programming FSQP method and for the Rosenbrock's hillclimb method. Rounded values are computed by a complementary special program (Chapter 6).

The ranges of unknowns are taken as follows (in mm): $t_f = 6$-20, $h_s = 84$-280, $t_s = 6$-25, $\varphi = 4$-15.

It can be seen that the minimum mass design ($k_f = 0$) results in much more stiffeners than the minimum cost design. The optimal plate dimensions depend on cost

Table 12.4. Optimal rounded sizes of a uniaxially compressed longitudinally stiffened plate, double fillet welds carried out by different welding methods, dimensions in mm.

Welding method	k_f/k_m	t_f	h_s	t_s	φ	A (mm^2)	K/k_m (kg)
SMAW	0.00	10	200	15	15	87000	2732
	0.10	13	210	17	11	91560	3516
	0.18	15	220	16	9	94680	3929
	0.20	16	220	16	8	95360	3945
	0.50	19	230	17	6	103260	5272
	1.00	19	230	17	6	103260	7301
	1.50	19	230	17	6	103260	9330
GMAW-C	0.0	10	200	15	15	87000	2732
	0.3	15	215	16	9	93960	3716
	0.5	16	220	16	8	95360	4146
	1.0	19	230	17	6	103260	5227
	1.5	19	230	17	6	103260	6220
SAW	0.0	10	200	15	15	87000	2732
	0.5	15	220	16	9	94680	3944
	1.0	19	230	17	6	103260	4767
	1.5	19	230	17	6	103260	5530

factors k_f/k_m and C_2, so the results illustrate the effect of the welding technology on the structure and costs.

It should be noted that, in the case of SMAW, the φ_{opt} values are very sensitive to k_f/k_m, so in Table 12.4 more k_f/k_m-values are treated.

For $k_f/k_m = 1.5$ the cost savings achieved by using SAW instead of SMAW or GMAW-C are $100*(9330-5530)/9330 = 41\%$ and $100 (6220-5530)/6220 = 11\%$.

In the case of SMAW and $k_f/k_m = 1.5$ the material cost component is $\rho LA = 103260*7.85*10^{-6}*4*10^3 = 3242$ kg, so the fabrication cost represents $100(9330-3242)/9330 = 65\%$ of the whole cost, this significant part of costs affects the dimensions and the economy of stiffened plates.

12.5 MINIMUM COST DESIGN OF RECTANGULAR CELLULAR PLATES

12.5.1 *The cost function*

The fabrication steps are described in Section 12.3. The optimization procedure described for square cellular plates in Section 12.3.3 is now generalized for rectangular plates.

The integer numbers of rib distances are φ_x and φ_y, respectively (Fig. 12.7). The number of ribs in x-and y direction is $\varphi_x + 1$ and $\varphi_y + 1$, resp. Assuming that all ribs have the same cross-sectional area, the whole volume of the cellular plate is

$$V = 2b_x b_y t_f + 2b_x ht_{rx} (\varphi_y + 1) + 2b_y h t_{ry} (\varphi_x + 1) \qquad (12.68)$$

where t_f is the thickness of face plates. Number of continuous ribs in y direction is

$\varphi_x + 1$, number of rib elements in x direction (internal intermittent, peripheral continuous, Fig. 12.7) is $\varphi_x (\varphi_y - 1) + 2$, number of face plate elements is $2\varphi_x \varphi_y$. Thus the number of all structural elements is

$$\kappa = 3 (\varphi_x \varphi_y + 1) \quad \text{and} \quad T_1 = 3 \sqrt{\rho V} \sqrt{3 (\varphi_x \varphi_y + 1)} \tag{12.69}$$

Note that here we use the fabrication time constants as orthogonally proposed by Pahl & Beelich (1982) and not on the basis of COSTCOMP software. Thus,

$$T_2 = \sum C_{2i} a_{wi}^{1.5} L_{wi} \tag{12.70}$$

$C_2' = 0.8 * 10^{-3}$ min/(mm$^{1.5}$mm) for manual arc-welding $C_2'' = 0.5 * 10^{-3}$ min/(mm$^{1.5}$ mm) for CO_2-welding a_w and L_w are the size and length of welds in mm, respectively.

It is assumed that the grid nodes are joined by manual-arc-welding with fillet welds and the face plate elements are connected to the grid by CO_2-welded fillet welds.

$$T_3 = \sqrt{Q} \sum C_{3i} a_{wi}^{1.5} L_{wi}, \quad C_3' = C_2', \quad C_3'' = C_2'' \tag{12.71}$$

The number of perpendicular joints of ribs is

$$2 (\varphi_x + 1) + 2 \varphi_x (\varphi_y - 1) = 2 (\varphi_x \varphi_y + 1)$$

It is assumed that the webs of ribs are welded with fillet welds of size $a_w = 0.7 \, t_{ry}$ in

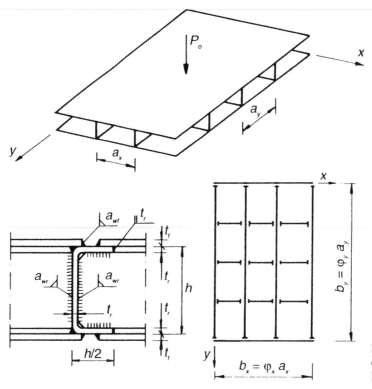

Figure 12.7. Details of a welded rectangular cellular plate.

the length of $2h$, and the flanges of ribs of length h are welded with welds of $a_w = t_{ry}$ (Fig. 12.7). Thus the $T_2 + T_3$ times for manual-arc-welded nodes are (in min)

$$T_2' + T_3' = (1 + \sqrt{Q})\, 0.8*10^{-3}* 2\,(\varphi_x \varphi_y + 1)\,[ht_{ry}^{1.5} + 2h\,(0.7t_{ry})^{1.5}] \qquad (12.72)$$

The total length of fillet welds for face plate elements is

$$2\,\varphi_x \varphi_y\,(2a_x + 2a_y) = 4\,\varphi_x \varphi_y\,(b_x/\varphi_x + b_y/\varphi_y)$$

and the fillet weld size is taken as $a_{wf} = 0.5\,t_f$, thus

$$T_2' + T_3' = (1 + \sqrt{Q})\, 0.5*10^{-3}\,(0.5t_f)^{1.5}\, 4\,\varphi_x \varphi_y\,(b_x/\varphi_x + b_y/\varphi_y) \qquad (12.73)$$

all sizes in mm.

12.5.2 *The design constraints*

Constraints on compressive elastic stresses in the central upper face plate element are as follows

$$\sigma_{x.\mathrm{max}} + \sigma_{xf.\mathrm{max}} \le \sigma_{\mathrm{adm}} \qquad (12.74)$$

$$\sigma_{y.\mathrm{max}} + \sigma_{yf.\mathrm{max}} \le \sigma_{\mathrm{adm}} \qquad (12.75)$$

where σ_{adm} is the admissible stress, $\sigma_{x,\mathrm{max}}$ and $\sigma_{y,\mathrm{max}}$ are caused by the bending of the whole plate, $\sigma_{xf,\mathrm{max}}$ and $\sigma_{yf,\mathrm{max}}$ are normal stresses due to the local bending of the face plate element.

It can be verified, similarly to the case of a square cellular plate (Section 12.1) that, because of the large torsional stiffness of cells, the whole rectangular cellular plate can be calculated as an isotropic one.

$$\sigma_{x,\mathrm{adm}} = M_{x\mathrm{max}}\, E_1\, e_1 / B, \quad \sigma_{y,\mathrm{adm}} = M_{y\mathrm{max}}\, E_1\, e_1 / B \qquad (12.76)$$

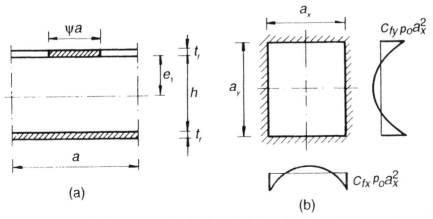

(a)

(b)

Figure 12.8. a) Effective cross-section for the calculation of bending stiffness, and b) Local bending of an upper plate element due to the uniformly distributed lateral load.

According to the isotropic plate theory (Timoshenko & Woinowsky-Krieger 1959)

$$M_{x\,\text{max}} = c_{mx}\, p\, b_x^2 \quad \text{and} \quad M_{y\,\text{max}} = c_{my}\, p\, b_x^2$$

there $p = 1.1\, p_0$, with the factor of 1.1 the self mass is considered, c_{mx} and c_{my} are given by Timoshenko & Woinowsky-Krieger (1959) for simply supported edges. The bending stiffness B is calculated considering the effective width of the compressed face plate element (Fig. 12.8)

$$B = E_1 I / a = E_1\, h^2 t_f \psi / (1 + \psi), \quad e_1 = h / (1 + \psi), \quad E_1 = E / (1 - \upsilon^2) \quad (12.77)$$

where E is the modulus of elasticity and υ is the Poisson's ratio.

We use here the effective width formula proposed by Usami & Fukumoto (1982)

$$\psi = 0.75 / \overline{\lambda}_p, \quad \overline{\lambda}_p = \frac{a}{t_f} \sqrt{\frac{12\,(1 - \upsilon^2)\sigma_{\max}}{4\pi^2 E}}$$

$$\text{or} \quad \psi = 1.426 / \lambda_p, \quad \lambda_p = \frac{a}{t_f} \sqrt{\frac{\sigma_{\max}}{E}} \quad (12.78)$$

Substitution of Equation (12.77) into Equation (12.76) yields

$$\sigma_{x.\text{max}} = c_{mx}\, p\, b_x^2 / (t_f h \psi_y) \quad \text{and} \quad \sigma_{y.\text{max}} = c_{my}\, p b_x^2 / (t_f h \psi_x) \quad (12.79)$$

Elimination of ψ from Equation (12.79) is performed using Equations (12.78) and (12.79) and yields

$$\sigma_{x.\text{max}} = c_{mx}^2 p^2 b_x^4 a_y^2 / (1.426^2 t_f^4 E h^2) \quad \text{and}$$
$$\sigma_{y.\text{max}} = c_{my}^2 p^2 b_x^4 a_x^2 / (1.426^2 t_f^4 E h^2) \quad (12.80)$$

Furthermore

$$\sigma_{xf.\text{max}} = 6 c_{fx} p_0 a_x^2 / t_f^2 \quad \text{and} \quad \sigma_{yf.\text{max}} = 6 c_{fy} p_0 a_x^2 / t_f^2 \quad (12.81)$$

where c_{fx} and c_{fy} are given in Table 12.5 according to Timoshenko & Woinowsky-Krieger (1959) for a uniformly loaded rectangular isotropic plate with clamped edges (Fig. 12.8b). Since c_{fx} and c_{fy} vary during the optimization procedure these values are calculated with approximate analytical formulae in a polynomial form

$$c_{fx} = c_0 + c_1 a_y/a_x + c_2\,(a_y/a_x)^2 + c_3\,(a_y/a_x)^3$$

Table 12.5. Coefficients for the calculation of bending moments and shear forces in a simply supported uniformly loaded rectangular isotropic plate.

b_y/b_x	1.0	1.1	1.2	1.3	1.4	1.5	1.6	1.7	1.9	2.0	3.0	4.0	∞
$10^4 c_{mx}$	479	554	627	694	755	812	862	908	985	1017	1189	1235	1250
$10^4 c_{my}$	479	493	501	503	502	498	492	486	471	464	496	384	375
$10^3 c_{ax}$	338	360	380	397	411	424	435	444	459	465	493	498	500
$10^3 c_{qy}$	338	347	353	357	361	363	365	367	369	370	372	372	372

Table 12.6. Coefficients for the calculation of maximum stresses in a uniformly loaded rectangular isotropic plate with clamped edges.

a_x/a_y	1.0	1.1	1.2	1.3	1.4	1.5	1.6	1.7	1.8	1.9	2.0	∞
$10^4 c_{fx}$	513	581	639	687	726	757	780	799	812	822	829	833
$10^4 c_{fy}$	513	538	554	563	568	570	571	571	571	571	571	571

On the contrary, values of c_{mx}, c_{my}, c_{qx} and c_{qy} are constant during an optimization procedure, because b_y and b_x are given in a numerical example.

Constraints on local buckling of rib webs due to bending

$$\sigma_{x.max} \leq \frac{23.9\pi^2 E_1}{12\gamma_b}\left(\frac{t_{rx}}{h}\right)^2 \quad \text{and} \quad \sigma_{y.max} \leq \frac{23.9\pi^2 E_1}{12\gamma_b}\left(\frac{t_{ry}}{h}\right)^2 \qquad (12.82a,b)$$

where γ_b is the safety factor for buckling.

Constraints on local buckling of rib webs due to shear

$$\tau_x = \frac{Q_x a_y}{h t_{rx}} = \frac{c_{qx} p b_x a_y}{h t_{rx}} \leq \frac{5.34\pi^2 E_1}{12\gamma_b}\left(\frac{t_{rx}}{h}\right)^2, \quad \tau_x \leq \tau_{adm} \qquad (12.83a)$$

$$\tau_y = \frac{Q_y a_x}{h t_{ry}} = \frac{c_{qy} p b_x a_x}{h t_{ry}} \leq \frac{5.34\pi^2 E_1}{12\gamma_b}\left(\frac{t_{ry}}{h}\right)^2, \quad \tau_y \leq \tau_{adm} \qquad (12.83b)$$

where $\tau_{adm} = \sigma_{adm}/\sqrt{3}$ is the admissible shear stress, c_{qx} and c_{qy} are given in Table 12.6 according to Timoshenko & Woinowsky-Krieger (1959).

Size constraints are the thickness limitations

$$t_{rx} \geq t_0, \quad t_{ry} \geq t_0 \quad \text{and} \quad t_f \geq t_0 \qquad (12.84)$$

where t_0 is the minimum thickness considering the welding technology. Note that the deflection constraint is not considered here because of the large stiffness of the whole cellular plate.

12.5.3 *The optimization procedure*

In a numerical example the values of p_0, b_y, b_x, σ_{adm}, E, υ, c_{mx}, c_{my}, c_{qx}, c_{qy}, t_0 are given, and the unknowns to be optimized for minimum cost K_{min} are as follows: φ_x, φ_y, h, t_f, t_{rx} and t_{ry}. In the cost function the k/k_m ratio is varied in the range of 0-1.5. For the purpose of comparison we have used here three mathematical programming methods.

– *The backtrack combinatorial method* is advantageous here, since the number of variables is only 6, φ_x and φ_y are integer numbers and the thicknesses should be commercially available, so the series of discrete values to be investigated can easily be defined. The starting point should be feasible.

– *The hillclimb method* is proposed by Rosenbrock (Chapter 6). No derivatives are required. The starting point should be feasible. We have supplemented this method with a secondary search for finding discrete values after having continuous ones.

– *FSQP (Feasible Sequential Quadratic Programming) method* (Chapter 6). The C version is a quite new development and we have worked also with the beta version on PC.

All the programs are written in C and run under Borland C++ on PC 486 type computer. These codes are quicker than the Fortran and Basic codes and are more transportable, we also could run them on workstation.

12.5.4 Numerical examples

Data: The intensity of the uniformly distributed normal load $p_0 = 5*10^{-3}$ N/mm^2, $p = 1.1*p_0 = 5*10^5$ N/mm^2, $t_0 = 2$ mm, E $= 2.1*10^5$ MPa, $\upsilon = 0.3$, $\rho = 7850$ kg/m^3, $\gamma_b = 2$, $\Theta = 3$, $b_x = 10$ m. To show the effect of yield stress of steel, calculations are made for steel Fe 360 with $\sigma_{adm} = 120$ MPa, and for steel Fe 510 with $\sigma_{adm} = 120*355/235 = 181$ MPa.

To show the effect of fabrication costs calculations are made for $k_f/k_m = 0$, 0.5, 1.0 and 1.5. The results are shown in Figures 12.9 and 12.10, and Table 12.7. Figure 12.9 shows the curves of the objective function as a function of x in the vicinity of the optimum value. It can be seen that for larger k_f/k_m values – larger fabrication costs – $\varphi_{x.opt}$ is smaller. Using higher-strength steel Fe 510, 4-12% cost savings can be achieved. The sensitivity of the objective function is small.

In Figure 12.10 the minimum K/k_m cost values are plotted in function of b_y/b_x for

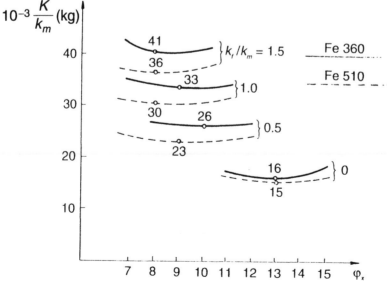

Figure 12.9. Results of a numerical example: Minimum costs for various k_f/k_m-ratios and the $\varphi_{x.opt}$ values.

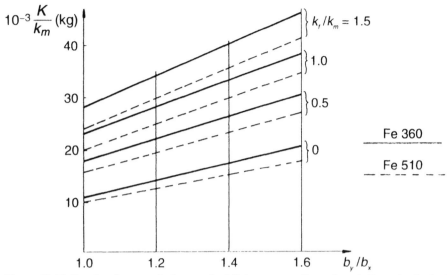

Figure 12.10. Results of a numerical example: Minimum costs for various k_f/k_m-ratios in the function of b_y/b_x.

Table 12.7. Results of a numerical example with $b_x = 10$ m, $b_y = 14$ m, steel Fe 360 obtained by three mathematical programming methods, dimensions in mm.

Method	k_f/k_m	φ_x	φ_y	t_f	t_{rx}	t_{ry}	h	K/k_m (kg)
CFSQP without	0	13.0	16.0	5.1	2.2.	2.0	278	14548
discretization	0.5	12.8	16.0	5.1	2.2	2.0	285	23627
	1.0	10.2	8.8	6.1	3.0	2.4	383	31743
	1.5	9.4	8.0	6.0	3.1	2.5	406	38705
Hillclimb without	0	13.1	15.7	5.0	2.3	2.0	299	14624
discretization	0.5	10.0	9.0	6.0	3.0	2.4	400	24609
	1.0	10.0	8.0	6.0	3.2	2.5	499	31888
	1.5	8.0	7.9	7.0	3.1	2.6	374	39094
Hillclimb with dis-	0	14	15	5	3	2	300	15229
cretization	0.5	10	9	6	3	3	400	25459
	1.0	10	8	6	4	3	450	33485
	1.5	8	8	7	4	3	375	40665
Backtrack	0	13	16	5	3	3	300	16162
	0.5	10	9	6	4	3	400	26149
	1.0	10	8	6	4	3	450	33485
	1.5	8	8	7	4	3	375	40665

steels Fe 360 and Fe 510. It can be seen that the K/k_m values vary with b_y/b_x approximately linearly.

These results are obtained by backtrack programming method.

In Table 12.7 the optimal dimensions obtained by three methods are given for a numerical example. It can be seen that the hillclimb and CFSQP methods resulted in very similar undiscretized optimal values. The results obtained by Hillclimb with discretization and by the discrete backtrack are also very similar.

12.5.5 *Conclusions*

Illustrative numerical examples show that, because of the large torsional stiffness of cellular plates, relatively large structures can be realized using thin plates. The optimal number of ribs decreases when the fabrication cost k_f/k_m increases. The sensitivity of the objective function is small. The use of Fe 510 instead of Fe 360 results in 4-12% cost savings. Active constraints are the normal stress limitation (Eq. 12.74) and the constraints on local shear buckling of rib webs (Eq. 12.83).

The comparison of the three mathematical programming methods shows that the hillclimb technique is quick but can result in local minima, the Backtrack is suitable for few variables defined by series of discrete values, the CFSQP method is very robust and the starting point can be unfeasible.

CHAPTER 13

Welded steel bridges

13.1 SURVEY OF SELECTED LITERATURE

A detailed survey of structural optimization studies published after 1960 is worked out by Cohn & Dinovitzer (1994). A catalogue is given which shows the main characteristics of examples which can be found in 44 publications, mainly books. The model beams, frames, trusses and plates treated in published works may be applied to bridges. The characteristics of cited examples are as follows: Sketch of the structural form, optimization level (section, member, structure, system, topological), loading (static or dynamic), materials (steel, concrete, composite steel-concrete), limit states (ultimate, serviceability), constraints (stress, deflection, cracking, fatigue, buckling, dynamics), objective function (single or multiple), computational method (e.g. analytic, branch and bound, dynamic programming, optimality criteria, gradient, sequential linear or quadratic programming, etc.). It is concluded that the number of actual engineering applications was very small and in order to broadening the practical range of application, more actual examples, easy-to-use software and expert systems should be available for designers.

Suruga & Maeda (1976) have compared the cost and weight of more floor systems of suspension bridges concluding that the conventional reinforced floor system is cheaper but heavier than the steel plate deck. Steel plate floor system is advantageous especially in aspect of erection and overall economy.

Konishi & Maeda (1976) have worked out the optimum design of simply supported welded I-section girders using the sequential linear programming method. In addition to material cost also the costs of drawing, machining, shop welding, shop assembly and shop painting have been considered. More welded splices of web and flanges have been calculated. Span lengths were between 16 and 30 m. The fabrication costs influence the number of different sections considerably. If the fabrication costs are higher, the number of different sections should be decreased.

In the Farkas' book (1984) the minimum cost design of a simply supported welded I-beam with one welded splice on the flanges is treated by a numerical example solved by the backtrack programming method. The beams are subject to a static, uniformly distributed load. The objective function expresses the cost of materials, welding and painting. Constraints on maximal bending stress and on local buckling of web and compressed flange are considered. Checks for shear stress are also performed. Hybrid I-beams are constructed from two types of steels with different yield stresses. Constraints on lateral buckling and deflection are not taken into account.

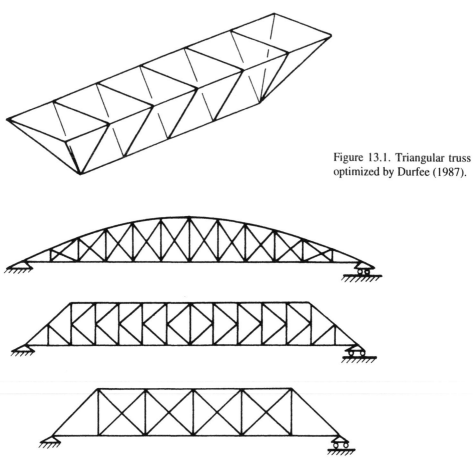

Figure 13.1. Triangular truss optimized by Durfee (1987).

Figure 13.2. Parker-, K- and Pratt-type trusses optimized by Adeli & Balasubramanyan (1988).

Ferscha (1987) has defined a cost function containing material and fabrication costs and, in addition, welding costs of butt welds of web splices and fillet welds connecting the flanges and the longitudinal stiffeners to the web. A systematic search method is used to obtain the optimal 4 variables (web height and thickness as well as cross-sectional areas of the two flanges).

According to Durfee (1987) the triangular cross-section truss is an effective structural alternate to a rectangular truss of four main chord members (Fig. 13.1). An actual numerical example of a highway bridge is investigated. The truss depth is optimized to minimize the weight of the structure. The specifications of the American Association of State Highway and Transportation Officials (AASHTO) are used. Square hollow sections (SHS) are applied. The span length is 45.7 m, the width is 9.14 m. Welded joint details are discussed.

Adeli & Balasubramanyan (1988) have developed an expert system for structural optimization and applied to different trusses. Three types of simply supported truss bridges have been treated (Fig. 13.2). Computerized determination of influence lines and their evaluation for AASHTO moving loads is worked out. Minimum weight design is treated considering constraints on stress, buckling, deflection and size limitations. Standard rolled and tubular profiles are applied. The mathematical method of feasible directions is used.

As mentioned above, the fabrication cost plays an important role in the optimization of bridge structures. There are very few publications about the costs of robotic welding, therefore the article of Touran & Ladick (1989) may be of interest. A detailed study on costs of robotic fabrication of an orthotropic steel bridge deck panel has been worked out on the basis of data obtained from five steel fabricators in USA. The investigated panel has had dimensions of 2.44*12.2 m and has been stiffened by trapezoidal ribs. The costs have been compared for conventional and robotic welding. Some characteristics are as follows:

- Traditional welding time/module: 26.4 man-hours,
- Robot welding time/module: 9.0 man-hours,
- Modules fabricated per year: 667 modules,
- Welding man-hours saved per year: 11606 hours,
- Labour rate including overhead and profit: $31.84/hour,
- Wage savings: $369535/year.

The welding operation, including consumables, makes up on average 14% of the total module fabrication cost, 38% is the material cost and 48% the additional labour cost (handling, preparation of material, stress relieving, grinding, coating, administration and equipment costs). Taking into account the robot investment cost, it was concluded that 5.6% reduction in the fabrication cost resulted from the use of robotics in welding operation as compared to conventional approach.

In the article of Memari et al. (1991) the finite element method (FEM) has been used to generate influence surfaces. The CONMIN non-linear programming software is used for optimization of continuous highway bridges. This software developed by Vanderplaats is based on the feasible direction method. As objective function only the material cost is defined. A numerical example of a three-span (30.5-42.7-30.5 m) highway bridge with three welded steel plate girders and with a concrete deck plate is treated. The cost of the conventionally designed continuous steel plate girder bridges can be decreased by 25%, i.e. 20% by the use of FEM and 5% by optimization, since the calculation using FEM does not need to use the wheel load distribution factors.

Ohkubo & Taniwaki (1991) have worked out the optimum design of cable-stayed bridges. The following parameters have been treated as constants: Span length, number of cables, height and width of the main box girder and pylon as well as materials. The variables have been as follows: Cross-sectional areas of cables, the reduced thicknesses of upper and lower flanges of each main girder element and pylon element, the distance from pylon to each cable anchor position in the main girder as well as the height of the lowest cable in the pylon from the axis of main girder. The reduced thicknesses consider also the longitudinal stiffeners. Design constraints on stresses in cables, in elements of the main girder and pylon have been considered according to the Japan Road Association specifications for highway bridges. The structure was analysed by FEM as a plane frame. The maximum and minimum bending moments, axial and shear forces due to traffic and impact loads have been calculated using influence lines. As objective function the material cost was defined.

The mathematical method was as follows: Constraints have been approximated by

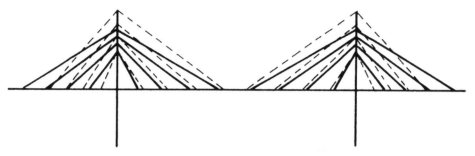

Figure 13.3. The initial (dashed) and optimum (continuous lines) cable topology of a cable-stayed bridge optimized by Negrao & Simoes (1994).

partial derivatives and reciprocal variables have been used. The approximate sub-problem was solved using dual method with Lagrangian functions developed by Fleury & Braibant (1986). As a numerical example a three-span bridge has been treated. The number of variables and constraints has been 67 and 158, respectively. It has been shown that the cable positions (topology) have a great influence on the minimum material cost.

In the study of Negrao & Simoes (1994) the dimensions of box-section pylons and I-section girders as well as the anchor positions of cables are selected as design variables. The optimum design is treated as a multiobjective optimization with goals of minimum cost of material, minimum stresses and displacements. The optimization method is combined with FEM code. Cable-stayed bridges are statically indeterminate and their structural behaviour is significantly affected by the cable arrangement and stiffness distribution among the cables, deck and pylons. Four numerical examples are worked out. It is shown that the minimum structural volume can be achieved by the modification of cable positions (Fig. 13.3).

In the article of Jármai & Farkas (1994) a parallel-chord belt-conveyor bridge is investigated to find the optimal topology for minimum volume (see Section 13.2).

Farkas & Jármai (1995) have worked out a detailed numerical example for the optimization of the height of a parallel-chord simply supported truss welded from circular hollow section (CHS) members with gap joints (see Section 11.2).

The minimum cost design of a simply supported SHS Vierendeel girder is worked out for uniformly distributed static loading by Farkas & Jármai (1996) (see Section 13.3).

13.2 MAIN TUBULAR TRUSS GIRDER OF A BELT-CONVEYOR BRIDGE

To illustrate numerically the effect of some structural parameters on the minimum weight design of tubular trusses for belt-conveyor bridges, the optimum topology is sought for a simply supported N-type truss (Fig. 13.4). The belt-conveyor is placed inside the bridge.

The total volumes of the planar main truss girder are calculated for three values of node distance 'a'.

The span length is kept constant $L_0 = 30$ m. For each 'a' the ratio $\omega = h/a$ is var-

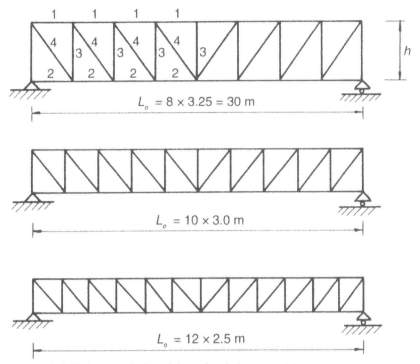

Figure 13.4. Various topologies of the main tubular truss girder.

ied and the optimum ω giving the minimum volume is determined.

The loads are as follows:

– Uniformly distributed vertical loads for a main truss girder:

Self-weight	p_G	= 3.7 kN/m,
Imposed load	p_{Q1}	= 0.5 kN/m,
Snow	p_{Q2}	= $1.0*s/2$ = 1.2 kN/m

where s is the width of the bridge.

Load on foot path for maintenance is p_{Q3} = 0.5 kN/m.

– Factored vertical load with safety factors according to the Eurocode 3 is

$$p_v = \gamma_G p_G + 0.9\ \gamma_Q (p_{Q1} + p_{Q2} + p_{Q3})$$
$$= 1.35*3.7 + 0.9*1.5\ (0.5 + 1.2 + 0.5) = 7.965 \text{ kN/m}$$

Horizontal wind load for one horizontal wind girder is p_{w0} = 0.8 $h/2$, the safety factor is γ_w = 1.5, the factor for simultaneous effects is 0.9, then the factored horizontal load is p_w = 0.9*1.5*0.8 $h/2$.

Four different square hollow sections are considered: 1. For upper chord, 2. Lower chord, 3. Inside columns, and 4. Diagonal braces. The outside columns are not treated, since they should be constructed as transverse frames and designed also for bending moments caused by the horizontal wind load.

The maximum forces in truss members are as follows:

Table 13.1. N_y forces from the vertical load p_y.

a	Lower chord	Inside columns	Diagonals
$L_0/8$	$7.5\,p_y a/\omega$	$2.5\,p_y a$	$4\,p_y a\sqrt{1+\omega^2}\,/\,\omega$
$L_0/10$	$12.0\,p_y a/\omega$	$3.5\,p_y a$	$5\,p_y a\sqrt{1+\omega^2}\,/\,\omega$
$L_0/12$	$17.5\,p_y a/\omega$	$4.5\,p_y a$	$6\,p_y a\sqrt{1+\omega^2}\,/\,\omega$

Table 13.2. Total volumes of a main girder of belt-conveyor bridge constructed from SHS members $10^{-6}\,V$ (mm^3).

$\omega = h/a$	0.8	1.0	1.2	1.4	1.6
$a = L_0/8$	144.1	137.7	138.6		
$a = L_0/10$		143.8	140.2	141.7	
$a = L_0/12$		153.9	146.1	144.1	148.0

– Upper chord: $N = N_v + N_h$ (compression), force from vertical load $N_v = p_v L_0^2\,/\,8h$, and from horizontal load $N_h = p_w L_0^2\,/\,8s$,

– Lower chord: $N = N_v + N_h$ (tension), and from the horizontal load $N_h = p_w L_0^2\,/\,8s$.

The forces from the vertical load are given in Table 13.1.

The compression members are designed for overall buckling according to EC3 using the buckling curve b for SHS struts and the limiting local slenderness according to CIDECT (Packer et al. 1992) $\delta_{SL} = (b/t)_{\lim} = 35$ for steel of yield stress $f_y = 235$ MPa. The optimum design of compressed SHS struts is treated in Section 9.1.

The end restraint factor for chord members is $K = 0.9$, for inside columns $K = 0.75$. For $f_y = 235$ MPa it is $\lambda_E = 93.91$. For a given compressive force N and strut

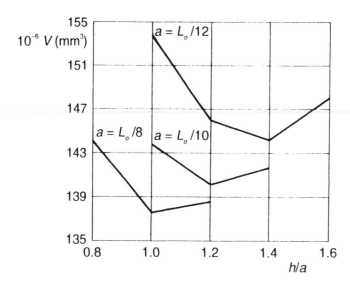

Figure 13.5. Total volumes of a main girder of belt-conveyor bridge constructed from SHS members.

length L (or x) the required cross-sectional area (or y) can be calculated by using a computer program.

The results of calculations are summarized in Table 13.2 and in Figure 13.5. It can be seen that the optimum $\omega = h/a$ ratios are different for various a-values. The absolute optimum is $\omega = 1$ for $a = L_0/8$.

Although the sensitivity of the volume function is small, the difference between the V-values for $\omega = 1$ is $100*(153.9-137.7)/153.9 = 10\%$, thus, 10% savings in weight can be achieved by using the optimum $a = L_0/8$ version instead of $a = L_0/12$ version. This optimum version is also advantageous regarding the fabrication costs, since it is constructed with less number of nodes.

13.3 MINIMUM COST DESIGN OF VIERENDEEL SHS TRUSSES

13.3.1 *Introduction*

Vierendeel trusses are used because of their aesthetic view and simple fabrication. On the contrary to triangulated trusses with diagonals, Vierendeel truss members should be designed for bending. The calculations show that the axial and shear forces can be neglected. In the CIDECT Design Guide (Packer et al. 1992) a numerical example has been worked out. In the book of Martin & Purkiss (1992) an illustrative numerical example is also treated.

Our aim is to apply the optimum design for Vierendeel trusses welded from square hollow section (SHS) members. In the cost function the material and fabrication costs are taken into account as in the authors' previous studies (e.g. Farkas & Jármai 1995). Design constraints relate to stresses due to bending, chord side wall failure, local buckling, deflection and prescription of the minimum height of a truss with parallel chords. The profiles of chords and verticals have the same outside dimensions, so the nodes can be regarded as fully rigid.

The thickness of all chords is t_0 and that of all verticals is t_1 so there are only three unknown profile dimensions. In addition the truss height is determined by service conditions (e.g. a closed footbridge should have a minimum height of 2.8 m). The only unknown to be optimized is the number of bays (spacings between verticals) in the case of a given span length of a simply supported girder. Thus, in an illustrative numerical example the optimum number of bays is sought to minimize the cost.

13.3.2 *The cost function*

The total cost contains the material and fabrication costs (see Chapter 5).
The volume of the structure shown in Figure 13.6 is expressed by

$$V = 2LA_0 + (\varphi+1)HA_1 \tag{13.1}$$

where A_0 and A_1 are the cross-section areas of the chords and verticals, respectively, $\varphi = L/a_0$ is the number of bays, L is the span length, H is the truss height. T_i are the fabrication times as follows:

– Time for preparation, assembly and tacking is

$$T_1 = C_1\Theta\sqrt{\kappa\rho V}, \quad C_1 = 1.0 \text{ min/kg}^{0.5} \tag{13.2}$$

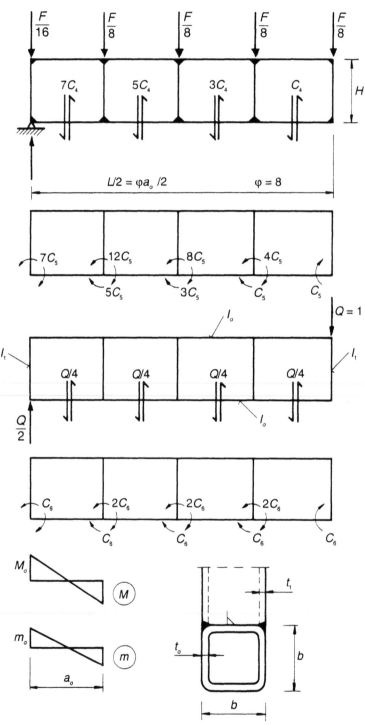

Figure 13.6. Half Vierendeel truss in the case of 8 bays. Shear forces and bending moments in lower nodes due to the external vertical load F and due to the virtual load $Q = 1$ for the calculation of maximum deflection. $C_4 = F/32$, $C_5 = FL/(32*16)$, $C_6 = QL/(4*16)$.

where Θ is a difficulty factor, for Vierendeel trusses it can be taken as 3, κ is the number of assembled structural elements, here $\kappa = \varphi + 3$.

– Time for welding

$$T_2 = \sum_i C_{2i} a_{wi}^n L_{wi} \tag{13.3}$$

where L_{wi} are the weld lengths, here for all welds

$$L_w = 8b\,(\varphi + 1) \tag{13.4}$$

$C_2 a_w^n$ are calculated according to the COSTCOMP software $a_w = t_1$ (see Fig. 13.6) and $C_2 a_w^n = 0.6*10^{-3} t_1^2$ for shielded metal arc welding (SMAW) of 1/2V butt welds and $0.8*10^{-3} t_1^2$ for SMAW of fillet welds (for $t_1 = 4\text{-}15$ mm), so we calculate with the mean value of

$$C_2 a_w^n = 0.7*10^{-3} t_1^2 \tag{13.5}$$

– Time for electrode changing, deslagging and chipping

$$T_3 = \sum_i C_{3i} a_{wi}^n L_{wi}$$

and

$$T_2 + T_3 = 1.3 \sum_i C_{2i} a_{wi}^n L_{wi} \tag{13.6}$$

The final form of the cost function to be minimized is as follows

$$\frac{K}{k_m} = \rho V + \frac{k_f}{k_m} \left[C_1 \Theta \sqrt{\kappa \rho V} + 1.3 * 0.7 * 10^{-3} t_1^2 L_w \right] \tag{13.7}$$

t_1 and L_w in mm, V in mm^3, $\rho = 7.85*10^{-6}$ kg/mm^3 (steel). For a wide range of cost factors we select the values of $k_m = 0.5\text{-}1.2$ \$/kg (steels) and $k_f = 0\text{-}45$ \$/manhour = 0-0.75 \$/min, thus $k_f/k_m = 0\text{-}1.5$ kg/min.

13.3.3 Design constraints

Constraint on chord side wall failure. According to Packer et al. (1992)

$$M_{max} \leq M^* = 0.5 f_y t_0\,(b_0 + 5t_0)^2 \tag{13.8}$$

Instead of the yield stress f_y we calculate with $f_{y1} = f_y / \gamma_{M1}$ where, according to EC3 $\gamma_{M1} = 1.1$. Furthermore, we use a reduction factor 0.9 for the effect of round-ings of SHS profiles, we introduce the notation $\delta_S = b/t$. Thus, Equation (13.8) can be written in the form

$$M^* = 0.45 f_{y1} b^3\,(\delta_S + 5)^2 / \delta_S^3 \tag{13.9}$$

According to Packer et al. (1992), to calculate with rigid joints, we take the limiting local plate slenderness of the chord connecting face

$$\delta_{S0L} = (b/t_0)_L = 16 \tag{13.10}$$

Thus

$$M_{max} \leq M^* = 0.04845 f_{y1} b^3 \tag{13.11}$$

Constraint on maximum elastic stress in verticals due to bending

$$M_{max} / W_x \leq f_{y1} \tag{13.12}$$

where the elastic section modulus of a SHS profile, considering also a reducing factor of 0.9 for roundings and a limiting plate slenderness for verticals δ_{S1L}

$$W_x = 0.9 * 4 (b - t_1)^2 t_1 / 3 = 1.2 b^3 (\delta_{S1L} - 1)^2 / \delta_{S1L}^3 \tag{13.13}$$

The local buckling constraint for verticals is given by CIDECT rules (Packer et al. 1992)

$$\delta_{S1L} = (b / t_1)_L = 1.1 \sqrt{E / f_y} \tag{13.14}$$

For $E = 2.1 * 10^5$ and $f_y = 355$ MPa it is $\delta_{S1L} = 26$, thus, Equation (13.12) can be written as

$$M_{max} \leq 0.042672 f_{y1} b^3 \tag{13.15}$$

It can be seen that Equation (13.15) gives larger b-values than Equation (13.11), thus, we calculate b from Equation (13.15), then t_1 using Equation (13.14) and t_0 with Equation (13.10).

13.3.4 *An illustrative numerical example*

The factored load of $F = 50$ kN is uniformly distributed along the upper nodes (Fig. 13.6), $L = 30$ m, steel Fe 510, $f_y = 355$ MPa, $f_{y1} = 355/1.1 = 323$ MPa. It can be seen from Figure 13.6 that the maximum bending moment acts in the verticals second from the outside post. The maximum bending moments and the selected profiles are given in Table 13.3.

The deflection constraint is defined by

$$w_{max} \leq w^* \tag{13.18}$$

Table 13.3. Maximum bending moments and selected profiles for different numbers of bays.

φ	8	10	12	14
M_{max}	FL/42.67	FL/50.00	FL/57.60	FL/65.33
b (mm) (Eq. 13.15)	137	130	124	119
$t_1 = b/26$ (mm)	5.3	5.0	4.8	4.6
$t_0 = b/16$ (mm)	8.5	8.1	7.7	7.4
Verticals $b_1 * t_1$	140*6.3	125*5	120*5	110*5
Chords $b_0 * t_0$	140*10	125*8.8	120*8	110*8
Modified verticals	140*6.3	140*6.3	120*5	120*5
Chords	140*10	140*10	120*8	120*8
A_1 (mm^2)	3230	3230	2210	2210
A_0 (mm^2)	4770	4770	3360	3360

Table 13.4. Maximum deflections for various numbers of bays calculated with Equation (13.21) with the data of $H = 2.8$ m, $L = 30$ m, $E = 2.1*10^5$ MPa and $F_0 = F/1.3 = 38462$ N.

ω	8	10	12	14
C_7	$3*32^2*16$	$3*40^2*20$	$3*48^2*24$	$3*56^2*28$
C_8	4	5	6	7
C_9	55	89	131	181
I_0*10^{-4} (mm⁴)	1268	1268	677	677
I_1*10^{-4} (mm⁴)	941	941	478	478
w_{max} (mm)	87	66	103	86

where w^* is the allowable deflection, according to EC3 $w^* = L/300$ and w_{max} should be calculated without multiplying the loads by load factors. The maximum deflection can be calculated by using the well-known principle of virtual work.

For a structure subject to bending

$$w_{max} = \sum_i \int \frac{Mmds}{EI_i} \qquad (13.19)$$

where M and m are the bending moments due to external load and a virtual force $Q = 1$ acting at the mid of the truss span (Fig. 13.6), respectively.

For a member with moment diagrams shown in Figure 13.6 it is

$$\int_0^{a_0} Mmds = M_0 m_0 a_0 / 3 \qquad (13.20)$$

The results of calculations are summarized in Table 13.4.

$$w_{max} = \frac{F_0 L^3}{C_7 E I_0} \left(C_8 + C_9 \frac{I_0 H}{I_1 L} \right) \qquad (13.21)$$

where $F_0 = F/\gamma$ is the load without safety factor $\gamma = 1.3$. Constants C_7, C_8 and C_9 are given in Table 13.4. I_0 and I_1 are the moments of inertia of chord and vertical profiles.

Size limitation for the truss height H. It can be seen from Equation (13.21) that the maximum deflection depends on the height so that a decrease of H decreases the w_{max}. If a minimum height is prescribed, then the limited deflection can be realized only by changing I_1 and I_0. In our numerical example the height limitation is

$$H \geq H_{min} = 2.8 \text{ m} \qquad (13.22)$$

We calculate the deflections with the moments of inertia of profiles determined using constraints of Equations (13.10), (13.14), (13.15) and (13.17) and then we modify the profile of verticals (I_1) to fulfill the deflection constraint (Eq. 13.18). The modified sections are given in Table 13.3. The maximum deflections calculated with the first and modified sections are summarized in Table 13.4. The results of cost calculations are shown in Table 13.5. It can be seen that the optimum number of bays is 12.

Table 13.5. Volumes and costs for different numbers of bays.

ω	8	10	12	14
$V*10^{-6}$ (mm³) (Eq. 13.1)	368	386	282	294
L_w (mm) (Eq. 13.4)	10080	11000	12480	13200
K/k_m (kg) (Eq. 13.7) for $k_f/k_m = 1.5$	4233	4516	3460	3654

13.3.5 Conclusions

The common width of chords and verticals can be calculated from the constraint on maximum elastic stress due to bending in the verticals second from outside posts. The thickness of verticals is obtained from the constraint on local buckling and on effective width.

The thickness of chords is calculated from the rule $b/t = 16$ so that the nodes can be treated as fully rigid ones. The final thickness of verticals is determined from constraints on deflection and on minimum truss height.

The optimum number of bays can be determined on the basis of cost calculations.

CHAPTER 14

Welded steel silos

14.1 INTRODUCTION

Silos are used for many engineering purposes. Several structural versions exist for ground or elevated silos, for storage and transportation of different materials such as coal, sand, cement, grains (wheat, peas etc.). Silos may be constructed from steel, aluminium or reinforced concrete. Steel silos can be welded or bolted. Corrugated plate elements are also used (Martens 1988).

A transit silo constructed from steel plate elements is investigated here (Fig. 14.1). This type consists of the following main structural parts: Roof, circular cylindrical bin, transition ringbeam, conical hopper and supporting columns. Our aim is to show the design procedure and fabrication cost calculations of these parts and to give designers aspects for the minimum weight and cost design.

Many articles can be found in the literature dealing with the stress and strength analysis of such silos (e.g. Gaylord 1984, Trahair et al. 1983, Teng & Rotter 1992), but the minimum cost design is not treated till now.

The main structural dimensions of a silo shown in Figure 14.1 are the height H and radius R of the cylindrical bin welded from horizontal courses of thin plates. For a given stored material, storage capacity of the bin and hopper, for a H/R ratio and H, the R value can be calculated and the structural dimensions of the silo parts can be designed on the basis of stress and buckling strength constraints.

The question of the optimum design is to determine the optimum H/R ratio for which the self weight of the structure and the cost is minimum. To illustrate the behaviour of theses objective functions a numerical example is selected and self weight and cost calculations are performed for various H/R ratios.

The slope angle of the hopper is determined by the friction angle of stored material, so it is not varied. The number of columns can be varied, but the maximum number is determined by the required minimum distance between columns to allow the emptying into lorries, and the minimum number is 6, required for the stability against horizontal action of wind and earthquake.

The design procedure and cost calculations are treated considering the practical ranges of dimensions regarding the numerical example. It should be noted that it is impossible to give general optimum design rules valid for all types of silos and all ranges of dimensions. For other types and dimensions a similar analysis and optimum design should be carried out.

The objective function is defined as the sum of material and fabrication costs as

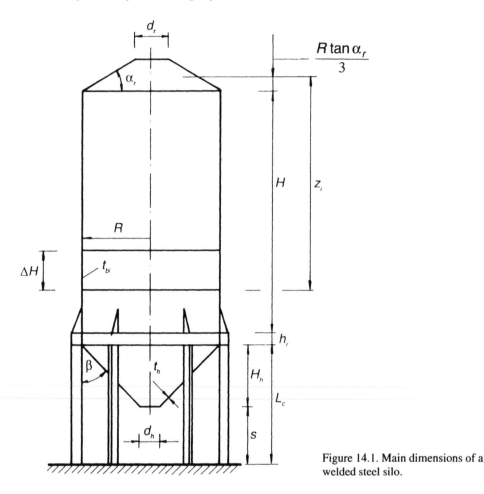

Figure 14.1. Main dimensions of a welded steel silo.

detailed in Chapter 5. We use here the fabrication time constants on the basis of COSTCOMP software.

14.2 DESIGN AND COST CALCULATION OF STRUCTURAL PARTS

14.2.1 *Roof*

In our numerical example treated in Section 14.2.4 the radius R is varied in the range of $R = 2.9$-4.25 m. For these radii a relatively simple roof structure can be used consisting of radial rafters of rolled I-section and trapezoidal plate segments welded to rafters and to inner and outer ring by fillet welds.

The snow load p_s (kN/m^2) acts on roof and the plate elements should be checked for bending. The number of rafters or plate elements can be determined from the restriction that the maximal plate width should not exceed 2.3 m to be transportable. Thus, the number of rafters is $n_r = 2R\pi/2.3$ rounded up to the next even number.

In an approximate calculation a plate strip can be designed for bending as a sim-

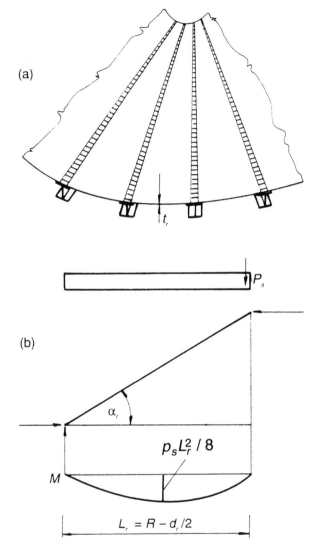

(a)

(b)

$$p_s L_r^2 / 8$$

M

$$L_r = R - d_r/2$$

Figure 14.2. Roof structure: a) Plate elements and rafters, and b) Bending moments in rafters due to snow load.

ply supported beam with a span length $L_p = 2R\pi/n_r$ and of thickness t_r (Fig. 14.2a)

$$\frac{M_{max}}{W_x} = \frac{6\gamma p_s L_p^2}{8 t_r^2} \le f_y \tag{14.1}$$

where $\gamma = 1.5$ is the safety factor, $f_y = 355$ MPa is the steel yield stress (for all parts of silo this yield stress is used). Calculating with $p_s = 1$ kN/m^2 = 10^{-3} N/mm^2 the required plate thickness from Equation (14.1) is

$$t_r = L_p \sqrt{\frac{3\gamma p_s}{4 f_y}} = L_p \sqrt{\frac{0.75 * 1.5 * 10^{-3}}{355}} = 1.78 * 10^{-3} L_p \ (L_p \text{ in mm}) \tag{14.2}$$

With the maximum value of $L_p = 2300$ mm, $t_r = 4$ mm. This thickness is used for all radii.

A rafter can be calculated approximately as a simply supported beam having span length of $L_r = R - d_r/2$, where $d_r = 800$ mm is the diameter of inner ring. The required section modulus is (Fig. 14.2b)

$$W_x = \gamma M_{max} / f_y = \gamma p_s L_p L_r^2 / (8 f_y) \tag{14.3}$$

where $\gamma = 1.5$, $f_y = 355$ MPa, $p_s = 10^{-3}$ MPa, $L_p = 2300$ mm.

The self mass of n_r plate elements can be calculated taking an element as a triangle with a width of $2R\pi/n_r$ and length $(R - d_r/2)/\cos \alpha_r$ ($\alpha_r = 30°$ is the roof slope angle):

$$G_p = \rho_s t_r \pi R (R - d_r / 2) / 0.866 \tag{14.4}$$

where $\rho_s = 7850$ kg/m^3 is the density of steel.

The self mass of n_r rafters with length $L_r/\cos 30°$ and cross-section area A (mm^2) is

$$G_{raf} = \rho_s n_r A L_r / \cos 30° \tag{14.5}$$

The total mass of a roof structure is

$$G_r = G_p + G_{raf} \tag{14.6}$$

The fabrication time of the roof, according to Chapter 5 is

$$\sum T_i = C_1 \Theta_r \sqrt{\kappa_r G_r} + 1.3 \sum C_{2ri} a_{wri}^n L_{wri} \tag{14.7}$$

where $C_1 = 1.0$ min/ kg$^{0.5}$, $\Theta_r = 3$, $\kappa_r = 2n_r$, according to COSTCOMP data, for $a_{wr} = 3$ mm and for GMAW-C welding method $C_{2r} = 1.7*10^{-3}$ min/mm^2, $n = 1$. The weld length is the sum of perimeters of the inner and outer ring as well as the length of longitudinal welds, approximately

$$L_{wr} = d_r \pi + 2R\pi + 2L_r n_r / \cos 30° \tag{14.8}$$

14.2.2 Bin

The circular cylindrical bin is loaded by the horizontal pressure of the stored material calculated with the Janssen's formula

$$p_h = p_0 \zeta, \quad p_0 = \frac{\rho R}{2\mu}, \quad \zeta = 1 - e^{-z/z_0}, \quad z_0 = \frac{R}{2\mu k} \tag{14.9}$$

where ρ is the density of stored material, μ is the friction coefficient of the material on the wall, z is the depth of stored material above the investigated section, k is the pressure coefficient. Note that the distance $R \tan \alpha_r / 3$ in Figure 14.1 is the possible height of the stored material in roof, this distance is in the following calculation neglected. With these coefficients the frictional stress on wall is

$$q = \mu p_h \tag{14.10}$$

and the vertical pressure in bin is

$$p_v = p_h / k \tag{14.11}$$

The circumferential membrane force in bin is

$$n_{\varphi b} = p_h R \tag{14.12}$$

and the meridional membrane force is

$$n_{zb} = \int_0^z p_v \, dz = \mu p_0 z_0 \left(\frac{z}{z_0} - \zeta \right) \tag{14.13}$$

The required bin thickness t_b can be calculated from the stress constraint

$$\sigma = \gamma_{red} n_{red} / t_b \leq f_y, \quad n_{red} = \sqrt{n_{\varphi b}^2 + n_{zb}^2 + \left| n_{\varphi b} n_{zb} \right|}$$

$$t_{b\max} = \frac{\gamma_{red} p_0 R}{f_y} \sqrt{\zeta_H^2 + \frac{1}{4k^2} \left(\frac{H}{z_0} - \zeta_H \right)^2 + \frac{\zeta_H}{2k} \left(\frac{H}{z_0} - \zeta_H \right)} \tag{14.14}$$

where $\zeta_H = 1 - e^{-H/z_0}$ and γ_{red} is the safety factor considering the approximate self weight (0.1*1.35), the dynamic effects of filling and emptying (1.2*1.5) as well as the sudden change of temperature (0.2)

$$\gamma_{red} = 0.1 * 1.35 + 1.2 * 1.5 + 0.2 = 2.135 \tag{14.15}$$

The bin thickness is limited to $t_{b\min} = 4$ mm by the fabrication requirements. When the t_b calculated from Equation (14.14) exceeds 4 mm, the thickness can be decreased step by step to 4 mm, since the bin is welded from shell courses. The width of courses is determined by the available plate width (e.g. 1500 mm).

The whole bin should be checked for local buckling due to wind acting on the empty silo. For a cylindrical shell with variable thickness the German standard for steel structures DIN 18800 Part 4 (1990) gives a complicated method. Instead of this method the API 650 formula can be used (Gaylord 1984)

$$\frac{H}{t_b} \leq 7200 \left(\frac{600 t_b}{R} \right)^{2/3} \tag{14.16}$$

where t_b is the average bin thickness.

The constraint on local buckling of bin courses due to the vertical pressure according to DIN 18800 Part 4 (1990) is expressed by

$$\sigma_z = 1.1 n_{zb} / t_b \leq \kappa_1 f_y / \gamma_M \tag{14.17}$$

where $\gamma_M = 1.1$ is a safety factor, κ_1 is the buckling coefficient

$$
\begin{aligned}
\kappa_1 &= 1 & \text{for} \quad \bar{\lambda}_s \leq 0.4 \\
\kappa_1 &= 1.274 - 0.686 \bar{\lambda}_s & \text{for} \quad 0.4 < \bar{\lambda}_s < 1.2 \\
\kappa_1 &= 0.65 / \bar{\lambda}_s^2 & \text{for} \quad 1.2 \leq \bar{\lambda}_s \\
\bar{\lambda}_s &= \sqrt{f_y / \sigma_{z.id}}, & \sigma_{z.id} = 0.605 E t_b / R
\end{aligned}
\tag{14.18}
$$

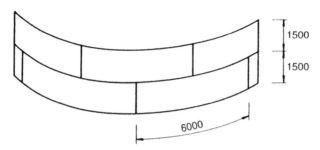

Figure 14.3. Courses of the bin welded to each other by horizontal and vertical butt welds.

With the value of the elastic modulus $E = 2.1*10^5$ MPa and for $f_y = 355$ MPa it is

$$\bar{\lambda}_s = \sqrt{\frac{R}{t_b}} \sqrt{\frac{f_y}{0.605E}} = \frac{1}{18.9179} \sqrt{\frac{R}{t_b}} \qquad (14.19)$$

In our numerical example treated in Section 4, the required maximum bin thickness is $t_b = 4$ mm, so all bin plate elements have the same thickness, 4 mm. Thus, the self mass of the bin is

$$G_b = 2R\pi H t_b \rho \qquad (14.20)$$

The difficulty factor is taken as $\Theta_b = 4$, since the bin is a spatial structure and needs a special erection method. The bin courses are welded from plate units having dimensions of 6000*1500 mm with horizontal and vertical double-sided I-welds with GMAW-C welding method (Fig. 14.3). Number of courses is $n_{c0} = H^{(m)}/1.5$, the length of circumferential welds is $2R\pi n_{c0}$, number of vertical welds is $n_v = 2R^{(m)}\pi/6.0$ rounded up to the next integer value, length of a vertical weld is 1.5 m. Number of assembled elements is $\kappa_b = 2R\pi n_{c0}/6$.

Thus, the total fabrication time for the bin is given by

$$\sum T_{bi} = \Theta_b \sqrt{\kappa_b G_b} + 1.3 C_{2b} a_{wb} L_{wb} \qquad (14.21)$$

where $a_{wb} = 4$ mm, $L_{wb} = 2R\pi n_{c0} + n_v H$. According to COSTCOMP (Chapter 5) for I-weld welded from both sides $10^3 C_{2b} a_{wb} = 2 (0.567 + 0.417*4) = 4.4$ min/mm.

For instance, the detailed cost calculation for a bin of $H = 12$ m, $R = 3.5$ m and $t_b = 4$ mm is given below.

$n_{c0} = 8$, $n_v = 4$, $\kappa_b = 8*4 = 32$, $\rho = 7850$ kg/m^3, with Equation (14.20) we get $G_b = 8286$ kg, using $k_f/k_m = 1.5$ with Equation (14.21) one obtains

$$K/k_m = 8286 + 1.5 (2060 + 1281) = 13298 \text{ kg}$$

and with

$$k_m = 1 \text{ \$/kg} \quad K = 13298 \text{ \$}$$

14.2.3 *Hopper*

The load component perpendicular to the conical hopper wall can be calculated according to DIN 1055 Part 6 (1987)

$$p' = \left(p_v c_b \sin^2 \beta + p_h \cos^2 \beta \right) \left[1 + \frac{\sin 2(90 - \beta)}{4\mu} \right] \tag{14.22}$$

where $c_b = 1.5$, β is the slope angle of hopper (Fig. 14.1).

The hoop (circumferential) membrane tension force is

$$n_h = p'R / \cos \beta \tag{14.23}$$

The meridional tension is given by Gaylord (1984)

$$n_m = \frac{p_v R}{2 \cos \beta} + \frac{Q_m}{2 R \pi \cos \beta} \tag{14.24}$$

where Q_m is the mass of stored material below the junction of hopper and ringbeam

$$Q_m = \frac{\rho \pi}{24} \left(8R^3 - d_h^3 \right) \cotan \beta \tag{14.25}$$

d_h is the diameter of the bottom ring of the hopper.

The required hopper wall thickness t_h can be obtained from the stress constraint

$$\sigma_h = \frac{\gamma_h \, n_{red.h}}{t_h} \leq f_y, \quad n_{red.h} = \sqrt{n_h^2 + n_m^2 - n_h n_m} \quad \gamma_h = 1.5 \tag{14.26}$$

The self weight of the hopper is expressed by

$$G_h = \frac{\rho_s \pi}{\sin \beta} \left[R^2 - \left(d_h / 2 \right)^2 \right] t_h \tag{14.27}$$

The number of plate elements is the same as in the case of roof $\kappa_h = 2R\pi / 2.3$. The difficulty factor is taken as $\Theta_h = 4$, since the hopper is fabricated from shell elements. The hopper thickness is for all silos treated in Section 14.3 $t_h = 4$ mm, so for I-welds between shell elements $10^3 C_{2h1} a_{wh} = 4.4$ min/mm. The length of a weld is $L_{wh1} = (R - d_h/2)/ \sin \beta$, the total length is $L_{wh1} \kappa_h$. The hopper is welded to the ringbeam by two circumferential fillet welds of size $a_w = 4$ mm, for which $C_{2h2} = 1.7*10^{-3}$ min/mm^2, and the length of welds is $L_{wh2} = 4R\pi$. Thus, the total fabrication time for hopper is given by

$$\sum T_{hi} = \Theta_h \sqrt{\kappa_h G_h} + 1.3 \left(C_{2h1} L_{wh1} \kappa_h + 4 C_{2h2} L_{wh2} \right) \tag{14.28}$$

14.2.4 *Columns*

Columns are loaded by self weight of roof, bin, hopper and ringbeam as well as by snow, wind and weight of the stored material.(The effect of earthquake load is treated in Section 14.4). In the case of more variable actions EC3 prescribes two combinations as follows: a) Considering only the most unfavourable variable action (Q) adding to the permanent actions (G): $\Sigma_j G_j + Q_1$, b) Considering all unfavourable variable actions multiplied by 0.9: $\Sigma_j G_j + 0.9 \, \Sigma_i Q_i$. The safety factor is $\gamma = 1.35$ for G and $\gamma = 1.5$ for Q.

The number of columns (n_{col}), as mentioned in the introduction, is limited by the transit function of the silo and the minimal number of columns is 6.

The snow load, according to Section 14.2.1 is $Q_{snow} = \pi R^2 p_s$. The mass of the stored material is $Q_{stor} = \rho V_{stor}$ the volume is given by

$$V_{stor} = \pi R^2 H + \pi H_h \left(\frac{d_h}{2}\right)^2 + \frac{\pi}{3} H_h \left(R - \frac{d_h}{2}\right)^2,$$

$$H_h = \left(R - \frac{d_h}{2}\right) \cotan \beta$$

(14.29)

In our numerical example the weight of stored material gives the leading combination, so the effect of snow and wind can be neglected.

The compressive force of a column is

$$N_c = \frac{1}{n_{col}} \left(1.35 \sum_j G_j + 1.5 Q_{stor}\right)$$

(14.30)

The overall buckling constraint is expressed by

$$N_c / A_c \leq \chi f_y$$

(14.31)

In the case of columns of square hollow section of width b_c and thickness t_c, $A_c = 4b_c t_c$. The length of a column with pinned ends is $L_c = H_h + 2$ m.

The buckling coefficient can be calculated on the basis of the JRA (Japanese Road Association) column curve which gives values similar to EC3 curve 'b':

$$\begin{aligned}
\chi &= 1 & \text{for} \quad \overline{\lambda}_c \leq 0.2 \\
\chi &= 1.109 - 0.545 \overline{\lambda}_c & \text{for} \quad 0.2 < \overline{\lambda}_c < 1 \\
\chi &= 1/\left(0.773 + \overline{\lambda}_c^2\right) & \text{for} \quad \overline{\lambda}_c \geq 1
\end{aligned}$$

(14.32)

The reduced slenderness is defined by

$$\overline{\lambda}_c = \frac{L_c \sqrt{6}}{b_c \lambda_E}, \quad \lambda_E = \pi \sqrt{\frac{E}{f_y}} = 76.41$$

The local buckling constraint according to EC3 is

$$\delta_c = b_c / t_c \leq \delta_{cL} = 42\varepsilon = 34, \quad \varepsilon = \sqrt{235 / f_y} \ (f_y \text{ in MPa})$$

(14.33)

Treating Equation (14.33) as active, the cross-sectional area can be expressed as

$$A_c = 4b_c t_c = 4b_c^2 / \delta_{cL}$$

(14.34)

Assuming that $0.2 < \overline{\lambda}_c < 1$, Equation (14.31) can be written as

$$\frac{N_c}{f_y} = \chi A_c = \left(1.109 - 0.545 \frac{L_c \sqrt{6}}{b_c \lambda_E}\right) \frac{4b_c^2}{\delta_{cL}}$$

(14.35)

which is a quadratic equation for b_c.

The self mass of columns is

$$G_{col} = \rho_s n_{col} A_c L_c$$

(14.36)

Since the sections are not welded, the fabrication cost of columns may be neglected. When $b_c > 400$ mm, welded box sections should be used.

It should be noted that, for the spatial stability of a silo, wind braces are needed between columns. The cost of these braces is neglected.

14.2.5 *Ringbeam*

The loads acting on the transition ringbeam cause compression, bending, shear and torsion. Since the open section beams have very small torsional stiffness, it is advantageous to use a welded box ringbeam (Farkas 1985).

The dimensions of the ringbeam can be calculated using the constraints on stress and local buckling of component plates. To construct a suitable connection between columns and ringbeam the distance between the two webs of the box beam should be equal to column width b_c thus, the flange width will be $b_r = b_c + 40$ mm (Fig. 14.4). Note that this is the reason why we treat the design of the ringbeam after the design of columns.

Since the horizontal component of the tensile force acting from the hopper causes compression in the ringbeam, it is necessary to use in the local buckling constraint of webs the same limiting plate slenderness $\delta_L = 42\varepsilon = 34$ as for flanges. Thus, the active local buckling constraints are as follows:

for webs $\qquad 2h_r / t_{wr} \leq \delta_L$ $\qquad\qquad\qquad\qquad$ (14.37)

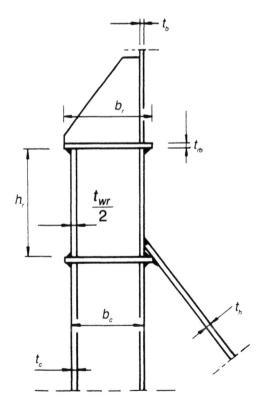

Figure 14.4. Welded box ringbeam connected with the bin, hopper and column.

for flanges $b_r / t_{rb} \leq \delta_L$ (14.38)

The vertical component of the tensile membrane force acting from the hopper is

$$y_r = \frac{\gamma Q_{\text{stor}}}{2R\pi}$$ (14.39)

This load causes bending and shear in vertical plane of the ringbeam

$$M_{r.\max} = y_r L_r^2 / 12, \quad L_r = 2R\pi / n_{\text{col}}$$ (14.40)

The horizontal component

$$x_r = y_r \tan \beta$$ (14.41)

causes compression

$$n_r = x_r R$$ (14.42)

The cross-sectional area of the ringbeam is

$$A_r = h_r t_{wr} + 2b_r t_{rb} = 2\left(h_r^2 + b_r^2\right) / \delta_L$$ (14.43)

and the section modulus is expressed by

$$W_{xr} = h_r^2 t_{wr} / 6 + h_r b_r t_{rb} = \left(h_r^3 + 3h_r b_r^2\right) / \left(3\delta_L\right)$$ (14.44)

In Equations (14.43) and (14.44) the only unknown is h_r which can be calculated from the stress constraint

$$\frac{M_{r.\max}}{W_{xr}} + \frac{n_r}{A_r} \leq f_y$$ (14.45)

Unfortunately, a closed formula cannot be derived from Equation (14.45).
The self mass of the ringbeam is

$$G_{rb} = \rho_s 2\pi \left(R + b_c / 2\right) A_r$$ (14.46)

The ringbeam is welded using four fillet welds of dimension a_w = 4 mm. The difficulty factor is Φ_{rb} = 4 since a curved beam should be fabricated.
κ_{rb} = 4, C_{2rb} = 1.7*10^{-3} min/mm^2. The whole weld length is

$$L_{wrb} = 4R\pi + 4\left(R + b_c\right)$$ (14.47)

The total fabrication time of a ringbeam is

$$\sum T_{rbi} = \Theta_{rb} \sqrt{\kappa_{rb} G_{rb}} + 1.3 * 4C_{2rb} L_{wrb}$$ (14.48)

14.3 NUMERICAL EXAMPLE

We select the extreme heights H = 7.5 and 18 m for integer numbers of courses n_{c0}. For a given constant storage capacity V_{stor} = 500 m^3 the required radius R can be calculated from Equation (14.29). The dimensions, self weights and costs are deter-

Table 14.1. Calculation of fabrication times T_1 and $T_2 + T_3$ according to Chapter 5 for structural parts of a welded silo. For all parts $C_1 = 1.0$ min/kg$^{0.5}$, $n = 1$.

Θ	3	4	4	4	
κ	$4R\pi/2300$	$n_{c0}n_v$	4	$4R\pi/2300$	
Structure	Roof	Bin	Ringbeam	Hopper	
G	Equation (14.6)	Equation (14.20)	Equation (14.46)	Equation (14.27)	
Welds a_w (mm)	Fillet, 3	I-weld, 4	Fillet, 3	I-weld, 4	Fillet, 4
C_2 (min/mm^2)	$1.7*10^{-3}$	$1.1*10^{-3}$	$1.7*10^{-3}$	$1.1*10^{-3}$	$1.7*10^{-3}$
L_w	Equation (14.8)	Equation (14.21)	Equation (14.47)	$(R - d_H/2)/\sin\beta$	$4R\pi$

Table 14.2. K/k_m (kg) values for four silos of equal storage capacity of 500 m^3.

k_f/k_m (kg/min)	R (m)	4.25	3.50	3.15	2.90
	H (m)	7.50	12.00	15.00	18.00
	H/R	1.76	3.43	4.76	6.20
0	Roof	2181	1449	1176	963
	Bin	6289	8286	9322	10299
	Ringbeam	4585	3653	3003	2521
	Hopper	2747	1855	1498	1266
	Columns	2681	2231	2068	1952
	Total	18483	17474	17067	17001
1.0	Roof	3769	2597	2073	1779
	Rin	8853	11627	13240	14295
	Ringbeam	6101	4943	4170	3597
	Hopper	4356	3065	2583	2169
	Columns	2681	2231	2068	1952
	Total	25760	24463	24134	23792
1.5	Roof	4563	3171	2589	2188
	Bin	10135	13297	15199	16293
	Ringbeam	6859	5888	4754	4135
	Hopper	5160	3670	3125	2620
	Columns	2681	2231	2068	1952
	Total	29398	28257	27735	27188
With earthquake load	Roof	4563	3171	2589	2188
1.5	Bin	10135	13297	15199	16293
	Ringbeam	6859	5888	4754	4135
	Hopper	5160	3670	3125	2620
	Columns	3506	2870	2809	2650
	Total	30223	28896	28476	27886

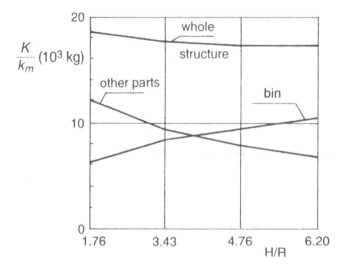

Figure 14.5. Cost of silo parts in the function of H/R for $k_f / k_m = 0$

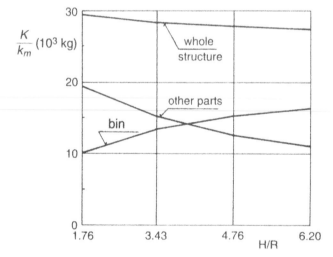

Figure 14.6. Cost of silo parts in the function of H/R for $k_f / k_m = 1,5$.

mined for four silos, i.e. for $H = 7.5$, 12, 15 and 18 m for storage of cement with a density of $\rho = 1600$ kg/m^3. For cement it is $\mu = 0.4$ and $k = 0.6$. The density of steel is $\rho_s = 7850$ kg/m^3. The data for the cost calculation are summarized in Table 14.1.

The results of the calculations are shown in Table 14.2. Figures 14.5 and 14.6 show the results as a function of H/R for $k_f/k_m = 0$ and 1.5.

14.4 CONSIDERATION OF EARTHQUAKE LOAD

When designing silos for earthquake risk zones, a horizontal force should be considered acting in the center of gravity, approximately in the height of $q = L_c + H/2$ (Fig. 14.1). According to Gaylord (1984) this force can be calculated for bins with aboveground bottoms by the formula

$$V = 0.2ZQ \tag{14.49}$$

where Z is the seismic zone coefficient. For instance, for zone 2 (moderate damage, with modified Mercalli scale intensity VII) $Z = 3/8$. Thus,

$$V = 0.075Q \tag{14.50}$$

In our numerical example $Q = Q_{stor} = \rho V_{stor} = 8000\text{kN}$, $V = 600\text{kN}$.

The combination of loads in Equation (14.30) should be modified, since the combination of

$$\Sigma G_j + 0.9 \, \Sigma Q_i$$

will be the leading one. The force acting on a column due to V in the case of 6 columns is

$$N_{cV} = \frac{Vq}{4R\cos 30°} \tag{14.51}$$

and the total compressive force acting on a column (number of columns $n_c = 6$)

$$N_c = 1.35 \, \Sigma G_j / 6 + 0.9 * 1.5 (Q_{stor} / 6 + N_{cV}) \tag{14.52}$$

The design of columns should be performed for this N_c. The calculations in our numerical example give the following results. The K/k_m (kg) values in Table 14.2 for columns should be modified. The other values can remain unchanged since the changes in column dimensions affect the other dimensions only very slightly. It can be seen that the tendency described below in conclusions does not change in our case when considering the earthquake load.

14.5 CONCLUSIONS

For a given storage capacity, including the hopper volume, and for given bin height H, the radius R can be determined. The aim was to find the optimal H/R ratio for a given storage capacity in the practical range of $H/R = 1.76$-6.20.

The calculations of an illustrative numerical example show the following.

1. When the H/R ratio increases *the self mass* of the bin increases but the self mass of the other parts of silo (roof, ringbeam, hopper and columns) decreases. The self mass of the whole structure has a minimum at the practical upper limit of $H/R = 6.20$. The difference between the self weights of the best and worst solution is $100(18483-17001)/17001 = 9\%$.

2. *The material and fabrication cost* of the whole structure decreases when H/R increases and reaches the minimum at the practical upper limit of H/R. Thus, designers have to choose the maximum H/R to achieve minimum costs. The cost difference between the best and worst solution is $100(29398-27188)/27188 = 8\%$. It should be noted that the optimum H/R- ratio depends on the characteristics of the stored material, thus, the optimization procedure should be performed for each type of stored material.

3. The number of columns should be minimum, the practical minimum number is

6. The slope angle of the hopper should be chosen in accordance with the friction angle of the stored material.

4. In the design of bin thickness the constraints on reduced stress and local buckling should be considered. The effect of a sudden temperature change as well as the dynamic filling and emptying effects are taken into account by multiplying factors. The thickness of the hopper is determined on the basis of the constraint on reduced stress.

5. The optimum dimensions of the ringbeam can be calculated from the stress and local buckling constraints. In the stress constraint the effect of compression and bending should be considered.

6. In the design of columns the effect of snow and wind can be neglected, the leading action is the weight of the stored material. This calculation should be modified when the effect of earthquake load is considered (see Section 14.4). Simple closed formulae can be derived for the optimum dimensions of columns of square hollow section.

Appendices

The lists of three computer programs are given in the Appendices A, B and C.

In Appendix A, the Rosenbrock's Hillclimb method for minimization of the objective function considering nonlinear inequality constraints is given in TURBO C.

In Appendix B, the SUMT method for minimization of the objective function considering non-linear inequality constraints is given in ANSI C.

In Appendix C, Backtrack method for minimization of the objective function with nonlinear inequality constraints is given in ANSI C.

The test problem in all cases is the so called 'post office problem'. Maximize or minimize the objective function, the volume of the parcel is

$$V = x_1 x_2 x_3$$

where x_i ($i = 1, 2, 3$) are the dimensions of the parcel.

There are explicit constraints

$$x_i^L \leq x_i \leq x_i^U, \quad x_1^L \quad \text{and} \quad x_i^U$$

are the lower and upper limits of variables.

In our case these limits are the same for every dimensions, 0.5 and 42 respectively. There is an implicit constraint x_4

$$x_4 = x_1 + 2x_2 + 2x_3$$

where the constraint value should be between the following limits:

$$x_4^L \leq x_4 \leq x_4^U$$

In our case these limits are 2.5 and 72 respectively.

The implicit constraint is active in every cases. The exact solutions of the problem are as follows:

– For minimization

$x_i = 0.5$

$x_i = 0.5$

$x_i = 0.5$

$V = 0.125$

– For maximization

$x_i = 24$

$x_i = 12$

$x_i = 12$

$V = 3456$

APPENDIX A

```
/* APPENDIX A */
/* Computer program of Rosenbrock's Hillclimb method for minimize objective function due to
nonlinear inequality constraints in TURBU C */
/* Prof. Jarmai, Karoly, University of Miskolc, Hungary */
/* H-3515 Miskolc, Egyetemvaros, Hungary, tel: +36/46/365111 ext 2028, fax: +36/46/367828
e-mail: jarmai@Kanga.alt.uni-miskolc.hu */
/* You can use the program in the education and academic research free, if you publish your re-
sults, please acknowledge the program */
/* Reference: Jarmai, K.: Single- and multivariable optimization techniques as a tools of decision
support system. Computers in Industry, 1989, Vol. 11, No. 3. pp. 249-266*/
/* Please send a copy for the authors when you use this program in your publication */

#include <stdio.h>
#include <conio.h>
#include <math.h>
#include <ctype.h>
#include <stdlib.h>

#define CR 13

/* UPPER LIMIT FOR CONSTRAINTS */
void upperl(z_1, z_2)
int     *z_1;
double  *z_2;
{
  switch (*z_1)
  {  case 1 :  *z_2 = 42.0 ;
              break;
     case 2 :  *z_2 = 42.0 ;
              break;
     case 3 :  *z_2 = 42.0 ;
              break;
     case 4 :  *z_2 = 72.0 ;
              break;
  }
}

/* LOWER LIMIT FOR CONSTRAINTS */
void lowerl(z_1, z_2)
int     *z_1;
double  *z_2;
{
  switch (*z_1)
  {  case 1 :  *z_2 = .5 ;
              break;
     case 2 :  *z_2 = .5 ;
              break;
     case 3 :  *z_2 = 0.5 ;
              break;
     case 4 :  *z_2 = 2.5 ;
              break;
```

```
  }
}
/* ACTUAL VALUES OF CONSTRAINTS */
void actualv(z_1, z_2, z_3, z_4, z_5)
int     *z_1;
double  *z_2, *z_3, *z_4, *z_5 ;
{
  switch (*z_1)
  {  case 1 : *z_2 = *z_3 ;
              break;
     case 2 : *z_2 = *z_4 ;
              break;
     case 3 : *z_2 = *z_5 ;
              break;
     case 4 : *z_2 = *z_3 + 2. * *z_4 + 2. * *z_5 ;
              break;
  }
}

/* FUNCTION VALUE */
void function(z_1, z_2, z_3, z_4, z_5)
int     *z_1;
double  *z_2, *z_3, *z_4, *z_5 ;
{
  *z_1 = *z_1 + 1 ;
  *z_2 = *z_3 * *z_4 * *z_5 ;
}

/* INPUT OF A DOUBLE VALUE */
void indou(f,px,py)
double *f;
int px,py;
{
  char c[15];
  int i,h;

  i = 0;
  h = 0;
  do {
    if (i == 0)
        gotoxy(px,py);
    c[i] = getch();
    cprintf("%c", c[i]);
    if (c[0] == CR)
        break;
    if ((c[0] == '0') && (c[1] == CR))
        break;
        if ((atof(c) != 0) && (c[i] == CR))
        h = 1;
    if (atof(c) == 0 && c[i] == CR)
        i = -1;
        i = i + 1;
```

```
   } while ( h == 0 );
  if ( i > 0 )
    *f = atof(c);
}

/* INPUT OF AN INTEGER VALUE */
void inint(fi,px,py)
int *fi;
int px,py;
{
  char c[15];
  int i,h;

  i = 0;
  h = 0;
  do {
    if (i == 0)
        gotoxy(px,py);
    c[i] = getch();
    cprintf("%c", c[i]);
    if (c[0] == CR)
        break;
    if ((c[0] == '0') && (c[1] == CR))
        break;
    if ((atoi(c) != 0) && (c[i] == CR))
        h = 1;
    if ((atoi(c) == 0) && (c[i] == CR))
        i = -1;
    i = i + 1;
  } while ( h == 0 );
  if ( i > 0 )
        *fi = atoi(c);
}

main()
/* SINGLE OPTIMIZATION BY THE ROSENBROCK'S HILLCLIMB METHOD */
{
  int    c, change, check1, dec, i, iq, ir, iss,i5, i7, j, jj, k, ka, ki, kj, kk, ko, kr, k4, k5, l, la, l1 ;
  int    ls, ls_start, lu, lx, lx_start, m, ns, ok, p, px, r, te, help ;
  double y[5][3] ;
  double al[5], bv[5], d[5], e[5], ei[5], h[5], ph[5], sb[5], x[5], x_start[5] ;
  double b[5][5], v[5][5], vv[5][5] ;
  double bb, bm, bw, cx, ch, cg, cv, de, de_start, uc, lc, xc, f, fd, f0, f1, ke, pw,  so, su ;
  char   ant, answ, ans ;

  char   MEM1[] = "1st variable" ;
  char   MEM2[] = "2nd variable" ;
  char   MEM3[] = "3rd variable" ;

/* SUBROUTINES FOR LIMITS AND VALUES OF UNKNOWNS */
  void upperl (int *z_1,double *z_2);
  void lowerl (int *z_1,double *z_2);
  void actualv (int *z_1,double *z_2,double *z_3,double *z_4,double *z_5);
```

```
    void function (int *z_1,double *z_2,double *z_3,double *z_4,double *z_5);

/* INPUT SUBROUTINES */
    void indou (double *f,int px,int py);
    void inint (int *f1,int px,int py);

    clrscr();
    window (1,1,80,4);
    textbackground(11);
    textcolor(0);
        clrscr();
    gotoxy(27,2);
    cprintf(" ROSENBROCK'S HILLCLIMB METHOD");
    gotoxy(22,3);
    cprintf(" CONSTRAINED SINGLE-OPTIMIZATION PROCEDURE ");
    x_start[0] = 0. ;
    x_start[1] = 5. ;              /****************************/
    x_start[2] = 10. ;            /*     STARTING  POINT    */
    x_start[3] = 10. ;            /****************************/
    de_start = 0.0000001 ;          /*  convergence criterion  */
    lx_start = 50 ;              /*  iteration limit      */
    ls_start = 10 ;             /*  step length / at start */
    m = -1 ;                   /*  MINIMUM (-1)  OR  MAXIMUM (1)  */
FIRST:j = 0 ;
    cx = ch = cg = 0. ;
    f = f0 = 0. ;
    ko = 0 ;

        p = 3 ;              /* number of independent variables */
        px = 1 ;              /* printing parameter */
        ns = 0 ;              /*  change step length    */
        l = 4 ;              /*  inequality constrains   */
        la = 0 ;
        lu = 0 ;
        iss = 0 ;
        iq = 0 ;
        te = 0 ;
        window(1,5,80,25);
        textbackground(1);
        textcolor(15);
        clrscr();
        gotoxy(56,2);
        cprintf("CURRENT      NEW");
        gotoxy(3,4);
        cprintf("SEARCH FOR MINIMUM (");
        textcolor(11);
        cprintf("-1");
        textcolor(15);
        cprintf(") OR SEARCH FOR MAXIMUM (") ;
        textcolor(11);
        cprintf("1");
        textcolor(15);
        cprintf(") :");
```

```
    gotoxy(58,4);
    cprintf("%d",m);
    gotoxy(3,5);
    cprintf("STEP LENGTH / STARTING VALUE :");
    gotoxy(58,5);
    cprintf("%d",ls_start);
    gotoxy(3,6);
    cprintf("UPPER LIMIT OF ITERATION :");
    gotoxy(58,6);
    cprintf("%d",lx_start);
    gotoxy(3,7);
    cprintf("VALUE OF CONVERGENCE FACTOR :");
gotoxy(56,7);
cprintf("%5.3e",de_start);
inint(&m,71,4);
gotoxy(71,4);
cprintf("%d",m);
inint(&ls_start,71,5);
ls = ls_start;
gotoxy(71,5);
    cprintf("%d",ls);
inint(&lx_start,71,6);
lx = lx_start ;
gotoxy(71,6);
cprintf("%d",lx);
indou(&de_start,69,7);
de = de_start ;
gotoxy(69,7);
cprintf("%5.3e",de);
    gotoxy(56,10);
cprintf("PREVIOUS    NEW ");
gotoxy(56,11);
cprintf("VALUES      VALUES");
gotoxy(3,11);
cprintf("STARTING POINT :");
textcolor(11);
gotoxy(32,10);
cprintf("LOWER    UPPER");
    gotoxy(32,11);
cprintf("LIMIT    LIMIT");
textcolor(15);
gotoxy(3,13);
cprintf("%s",MEM1);
gotoxy(3,14);
cprintf("%s",MEM2);
gotoxy(3,15);
cprintf("%s",MEM3);
for(i = 1 ; i <= p ; i++)
{
    upperl(&i,&ch);
    lowerl(&i,&cg);
  textcolor(11);
    gotoxy(40,12+i);
```

```
        cprintf("%6.2f", ch);
        gotoxy(31,12+i);
        cprintf("%6.2f", cg);
        textcolor(15);
        gotoxy(56,12+i);
        cprintf("%6.2f", x_start[i]);
        indou(&x_start[i],70,12+i);
        x[i] = x_start[i];
        gotoxy(69,12+i);
        cprintf("%6.2f", x[i]);
}
window (1,25,80,25);
textbackground(15);
textcolor(0);
clrscr();
gotoxy(52,1);
cprintf("PUSH ANY KEY TO CONTINUE !");
getch();
window(1,5,80,25);
textbackground(1);
textcolor(11);
    clrscr();
gotoxy(3,2);
cprintf("SEARCH FOR MINIMUM (-1)  OR SEARCH FOR MAXIMUM (1) :");
gotoxy(60,2);
cprintf("%d" ,m);
gotoxy(3,3);
cprintf("NUMBER OF UNKNOWNS :");
gotoxy(60,3);
cprintf("%d", p);
gotoxy(3,4);
cprintf("NUMBER OF INEQUALITY CONSTRAINS :");
gotoxy(60,4);
cprintf("%d", l);
gotoxy(3,5);
cprintf("UPPER LIMIT OF ITERATION :");
gotoxy(60,5);
cprintf("%d", lx);
gotoxy(3,6);
cprintf("STEP LENGTH / STARTING VALUE :");
gotoxy(60,6);
cprintf("%d", ls);
gotoxy(3,7);
cprintf("VALUE OF CONVERGENCE FACTOR :");
gotoxy(60,7);
cprintf("%e", de);
textcolor(15) ;

for (j = 1 ; j <= p ; j++ )
    e[j] = x[j]/ls ;
function(&ko, &f, &x[1], &x[2], &x[3] );
    f = m*f ;
    f1 = 0. ;
```

```
            for (j = 1 ; j <= l ; j++ )
            {
            upperl(&j,&ch);
            lowerl(&j,&cg);                    /*    computation of    */
            al[j] = ( ch -cg ) * 0.001 ;       /*     boundary zone     */
            }
            for (i = 1 ; i <= p ; i++ )
            for (j = 1 ; j <= p ; j++ )
            {
               if (i == j)
                    v[i][j] = 1. ;
               else
                    v[i][j] = 0. ;
            }
            for (i = 1 ; i <= p ; i++ )          /*  the step length is  */
            ei[i] = e[i] ;                     /*   the original size   */
BEGIN:for (j = 1 ; j <= p ; j++ )
            {
            if (ns == 0 )
               e[j] = ei[j] ;                  /*  take the original  */
            sb[j] = 2. ;                       /*    step length    */
            d[j] = 0. ;
            }
            fd = fl ;
            i = 1 ;
            if (iq != 0)
            {
            for (k = 1 ; k <= p ; k++ )
               x[k] = x[k] + e[i] * v[i][k] ;       /*  move with the step length  */
            for (k = 1 ; k <= l ; k++ )
               h[k] = f0 ;
            }
START:function(&ko, &f, &x[1], &x[2], &x[3] );
            fl = m * f ;
            if ( iss == 0 )
            f0 = fl ;
            iss = 1 ;
            if ( (double) fabs(fd - fl) <= de * (double) fabs(fl) )  /* CONVERGENCE CRITERION */
            {
            te = 1 ;
            goto PRINT ;
            }
            j = 1 ;
            do
            {
            upperl(&j, &ch) ;
            lowerl(&j, &cg) ;
            actualv(&j, &cx, &x[1], &x[2], &x[3]) ;
            uc = ch ;
            lc = cg ;
            xc = cx ;
            if ((xc < lc) || (xc > uc) || (fl < f0))
            {
```

```
         if (iq == 0 )
            goto PRINT ;
         for (ll = 1 ;ll <= p ; ll++ )
            x[ll] = x[ll] - e[i] * v[i][ll] ;          /*  move back  */
         e[i] = -0.5 * e[i] ;                    /* reduce step length  */
         if (sb[i] < 1.5 )
               sb[i] = 0. ;
         goto LABEL ;
      }
      else
        if ((xc < lc + al[j]) || (xc > uc - al[j]))
        {
           bw = al[j] ;
           help = 0 ;
           ph[j] = 1.0 ;
           if ((xc < lc) || (uc < xc))
           {
               ph[j] = 0. ;
               help = 1 ;
           }
               else
               if ((lc < xc) && (xc < lc + bw))
               {
                  pw = (lc + bw - xc) / bw ;
                  ph[j] = 1.-3.*pw + 4. *pw*pw -2. *pw*pw*pw ;  /* function modification */
                  help = 1 ;
               }
               else
                 if ((uc - bw < xc) && (xc < uc))
                 {
                    pw = (xc - uc + bw) / bw ;
                    ph[j] = 1 - 3. * pw + 4 * pw*pw - 2. * pw*pw*pw ;
                    help = 1 ;
                 }
           if (help == 1)
               f1 = h[j] + (f1 - h[j]) * ph[j] ;           /*  MODIFIED FUNCTION  */
        }
        else
            h[j] = f0 ;
j = j + 1 ;
}
while (j <= l);
iq = 1 ;
if (f1 < f0)
{
    if (iq == 0 )
     goto PRINT ;
    for (ll = 1 ;ll <= p ; ll++ )
     x[ll] = x[ll] - e[i] * v[i][ll] ;          /*  wrong direction    */
    e[i] = -0.5 * e[i] ;                    /*   step = -0.5 * step */
    if (sb[i] < 1.5 )
      sb[i] = 0. ;
}
```

```
    else
    {
        d[i] = d[i] + e[i] ;                    /*  good  direction    */
        e[i] = 3. * e[i] ;                      /*  step = 3 * step    */
        f0 = f1 ;
        if (sb[i] >= 1.5 )
            sb[i] = 1. ;
    }
LABEL:for (jj = 1 ; jj <= p ; jj++ )
    {
        if (sb[jj] >= 0.5 )
        {
         if ( i == p )
         {
           i = 1 ;
           if (iq != 0 )
           {
              for (k = 1 ; k <= p ; k++ )
              x[k] = x[k] + e[i] * v[i][k] ;        /*** step ***/
              for (k = 1 ; k <= l ; k++ )
              h[k] = f0 ;
              }
          goto START ;
          }
          else
          {
          i = i + 1 ;
          for (k = 1 ; k <= p ; k++ )
            x[k] = x[k] + e[i] * v[i][k] ;          /*** step ***/
          for (k = 1 ; k <= l ; k++ )
            h[k] = f0 ;
          goto START ;
          }
        }
      }
    for (r = 1 ; r <= p ; r++ )                  /************************/
       for (c = 1 ; c <= p ; c++ )
          vv[r][c] = 0. ;                        /*                */
    for ( r = 1 ; r <= p ; r++ )
       {                                         /*                */
       kr = r ;
       for ( c = 1 ; c <= p ; c++ )             /*                */
       {
        for ( k = kr ; k <= p ; k++ )           /*   ROTATING  OF  */
        vv[r][c] = d[k] * v[k][c] + vv[r][c] ;
        b[r][c] = vv[r][c] ;                    /*                */
       }
     }                                          /*                */
    bm = 0. ;
    for ( c = 1 ; c <= p ; c++ )                /*                */
       bm = bm + b[1][c] * b[1][c] ;
    bm = sqrt(bm) ;                             /*   COORDINATE    */
    for ( c = 1 ; c <= p ; c++ )
```

```
        v[1][c] = b[1][c] / bm ;                    /*              */
    for ( r = 2 ; r <= p ; r++ )
    {                                       /*                */
        ir = r - 1 ;
        for ( c = 1 ; c <= p ; c++ )               /*              */
        {
            su = 0. ;                       /*     SYSTEM      */
            for ( kk = 1 ; kk <= ir ; kk++ )
            {                               /*              */
                so = 0. ;
                for ( kj = 1 ; kj <= p ; kj++ )          /*              */
                    so = so + vv[r][kj] * vv[kk][kj] ;
                su = so * v[kk][c] + su ;            /*              */
            }
            b[r][c] = vv[r][c] - su ;            /*              */
        }
    }                                       /*              */
    for ( r = 2 ; r <= p ; r++ )
    {                                       /*              */
        bb = 0. ;
        for ( k = 1 ; k <= p ; k++ )               /*              */
            bb = bb + b[r][k] * b[r][k] ;
        bb = sqrt(bb) ;                     /*              */
        for ( c = 1 ; c <= p ; c++ )
            v[r][c] = b[r][c] / bb ;               /***************************/
    }
    lu = lu + 1 ;
    la = la + 1 ;
    if (la != px)
        goto BEGIN ;
PRINT:gotoxy(3,9) ;
    cprintf("STAGE :");
    gotoxy(46,9);
    cprintf("%d",  lu);
    gotoxy(3,10);
    cprintf("OBJECTIVE FUNCTION :");
    gotoxy(46,10);
    cprintf("%f", f0);
    gotoxy(3,11);
    cprintf("PROGRESS :");
    gotoxy(46,11) ;
    cprintf("%f", bm);
    gotoxy(3,12) ;
    cprintf("NUMBER OF OBJECTIVE FUNCTION COMPUTATION : ");
    gotoxy(46,12);
    cprintf("%d", ko);
    gotoxy(3,13);
    cprintf("VALUE OF X IN THIS STAGE :") ;
    for ( j = 1 ; j <= p ; j++ )
    {
        gotoxy(37,12+j) ;
        cprintf(" x[%d] =  %f \n", j, x[j]) ;
    }
```

```
/*    if (lu == 1)
    {
        gotoxy(2,21) ;
        cprintf("                              ");
    }          */
        la = 0 ;
    if ((iq == 0) || (lu == 0))
    {
        window(1,21,80,23);
        textbackground(4);
        textcolor(14);
        clrscr();
        gotoxy(5,2);
         cprintf("***** THE STARTING POINT DOES NOT SATISFY THE CONSTRAINTS
*****");
    }
    else if ((te == 1) || (lu >= lx))          /*   ITERATION  LIMIT  */
        {
          gotoxy(3,17);
          cprintf(" FINAL STEP LENGTH :");
          for (j = 1 ; j <= p ; j++ )
          {
              gotoxy(46,16+j);
              cprintf("%f", e[j] );
              }
        }
        else goto BEGIN ;
  do
  {
  window (1,24,80,25);
    textbackground(15) ;
    textcolor(0) ;
    clrscr();
    gotoxy(3,1) ;
    cprintf("(");
    textcolor(4);
    cprintf("O");
    textcolor(0);
    cprintf(") CONTINUE WITH OPTIMIZATION ");
    gotoxy(50,1) ;
    cprintf("(");
    textcolor(4);
        cprintf("E");
    textcolor(0);
    cprintf(") END OF COMPUTATION");
    gotoxy(3,2);
    cprintf("(");
    textcolor(4);
    cprintf("D");
    textcolor(0);
    cprintf(") CONTINUE WITH FINDING DISCRETE VALUES ");
    gotoxy(50,2);
    textcolor(4);
```

```
        cprintf("SELECT CODE ---> ");
        textcolor(0);
        ans = getch();
        switch(ans)
        {
            case 'o' :
            case 'O' :  dec = 0 ;
                        goto FIRST ;
            case 'd' :
            case 'D' :  dec = 1 ;
                        goto DISCR ;
            case 'e' :
            case 'E' :  dec = 2 ;
                        break ;
            default  :  dec = 3 ;
        }
        }
    while(dec == 3) ;
DISCR:  if (( ans == 'd') || ( ans == 'D' ))
    {
    window(1,5,80,25);
    textbackground(1);
    textcolor(11);
    clrscr();
    gotoxy(10,3);
    cprintf("---------------- COMPUTE DISCRETE VALUES ------------------ ");
    textcolor(15);
    do
    {
        change = 0 ;
        window(1,10,80,25);
        textbackground(1);
        textcolor(15);
        clrscr();
        gotoxy(53,1);
        cprintf("LOWER          UPPER ");
        gotoxy(53,2);
        cprintf("LIMIT          LIMIT ");
        gotoxy(3,4);
        cprintf("%s", MEM1);
        gotoxy(3,5);
        cprintf("%s", MEM2);
        gotoxy(3,6);
        cprintf("%s", MEM3);
        textcolor(11);
        gotoxy(34,1);
        cprintf("CONTINUOUS");
        gotoxy(34,2);
        cprintf("VALUES");
        for (i = 1 ; i <= p ; i++ )
        {
            textcolor(11);
            gotoxy(31,3+i);
```

```
  cprintf("%9.4f",x[i]);
upperl(&i,&ch);
lowerl(&i,&cg);
gotoxy (7,13);
cprintf("RANGES FOR LOWER LIMIT %d-TH VARIABLE :  ", i) ;
cprintf("%10.6f  --> %10.6f ", cg, x[i]);
gotoxy (7,14);
cprintf("RANGES FOR UPPER LIMIT %d-TH VARIABLE :  ", i) ;
cprintf("%10.6f  --> %10.6f ", x[i], ch);
textcolor(15);
check1 = 0 ;
do
{
     if(check1 != 0)
  {
     textcolor(13);
     gotoxy (12,11);
     cprintf("     LOWER LIMIT IS WRONG  -->   CHECK RANGES !     ");
     textcolor(15);
  }
  gotoxy(51,3+i);
  cprintf("           ");
  gotoxy (53,3+i);
  cscanf("%lf",&y[i][1]) ;
  check1 = check1 + 1 ;
  gotoxy (5,11);
     cprintf("                                 ") ;
  textcolor(15);
}
while((y[i][1] > x[i]) || (y[i][1] < cg));
gotoxy(51,3+i);
cprintf("%7.3f",y[i][1]);
check1 = 0 ;
do
{
if(check1 != 0)
  {
     textcolor(13);
     gotoxy (12,11);
     cprintf("     UPPER LIMIT IS WRONG  -->   CHECK RANGES !     ");
     textcolor(15);
  }
  gotoxy(68,3+i);
  cprintf("           ");
     gotoxy (70,3+i);
  cscanf("%lf",&y[i][2]) ;
  check1 = check1 + 1 ;
  gotoxy (5,11);
  cprintf("                                 ") ;
     textcolor(15);
}
while((y[i][2] < x[i]) || (y[i][2] > ch)) ;
gotoxy(68,3+i);
```

```
        cprintf("%7.3f",y[i][2]);
    }
    window (1,25,80,25);
    textbackground(15);
    textcolor(0);
    clrscr();
    gotoxy(52,1);
    cprintf("PUSH ANY KEY TO CONTINUE !");
    getch();
        getch();

/***** FIND DISCRETE VALUES *****/

    cv = 1e+25 ;
        for ( i = 1 ; i <= p ; i++)
        bv[i] = 0. ;
    ke = pow(2.,p);
    for ( ka = 1 ; ka <= ke ; ka++ )
    {
        k5 = ka - 1 ;
        for ( i5 =1 ; i5 <= p ; i5++ )
        {
          ki = k5 - (k5/2) * 2 ;
          if (k5 < 1)
            ki = 0 ;
          k5 = (k5 - ki) / 2 ;
          x[i5] = y[i5][ki+1] ;
        }
        j = 1 ;
        while(j <= l)
        {
          upperl(&j, &ch) ;
          lowerl(&j, &cg) ;
          actualv(&j, &cx, &x[1], &x[2], &x[3]) ;
          xc = cx ;
          lc = cg ;
          uc = ch ;
          if((xc > uc) || (xc < lc))
            break;
          j = j + 1;
        }
        if((xc <= uc) && (xc >= lc))
        {
          function(&ko, &f, &x[1], &x[2], &x[3]);

          /***** FIND MINIMAL FUNCTION *****/

          if (f < cv)
          {
            cv = f ;
                for (i7 =1 ;i7 <= p ; i7++ )
                bv[i7] = x[i7] ;
              }
```

```
            }
        }
        if (bv[p] == 0)
        {
            window(1,21,80,23);
                textbackground(4);
                textcolor(14);
                clrscr();
                gotoxy(5,2);
                cprintf("*****  THERE IS NO RESULT WITH DISCRETE , CHANGE (Y/N) ?
*****");
                answ = getch();
                if ((answ == 'y') || (answ == 'Y'))
                    change = 1 ;
        }
    }
    while(change == 1);
    window(3,8,77,25);
    textbackground(1);
    textcolor(15);
    clrscr();
    gotoxy(6,4);
    cprintf("THE FINAL DISCRETE X VALUES ARE :");
    gotoxy(6,6);
    cprintf("%s", MEM1);
    gotoxy(6,7);
    cprintf("%s", MEM2);
    gotoxy(6,8);
    cprintf("%s", MEM3);
    for (i = 1 ; i <= p ; i++)
    {
        gotoxy(35,5+i);
        cprintf("%6.2f", bv[i]);
    }
    gotoxy(6,12);
    cprintf("FINAL OBJECTIVE FUNCTION : ");
    gotoxy(35,12);
    cprintf("%f",cv);
    do
    {
    window (1,24,80,25);
        textbackground(15) ;
        textcolor(0) ;
        clrscr();
        gotoxy(3,1) ;
        cprintf("(");
        textcolor(4);
            cprintf("O");
        textcolor(0);
        cprintf(") CONTINUE WITH OPTIMIZATION ");
        gotoxy(50,1) ;
        cprintf("(");
            textcolor(4);
```

```
        cprintf("E");
        textcolor(0);
        cprintf(") END OF COMPUTATION");
        gotoxy(50,2);
        textcolor(4);
        cprintf("SELECT CODE ---> ");
        textcolor(0);
        ans = getch();
        switch(ans)
        {
        case 'o' :
        case 'O' :  dec = 0 ;
                        goto FIRST ;
        case 'e' :
        case 'E' :  dec = 2 ;
                        break ;
        default  :  dec = 3 ;
        }
        }
        while(dec == 3) ;
        }
}
```

Note: In Win32s applications the cprintf, cscanf should be replaced by printf and scanf, respectively and leave out window (··), textbackground (·), textcolor (·) statements and somewhere cltscr ( ) statement too.

APPENDIX B

```
/* APPENDIX B */
/* Computer program of SUMT method for minimize objective function due to nonlinear inequality constraints in ANSI C */
/* Prof. Jarmai, Karoly, University of Miskolc, Hungary */
/* H-3515 Miskolc, Egyetemvaros, Hungary, tel. +36/46/365111 ext 2028, fax +36/46/367828 e-mail jarmai@Kanga.alt.uni-miskolc.hu */
/* You can use the program in the education and academic research free, if you publish your results, please acknowledge */
/* Please send a copy for the authors when you use this program in your publication */

#include <stdio.h>
#include <math.h>
#include <conio.h>

#define      N    3     /* Number of variables */
#define      M    7     /* Number of constraints */
#define      E    1e-3  /* Convergence parameter */
#define      H    1e-6  /* Step length for gradient computation */
#define      EP   0.001 /* See penalty function */
#define      T    10    /* Multiplier of 'ra' */

double  celf(double *x, double ra);
double  bunt_f(double *x);
int     felt(double *x, double *c);
```

```c
int    grad(double *x, double ra, double *g);

int    cfvh,ffvh;

int  main(void)
{
double  x[N],        /* Coordinates of new point      */
   y[N],         /* Coordinates of previous point */
   g[N],         /* Gradient of objective function */
   d[N],         /* Direction in the actual point  */
   u[N],         /* Direction in the previous point */
   c[M],         /* Constraints               */
   v[N],m[N];    /* vectors for calculations      */
double  ra;      /* coefficient of penalty function */
double  h[N][N];       /* approximation of Hesse-matrix  */
double  ff,fq,gp,gr,gq,z,zz,w,dd,l,kk,wk,a,b,s1,s2;
int    i,j,kod2,cc;

clrscr();
printf("    CONSTRAINED NONLINEAR OPTIMIZATION WITH SUMT TECHNIQUE\n\n");
for(i = 0; i < N; i++)
      {
      printf("%d VARIABLE STATING VALUE : ",(i+1));
      scanf("%lf",&(x[i]));
      }

ffvh = 0;
cfvh = 0;

if(felt(x,c))
      {
      printf(" THE STARTING POINT VIOLATED THE CONSTRAINTS\n");
      return(0);
      }

cc = 0;

ra = 0.0;
grad(x,ra,g);
s1 = 0.0;
s2 = 0.0;
for(i = 0; i < N; i++)
      {
      x[i] -= H;
      a = bunt_f(x);
      x[i] += 2*H;
      b = bunt_f(x);
      s1 += ((b-a)/(2*H))*((b-a)/(2*H));
      x[i] -= H;
      s2 -= g[i]*(b-a)/(2*H);
      }
ra = s2/s1;
if(ra < 1)
```

```
        ra = 1.0;

do                  /* starting sumt iteration */
        {
        for(i = 0; i < N; i++)
                {
                for(j = 0; j < N; j++)
                        h[i][j] = 0.0;
                h[i][i] = 1.0;
                }

        do          /* starting DFP technique */
                {
                grad(x,ra,g);
                for(i = 0; i < N; i++)
                        {
                        y[i] = x[i];
                        u[i] = g[i];
                        d[i] = 0.0;
                        }
                ff = celf(x,ra);

                for(i = 0; i < N; i++)
                        for(j = 0; j < N; j++)
                                d[i] -= h[i][j]*g[j];

                for(i = 0, gp = 0; i < N; i++)
                        gp += g[i]*d[i];

                if(gp > 0)
                        {
                        printf("peak point! \n");
                        felt(x,c);
                        fq = celf(x,0.0);
                        printf("VALUE OF THE OBJECTIVE FUNCTION : %f\n",fq);
                        for(i = 0; i < N; i++)
                                printf("X%d VARIABLE VALUE : %f\n",i,x[i]);
                        printf("COEFFICIENT OF PENALTY FUNCTION (ra) : %g\n",ra);
                        printf("NUMBER OF ITERATIONS : %d\n",cc);
                        for(i = 0; i < M; i++)
                                printf("X%d VALUE OF CONSTRAINTS : %f\n",i,c[i]);
                        printf("NUMBER OF CONSTRAINTS CALL : %d \n",ffvh);
                        printf("NUMBER OF OBJECTIVE FUNCTION CALL : %d \n",cfvh);
                        return(0);
                        }

                l = 1.0;
                for(i = 0; i < N; i++)
                        x[i] = y[i]+l*d[i];
                while(felt(x,c))
                        {
                        l = l/2;
                        for(i = 0; i < N; i++)
```

```
                    x[i] = y[i]+l*d[i];
        }

grad(x,ra,g);

for(i = 0, gq = 0; i < N; i++)
        gq += g[i]*d[i];

fq = celf(x,ra);

if((gq > 0) || (fq > ff))
        {
        kod2 = 1;              /* x not an optimum */
        while(kod2 && (l != 0)) /* cubic interpolation */
                {
                zz = 3*(ff-fq)/l+gp+gq;
                w = zz*zz-gp*gq;
                if(w < 0)
                        w = 0;
                w = sqrt(w);
                dd = l*(1-(gq+w-zz)/(gq-gp+2*w));
                for(i = 0; i < N; i++)
                        x[i] = y[i]+dd*d[i];
                grad(x,ra,g);
                for(i = 0, gr = 0; i < N; i++)
                        gr += g[i]*d[i];
                z = celf(x,ra);
                if((z > ff) || (z > fq))
                        {
                        if(gr > 0)
                                {
                                l = dd;
                                fq = z;
                                gq = gr;
                                }
                        else
                                {
                                l = l-dd;
                                ff = z;
                                gp = gr;
                                }
                        }
                else
                        kod2 = 0;          /* x optimum */
                }           /* end od interpolation */
        for(i = 0; i < N; i++)
                {
                u[i] = g[i]-u[i];
                v[i] = x[i]-y[i];
                }
        for(i = 0, kk = wk = 0; i < N; i++)
                {
                m[i] = 0;
```

```
                        for(j = 0; j < N; j++)
                                m[i] += h[i][j]*u[j];
                        kk += m[i]*u[i];
                        wk += v[i]*u[i];
                        }
                if((kk != 0) && (wk != 0))
                        for(i = 0; i < N; i++)
                                for(j = 0; j < N; j++)
                                        h[i][j] += (-m[i]*m[j]/kk+v[i]*v[j]/wk);
                } /* end if */
        else
                {
                for(i = 0; i < N; i++)
                        for(j = 0; j < N; j++)
                                h[i][j] -= d[i]*d[j]/gp;
                }
        cc++;
        fq = celf(x,ra);
        } while(fabs(ff-fq) > E );

    ra = ra/T;
    } while(fabs(T*ra*bunt_f(x)) > E );

felt(x,c);
fq = celf(x,0.0);
printf("VALUE OF THE OBJECTIVE FUNCTION : %f\n",fq);
for(i = 0; i < N; i++)
        printf("X%d VARIABLE VALUE : %f\n",i,x[i]);
printf("COEFFICIENT OF THE PENALTY FUNCTION (ra) : %g\n",(ra*T));
printf("NUMBER OF ITERATIONS : %d\n",cc);
for(i = 0; i < M; i++)
        printf("X%d VALUE OF CONSTRAINT : %f\n",i,c[i]);
printf("NUMBER OF CONSTRAINTS CALL : %d \n",ffvh);
printf("NUMBER OF OBJECTIVE FUNCTION CALL : %d \n",cfvh);
return(0);
}                       /* END of program */

/* ++++++++++++++++++++++++++++++++++++++++++++++++++++++++++++++++++++++++ */

int     grad(double *x, double ra, double *g)
{
int        i;
double  a,b;

for(i = 0; i < N; i++)
        {
        x[i] -= H;
        a = celf(x,ra);
        x[i] += 2*H;
        b = celf(x,ra);
        g[i] = (b-a)/(2*H);
        x[i] -= H;
        }
```

```
return(0);
}

/* +++++++++++++++++++++++++++++++++++++++++++++++++++++++++++++++++++++++ */

double bunt_f(double *x)
{
double  c[M];
double  sum;
int         i;

felt(x,c);
sum=0;
for(i=0; i<M; i++)
        if( c[i] < EP )
                sum += (2*EP-c[i])/(EP*EP);
        else
                sum += 1/(c[i]);
return(sum);
}

/* +++++++++++++++++++++++++++++++++++++++++++++++++++++++++++++++++++ */

int felt(double *x, double *c)
{
int    i;
int  kod = 0;

ffvh++;
c[0] = 72-(x[0]+2*x[1]+2*x[2]);   /*   constraints, both explicit and implicit   */
c[1] = x[0];
c[2] = x[1];
c[3] = x[2];
c[4] = 42-x[0];
c[5] = 42-x[1];
c[6] = 42-x[2];
for(i = 0; i < M ; i++)
        if(c[i] < 0)
                kod = 1;
return(kod);        /* if kod=1 the constraints are violated */
}

/* +++++++++++++++++++++++++++++++++++++++++++++++++++++++++++++++++++++++ */

double celf(double *x, double ra)
{
double  z;

cfvh++;
z = -x[0]*x[1]*x[2];     /*  objective function  */
return(z+ra*bunt_f(x));
}
```

```
/* +++++++++++++++++++++++++++++++++++++++++++++++++++++++++++++++++++++ */
```

APPENDIX C

```
/*  APPENDIX C  */
/* Computer program of Backtrack method for minimize objective function due to nonlinear ine-
quality constraints in ANSI C */
/* Prof. Jarmai, Karoly, University of Miskolc, Hungary */
/* H-3515 Miskolc, Egyetemvaros, Hungary, tel. +36/46/365111 ext 2028, fax +36/46/367828
e-mail jarmai@Kanga.alt.uni-miskolc.hu */
/* You can use the program in the education and academic research free, if you publish your re-
sults, please acknowledge the program */
/* Reference: Farkas, J., Jarmai, K.: Optimum design of steel structures with DSO. CISM Course
on Discrete Structural Optimization, */
/* Udine, Italy, June 17-21, 1996. Springer Verlag, Heidelberg, New York, 65 p. (under publica-
tion) */
/* Please send a copy for the authors when you use this program in your publication */


#define M              2    /* number of constraints */
#define N              3    /* number of variables  */
#define EM             14   /* number of possible discrete sizes for one variable */
#define XA(i)     x[i][av[i]]


double  x[N][EM] =       {
                     { 0.5,1,2,4,6,8,10,12,14,16,18,20,22,24 }, /* these are the discrete sizes
for one variable: first */
                     { 0.3,0.5,1,2,4,6,8,10,12 },  /* these are the discrete sizes for one variable:
second */
                     { 0.3,0.5,1,2,6,8,10,12 },    /* these are the discrete sizes for one variable:
third */
                     };
          /* maximum number of discrete sizes for every variable */
int           esz[N] = { 13,7,7 };   /* maximal No. of index = elem_szam - 1 */
int           av[N];     /* actual index */
int           opt[N];    /* optimum index */
double  kmin;            /* optimum of the objective function */

double        celf();
double        xend();

int      main(void)
{
int           i;
int           j;
      printf("-------------------------------------------------\n");
      for(j=0;j<N;j++)               /* starting from the maximal discrete values of variables */
      opt[j] = av[j] = esz[j];
      kmin = celf();

if(!felt())
      {
```

```
                printf("The maximal values do not satisfy the constraints\n");
                return;
                }

felezo(0);
i=1;

do

        {
        while(i < (N-1))
                {
                felezo(i);       /* finding the minimum value of one variable */
                i++;             /* using halving procedure */
                }
        val();              /* computation of the two last variables */
        j=i-2;
        av[j]++;
        while( (av[j] > esz[j]) && !(j < 0))
                {
                j--;
                if(j>=0)
                        av[j]++;
                }
        i=j+1;
        } while(i > 0);       /* end of do cycle */

        printf("Objective function minimum: %g\n",kmin);
for(j=0;j<N;j++)
printf("Values of the XA(%d) variables: %g\n",j,x[j][opt[j]]);
        printf("+++++++++++++++++++++++++++++++++++++++++++++++++++++++++++++++\n");
}

/* +++++++++++++++++++++++++++++++++++++++++++++++++++++++++++++++++++++++++++++ */

double  celf()          /* computation of the objective function */
{
        return (XA(0)*XA(1)*XA(2));
}

/* +++++++++++++++++++++++++++++++++++++++++++++++++++++++++++++++++++++++++++++ */

double  xend() /* computation of the last variable from the objective function */
{
        return(kmin / (XA(0)*XA(1)));
}

/* +++++++++++++++++++++++++++++++++++++++++++++++++++++++++++++++++++++++++++++ */

int     felt()          /* checking the constraints (ff >= 0) ? */
{
        double  ff[M];
        int             i;
```

```
        ff[0] = 72.-(XA(0)+2.*XA(1)+2.*XA(2)); /* first constraint: upper limit */
        ff[1] = (XA(0)+2.*XA(1)+2.*XA(2))-2.5; /* second constraint: lower limit */

        for(i=0; i<M; i++)
                if(ff[i] < 0)
                        return 0;
        return 1;       /* satisfy the constraints */
}

/* +++++++++++++++++++++++++++++++++++++++++++++++++++++++++++++++++++++ */

int     val()   /* computation of the last two variables */
{
int             i,j;
int             NI;
double  k,xn;

do
        {
        NI=N-1;                         /* index of the last variable */
        xn=xend();
        if(xn < x[NI][0])
                return(0);
        if(felt())
                {
                j=0;
                while((xn > x[NI][j]) && (j<esz[NI]))
                        {
                        j++;
                        }
                av[NI]=j;
                do
                        {
                        av[NI]--;
                        } while(!(av[NI]<0) && felt());
                av[NI]++;
                if(((k=celf()) < kmin) && felt())
                        {
                        kmin=k;
                        for(i=0;i<N;i++)
                                opt[i]=av[i];
                        }
                }
        av[NI-1]++;
        }while(av[NI-1]<=esz[NI-1]);
return(0);
}

/* +++++++++++++++++++++++++++++++++++++++++++++++++++++++++++++++++++++ */

int felezo(i)   /* halving procedure */
int             i;
{
```

```
int  fh,ah,kp;
int  ii;

av[i]=0;
for(ii = i+1; ii < N; ii++)
        av[ii] = esz[ii];
if(felt())
        return(0);
fh=esz[i];
ah=0;
kp = (fh+ah+1) >> 1;
while((fh-ah)!=1)
        {
        av[i]=kp;
        if(felt())
                {
                fh=kp;
                }
        else
                {
                ah=kp;
                }
        kp= (fh+ah+1) >> 1;
        }
return(av[i] = kp);
}
```

/* +++ */

References & bibliography

REFERENCES TO CHAPTER 1

Barsom, J.M. 1971. Fatigue-Crack Propagation in Steels of Various Yield Strengths. *Trans. ASME, J. Eng. Ind., Ser. B* 4: 1190.

Boley, B.A. & Weiner, J.H. 1960. *Theory of thermal stresses*. New York: Wiley.

Chang Doo Jang & Seung Il Seo 1995. Basic studies for automatic fabrication of welded built-up beams. *Journal of Ship Production* 11(2): 111-116.

Chapetti, M.D. & Otegui, J.L. 1995. Importance of Toe Irregularity for Fatigue Resistance of Automatic Welds. *Int. J. of Fatigue* 17(8): 531.

Farkas, J. 1969. Analytische Betrachtung von Methoden zur Verminderung der Krümmung infolge Längsschrümpfung. *Schweisstechnik Berlin* 66: 406-408.

Grimme, D. et al. 1984. Untersuchungen zur Betriebsfestigkeit von geschweissten Offshore-Konstruktionen in Seewasser, ECSC Agreement 7210 KG/101 Final Report.

Godfrey, P.S. & Hicks, J.G. 1987. Control of fatigue performance of welded joints by attention to profile and various post weld treatments, TS27. *Steel in Marine Structures, Int. Conf. Delft*, Elsevier.

Gurney, T.R. 1979. *Fatigue of welded structures*. 2nd ed. Cambridge, Univ. Press.

Haagensen, P.J. 1985. Improving the fatigue strength of welded joints, Fatigue Handbook. *Offshore Steel Structures*. Tapir: Ed. A. Almar Naess.

Haagensen, P.J. 1992. Weld Improvement Methods for Increased Fatigue Strength. *Int. Conf. on Engineering Design in Welded Constructions, Madrid*, Spain, 7-8 Sept. 1992, International Institute on Welding, Pergammon Press, Oxford, pp. 73-92.

Haagensen, P.J. 1996. Weld improving methods – Applications and implementations in design codes. *International Conference on Fatigue of Welded Components and Structures, Senlis*: 1-24.

Josefson, B.L. 1993. Prediction of residual stresses and distortion in welded structures. *Transactions of the ASME* 115: 52-57.

Kado, S. et. al. 1988. The improvement of fatigue strength in welded high tensile strength steels by additional weld run with coated electrodes. *IIW Doc* XIII-1289-88.

Lieurade, H.P. et al. 1992. Efficiency of improvement techniques on the fatigue strength as a function of the type of welded joint. *IIW Doc* XIII-1467-92.

Lopez Martinez & Blom, A. 1995. Influence of life improvement techniques on different steel grades under fatigue loading. *Fatigue Design'95*. VTT, Espoo, Finland.

Maddox, S.J. 1991. *Fatigue Strength of Welded Structures*. 2nd ed, Abington Publishing.

Ohta, S. et al. 1988. Methods of improving the fatigue strength of welded joints by various toe treatments. *IIW Doc* XIII-1289-88.

Okerblom, N.O. 1955. *The calculations of deformations of welded metal structures*, (translated by the Dept. of Scientific and Industrial Research, 1958) London: HMSO.

Okerblom, N.O., Demyantsevich, V.P. & Baikova, I.P. 1963. *Design of fabrication technology of welded structures Leningrad* (in Russian). Sudbromgiz.
Rykalin, N.N. 1951. *Calculation of heat processes in welding* (in Russian). Masgiz.
Trufiakov, V.I. et al. 1995. Ultrasonic impact treatment of welded joints. *IIW Doc* XIII-1609-95.
White, J.D. 1977a. Longitudinal shrinkage of a single pass weld. *CUED/C-Struct./TR.57*. Univ. of Cambridge, England, Department of Engineering.
White, J.D. 1977b. Longitudinal stresses in a member containing noninteracting welds. *CUED/C-Struct./TR.58*. Univ. of Cambridge, Dept. of Engineering.
White, J.D. 1977c. Longitudinal shrinkage of multi-pass welds. *CUED/C-Struct./TR.59*. Univ. of Cambridge, Dept. of Engineering.
Winkander, L. et al. 1994. Finite element simulation and measurement of welding residual stresses. *Modelling Simul. Mater. Sci. Eng.* 2: 845-864.
Wozney, G.P. & Crawmer, G.R. 1968. An investigation of vibrational relief in steel. *Weld, J.* 47(9): 411-419.

REFERENCES TO CHAPTER 2

Chen, W.F. & Atsuta, T. 1976-77. *Theory of beam-columns.* Volumes 1-2. New York: McGraw Hill..
German Design Rules DASt-Richtlinie 016. 1986. *Entwurf. Bemessung und konstruktive Gestaltung von Tragwerken aus dünnwandigen kaltgeformten Bauteilen.* Köln: Deutscher Ausschuss für Stahlbau.
Hancock, G.J. 1994. *Design of cold-formed steel structures.* 2nd ed. Australian Institute of Steel Construction, Sydney.
Murray, N.W. 1984. *Introduction to the theory of thin-walled structures.* Oxford: Clarendon Press.
Rhodes, J. (ed.) 1991. *Design of cold-formed steel members.* Amsterdam: Elsevier.
Umanskiy, A.A. 1961. *Structural mechanics of aircrafts* (in Russian). Moscow: Oborongiz.
US Manual 1986. American Iron and Steel Institute, Washington, DC. Parts I-VII.
Vlasov, V.Z. 1963. *Thin-walled elastic rods* (in Russian). Moscow: Izd. Akad. Nauk SSSR, Moscow.
Yu, Wei-Wen 1991. *Cold-formed steel design.* 2nd ed. New York: Wiley & Sons.

REFERENCES TO CHAPTER 3

Ayrton, W.E. & Perry, J. 1886. On struts. *The Engineer* 62: 464.
Beer, H. & Schulz, G. 1970. Bases théoriques des courbes européennes de flambement. *Construction Métallique* 3.
Beedle, L.S. (ed.) 1991. *Stability of metal structures. A world view.* 2nd ed. Structural Stability Research Council, Bethlehem, USA.
Braham, M., Grimault, J.P. et al. 1980. Buckling of thin-walled hollow sections. Cases of axially-loaded rectangular sections. *Acier-Stahl-Steel* 45: 30-36.
Chen, W.F. & Atsuta, T. 1976-77. *Theory of beam-columns,* Vols 1-2. New York: McGraw Hill.
Chen, W.F. & Lui, E.M. 1987. *Structural stability.* New York: Elsevier.
Chen, W.F. & Lui, E.M. 1991. *Stability design of steel frames.* Boca Raton: Lewis Publ.
Duan, L. 1990. Stability analysis and design of steel structures. PhD Dissertation, Purdue University.
Duan, L. & Chen, W.F. 1989. Design interaction equation for steel beam-columns. *J. Struct. Eng ASCE* 115: 1225-1243.
Ellinas, C.P., Supple, W.J. & Walker, A.C. 1984. *Buckling of offshore structures.* London: Granada.

Euler, L. 1776. Determinatio onerum, quae columnae gestare valent. *Acta Acad. Sci. Petrop.* 2.

Farkas, J. 1977. The effect of residual welding stresses on the buckling strength of compressed plates. *Proc. Regional Colloquium on Stability of Steel Structures*, Budapest: 299-306.

Faulkner, D., Adamczak, J.C. et al. 1973. Synthesis of welded grillages to withstand compression and normal loads. *Computers and Struct.* 3: 221-246.

Handbook of structural stability 1971. Column Research Council of Japan, Tokyo, Corona Publ.

Kollár, L. & Dulácska, E. 1984. *Buckling of shells for engineers*. Budapest, Akadémiai Kiadó.

Maquoi, R. & Rondal, J. 1978. Mise en équation des nouvelles courbes européennes de flambement. *Construction Métallique* 15(1): 17-30.

Petersen, Ch. 1980. *Statik und Stabilitat der Baukonstruktionen*. Braunschweig-Wiesbaden, Vieweg & Sohn.

Rondal, J., Würker, K.-G. et al. 1992. *Structural stability of hollow sections*. Köln: Verlag TÜV Rheinland.

Sohal, I.S., Duan, L. & Chen, W.F. 1989. Design interaction equations for steel members. *J. Struct. Eng ASCE* 115: 50-1665.

Timoshenko, S.P. & Gere, J.M. 1961. *Theory of elastic stability*. 2nd ed. New York: Mc Graw Hill.

Trahair, N.S. 1993. *Flexural-torsional buckling of structures*. London: E. & FN Spon.

Usami, T. & Fukumoto, Y. 1982. Local and overall buckling of welded box columns. *J. Struct. Div. Proc. ASCE* 108: 525-541.

Vol'mir, A.S. 1967. *Stability of deformable systems* (in Russian). Moscow: Nauka.

Waszczyszyn, Z., Cichon, Cz. & Radwanska, M. 1994. *Stability of structures by finite element methods*. Amsterdam: Elsevier.

REFERENCES TO CHAPTER 4

Allen, H.G. 1969. *Analysis and design of structural sandwich panels*. Oxford: Pergamon Press.

Bert, C.W. & Francis, P.H. 1974. Composite materials Mechanics: Structural mechanics. *AIAA Journal* 12(9): 1173-1186.

Dwyer, F.J. 1976. A review of factors affecting durabilitþy characteristics of flexible urethane foams. *Journal of Cellural Plastics* 2: 104-113.

Earles, S.W.E. 1966. Theoretical estimation of the frictional energy dissipation in simple lap joints. *Journal of Mechanical Eng. Sci.* 8(2): 207-214.

Emerson, P.D. 1974. Application of constrained-layer damping to control noise in machine parts. *Trans. ASME. Journal of Engineering for Industry* 2: 299-303.

Farah, A., Ibrahim I.M. & Green R. 1977. Damping of floor vibrations by constrained viscoelastic layers. *Canadian Journal of Civil Engineering* 4: 405-411.

Farkas, J. & Jármai, K. 1982. Structural synthesis of sandwich beams with outer layers of box section. *Journal of Sound and Vibration* 84(1): 47-57. London: Academic Press Inc. Limited.

Grosskopf, P. & Winkler, Th. 1973. Auslegung von GPK/Hartschaum – Verbundwerkstoffen. *Kunststoff* 63(12): 881-888.

Hammwill, W.J. & Andrew, C. 1974. Vibration reduction with continuous damping inserts, *Proceedings 14th Machine Tool Design and Research Conference*: 455-502.

Jármai, K. 1989c. Application of decision support system on sandwich beams, verified by experiments. *Computers in Industry* 11(3): 267-274. Elsevier Applied Science Publishers.

Jones, D.I.G. & Parin, M.L. 1972. Technique for measuring damping properties of thin viscoelastic layers. *J. of Sound and Vibration* 24(2): 201-210.

Katzenschwanz, N. 1967. Reibungsdampfung an Schweissverbindungen. Diss. TH München.

Kerwin, E.M. 1959. Damping of flexural waves by a constrained viscoelastic layer. *Journal of the Acoustical Society of America* 31(7): 952-962.

Kerwin E.M. 1965. Macromechanisms of damping in composite structures. *Internal friction damping and cyclic plasticity ASIM Spec. Techn. Publ.* No. 378. Philadelphia: 125-147.

Kronenberg, M. et al. 1956. Practical design technique for controlling vibration in welded machines. *Machine design* 28(14): 103-109.

Lazan, B.J. 1968. *Damping of materials and members in structural mechanics*. Oxford: Pergamon Press.

Markus, S., Oravsky, V. & Simkova, O. 1974. Philosophy of optimum design of damped sandwich beams. *Acta Technica* CSAV 6: 647-661.

Mead, D.J. 1976. Loss factors and resonant frequencies of periodic damped sandwich plates, Trans. ASME, *Journal of Engineering for Industry* 2: 75-80.

Mead, D.J. & Markus, S. 1969. The forced vibration of a three-layer, damped sandwich beam with arbitrary boundary conditions. *Journal of Sound and Vibration* 10: 163-175.

Nashif, A.D., Jones, D.I.G. & Henderson, J.P. 1985. *Vibration Damping*. New York: John Wiley & Sons, Inc.

Oberst, H. 1952. Über die Dämpfung der Biegeschwingungen Dünner Bleche durch Fest Haftende Beläge. *Akustische Beihefte* 4: 181-194.

Pearce, T.R.A. & Webber J.P.H. 1971. Buckling of sandwich panels with laminated face plates. *Aeronautical Quarterly*, Oct., pp. 148-160.

Plunkett, R. 1981. Friction damping. In P.J. Torvik (ed.), *Damping applications for vibration control*. ASME Publication. AMD 38: 65-74.

Rangajanan, R. & Rao, B.V. 1968. Effect of weld length on the stiffness and structural damping of built-up structures. *Vishwakarma* 8(10): 28-32.

Rao, D.K. 1978. Frequency and loss factors of sandwich beams under various boundary conditions. *Journal of Mechanical Engineering Science, IMechE* 20(5): 271-282.

Rao, S.S. 1977. Vibration of short sandwich beams. *Computers and Structures* 9(3): 185-191.

Rao, S.S. 1990. *Mechanical Vibration*. 2nd ed. Addison-Wesley Publishing.

Ross, D, Ungar, E.E. & Kerwin, E.M. Jr. 1959. Damping of plate flexural vibrations by means of viscoelastic laminate. In ASME (ed.), *Structural damping*: 49-88. New York: ASME.

Ruzicka, J.E. 1961a. Damping structural responses using viscoelastic shear-damping mechanisms. Part I. Design configurations. *Journal of Eng. for Industry* 83: 403-413.

Ruzicka, J.E. 1961b. Damping structural responses using viscoelastic shear-damping mechanisms. Part II. Experimental results. *Journal of Eng. for Industry* 83: 414-424.

Snowdon, J.C. 1968. *Vibration and Shock in Damped Mechanical Systems*. New York: John Wiley & Sons, Inc.

Schlesinger, A. 1979. Vibration isolation in the presence of Coulomb friction. *Journal of Sound and Vibration* 63(2): 213-224.

Stamm, K. & Witte, H. 1974. *Sandwichkonstruktionen,* pp. 338. Wien, New York: Springer-Verlag.

Soyfer, A.M. & Filekin, V.P. 1958. Konstruction, damping the vibration of a part of driving mechanism GTD, BUZ Aviac. *Technika* 1: 158-164.

Sun, C.T. & Lu, Y.P. 1995. *Vibration damping of structural elements,* 372 p. Englewood Cliffs, New Jersey: Prentice Hall.

Ungar, E.E. 1962. Loss factors of viscoelastically damped beam structures. *Journal of the Acoustical Society of America* 34: 1082-1089.

Ungar, E.E. & Kerwin 1959. Loss factors of viscoelastic systems in terms of energy concepts. *Journal of the Acoustical Society of America* 34(7): 954-957.

Warnaka, G.E. & Miller, H.T. 1968. Strain-frequency temperature relationships in polymers. *Journal of Engineering Industry* B90(3): 491-498.

Yan, H.J. & Dowell, E.M. 1974. Elastic sandwich beam or plate equations equivalent to classical theory. *Journal of Applied Mechanics*, Tranaction of the ASME, June, pp. 526-528.

Yin, T.P., Kelly, T.J. & Barry, J.B. 1967. Quantitative evaluation of constrained-layer damping. Trans. of ASME, *Journal of Engineering for Industry* 11: 773-784.

REFERENCES TO CHAPTER 5

Aichele, G. 1985. *Kalkulation und Wirtschaftlichkeit in der Schweisstechnik* (Calculation and economy in welding technology). Düsseldorf: Deutscher Verlag für Schweisstechnik.

Bodt, H.J.M. 1990. *The global approach to welding costs*. The Netherlands Institute of Welding, The Hague.

Chalmers, D.W. 1986. Structural design for minimum cost. In *Advances in Marine Structures. Proceedings of Internat. Conference Dunformline*: 650-669. London: Elsevier.

COSTCOMP 1990. *Programm zur Berechnung der Schweisskosten* (Program for the calculation of welding costs). Düsseldorf: Deutscher Verlag für Schweisstechnik.

Farkas, J. 1991. Techno-economic considerations in the optimum design of welded structures. *Welding in the World* 29(9-10): 295-300.

Fern, D.T. & Yeo, R.B.G. 1990. Designing cost effective weldments. In *Welded Structures '90. Int. Conference, London*: 149-158. Welding Institute, Abington Publ.

Horikawa, K., Nakagomi, T. et al. 1992. The present position in the practical application of arc welding robots. *Welding in the World* 30(9-10): 256-274.

Likhtarnikov, Ya.M. 1968. *Metal Structures* (in Russian). Moskva: Stroyizdat.

Malin, V. 1986. Designer's guide to effective welding automation. Part 1-2. *Welding Journal*, Vol. 64 (1985) November, pp. 17-29, Vol. 65 (1986) June, pp. 43-52.

Ntuen, C.A. & A.K. Mallik 1987. Applying artificial intelligence to project cost estimating. *Cost Engineering* 29(5): 8-13.

Ott, H.H. & Hubka, V. 1985. Vorausberechnung der Herstellkosten von Schweisskonstruktionen. In *Proc. Int. Conference on Engineering Design ICED Hamburg*: 478-487. Zürich: Edition Heurista.

Pahl, G. & Beelich, K.H. 1982. Kostenwachstumsgesetze nach Ahnlichkeitsbeziehungen für Schweiss-verbindungen. VDI-Bericht Nr.457. Düsseldorf, pp. 129-141.

REFERENCES TO CHAPTER 6

Annamalai, N. 1970. Cost optimization of welded plate girders. Dissertation, Purdue Univ. Indianapolis, Ind.

Azarm, S. & Eschenauer, H. 1993. A minimax reduction method for multiobjective decomposition based design optimization. *Struct. Optim.* 6: 94-98.

Bitner, J.R. & E.M. Reingold 1975. Backtrack programming techniques. *Communications of ACM* 18: 651-656.

Box, M.J. 1965. A new method of constrained optimization and a comparison with other methods. *Computer Journal* 8: 42-52.

Cohon, J.L. 1978. *Multi-objective programming and planning*. New York: Academic Press.

Davidon, W.C. 1959. *Variable metric method for minimization*. Argonne Nat. Lab. Illinois.

Eschenauer, H., Koski, J. & Osyczka, A. (eds) 1990. *Multicriterion design optimization: Procedures and applications*. Berlin, Heidelberg, New York: Springer.

Eschenauer, H.A. 1995. Multicriterion structural optimization as a technique for quality improvement in the design process. *Microcomputers in Civil Engineering* 10(4): 257-267.

Farkas, J. 1984. *Optimum Design of Metal Structures*. Budapest: Akadémiai Kiadó, Chichester: Ellis Horwood.

Farkas, J. & K. Járma 1995. Multiobjective optimal design of welded box beams. *Microcomputers in Civil Engng.* 10: 249-255.

Farkas, J. & L. Szabó 1980. Optimum design of beams and frames of welded I-sections by means of backtrack programming. *Acta Techn. Hung.* 91: 121-135.

Fiacco, A.V. & McCormick, G.P. 1968. *Nonlinear sequential unconstrained minimization technique*. New York: John Wiley & Sons, Inc.

Fletcher, R. & Powell, M.J.D. 1963. A rapidly convergent descent method for minimization. *Computer Journal* 6: 163-168.

Ghani, S.M. 1972. An improved complex method of function minimization. *Computer Aided Design*: 71-78.

Gill, P.E., Murray, W. & Wright, W.H. 1982. *Practical optimization*. London: Academic Press.

Golomb, S.W. & Baumert, L.D. 1965. Backtrack programming. *J. Assoc. Computing Machinery* 12: 516-524.

Himmelblau, D.M. 1972. *Applied nonlinear programming*. New York: McGraw-Hill.

Hock, W. & Schittkowski, K. 1981. Test examples for nonlinear programming codes. *Lecture Notes in Economics and Mathematical Systems*. Berlin: Springer.

Hooke, R. & Jeeves, T.A. 1961. Direct search solution of numerical and statistical problems. *J. Assoc. Comp. Machinery* 8: 212-229.

Hwang, C.L. & Masud, A.S.M. 1979. Multiple objective decision making: Methods and applications. Lecture notes in economics and mathematical systems, 164. Berlin, Heidelberg, New York: Springer.

Jármai, K. 1982. Optimal design of welded frames by complex programming method. Publ. Techn. Univ. Heavy Ind. *Ser. C. Machinery* 37: 79-95.

Jármai, K. 1989. Single- and multicriteria optimization as a tool of decision support system. *Computers in Industry* 11: 249-266.

Jármai, K. 1989. Application of decision support system on sandwich beams verified by experiments. *Computers in Industry* 11: 267-274.

Jármai, K. 1990. Decision support system on IBM PC for design of economic steel structures, applied to crane girders. *Thin-walled Structures* 10: 143-159.

Jendo, S. 1990. Multiobjective optimization. In Save, M. & Prager, W. (eds), *Structural optimization, Volume II: Mathematical Programming*: 311-342. New York: Plenum Press.

Jendo, S. & Paczkowski, W.M. 1992. Multi-criterion discrete optimization of large scale frame structures. In Phua, K.H. et al., *Proc. ICOTA'92*. Singapore: World Scientific Publ.

Karczewski, J.A., Niczyj, J. & Paczkowski, W.M. 1991. Discrete optimization of the space truss being controlled by an expert system. In Wester, T., Medwadowski, S.J. & Morgensen, I. (eds), *Proc. Int. IASS Symp.* 3: 191-197. Copenhagen: Kunstakademiets Forlag Arkitekskolen.

Keeney, R.L. & Raiffa, H. 1976. *Decision with multiple objectives: Preferences and value trade-offs*. New York: John Wiley & Sons.

Kirsch, U. 1975. Multilevel approach to optimum structural design. *ASCE J. Struct. Div.* 101: 957-974.

Kirsch, U. 1993. *Structural optimization, Fundamentals and application*. Heidelberg: Springer Verlag.

Kiusalaas, J. & Reddy, G.B. 1977. DESAP1 – A structural design program with stress and displacement constraints. Volumes I, II, III. NASA Contractor Report, NASA CR-2794-95-96. Washington, D.C.

Khot, N.S. & Berke, L. 1984. Structural optimization using optimality criteria methods. In Atrek, E., Gallagher, R.H. et al. (eds), *New directions in optimum structural design*: 47-74. Chichester, New York: Wiley & Sons.

Knuth, D.E. 1975. Estimating the efficiency of backtrack programs. *Mathematics of Computation* 29: 121-136.

Koski, J. 1984. Bicriterion optimum design method for elastic trusses. *Acta Polytechnica Scandynavica, Mech. Engng. Series 86*.

Lewis, A.D.M. 1968. Backtrack programming in welded girder design. In *Proc. 5th Annual SHARE-ACM-IEEE Design Automation Workshop*, Washington, 28/1-28/9.

Lin, J.G. 1976. Maximal vector and multi-objective optimization. *JOTA* 18: 41-63.

Lounis, Z. & Cohn, M.Z. 1995. An engineering approach to multicriteria optimization of bridge structures. *Microcomputers in Civil Engineering* 10(4): 233-238.

Makowski, Z.S. (ed.) 1984. *Analysis, design and construction of braced domes*. London: Granada.

Michell, A.G.M. 1904. The limits of economy of material in framed structures. *Phil. Mag. Series 6* 8: 589-597.

Nelder, J.A. & Mead, R. 1964. A simplex method for function minimization. *Computer Journal* 7: 308-313.

Von Neumann, J. & Morgenstern, O. 1947. *Theory of game and economic behaviour.* Princeton: Princeton University Press.

Niczyj, J. & Paczkowski, W.M. 1992. An expert system of controlling a program for discrete structural optimization. In Phau, K.H. et al. (eds), *Int. Conf. Optim. Tech. Appl. (ICOTA'92)* 2: 835-842. Singapore: World Scientific Publ.

Osyczka, A. 1984. *Multicriterion Optimization in Engineering.* Chichester: Ellis Horwood.

Osyczka, A. 1992. *Computer Aided multicriterion optimization system.* Krakow: International Software Publishers.

Pareto, V. 1986. *Cours d'economie politique.* Volumes 1 and II. Lausanne: F. Rouge.

Peschel, M. & Riedel, C. 1976. *Polioptimierung eine Entscheidurcgshilfe für ingenzeurtechnische Compromisslösungen.* Berlin: VEB Verlag Technik.

Qian, L.X. et al. 1984. An approach to structural optimization-SQP. *Eng.Opt.* 8: 83-100.

Rosenbrock, H.H. 1960. An automatic method for finding the greatest or least value of a function. *Computer Journal* 3: 175-184.

Rozvany, G.I.N. 1989. *Structural optimization via optimality criteria.* Dordrecht: Kluwer.

Schittkowski, K., Zillober, C. & Zotemantel, R. 1994. Numerical comparison of nonlinear programming algorithm for structural optimization. *Journal of Structural Optimization* 7(1/2): 1-19.

Schmit, L.A. jr. 1960. Structural design by systematic synthesis. *Proc. 2nd Conference Electronic Computation.* ASCE, New York, 105-132.

Sobieszczanski-Sobieski, J., James, B.B. & Dori, A.R. 1985. Structural optimization by multilevel decomposition. *AIAA J.* 23: 1775-1782.

Spendley, W., Hext, G.R. & Himsworth, F.R. 1962. The sequential application of designs in optimization and evaluationary operation. *Technometrics* 4: 441.

Stadler, W. 1980. Preference optimality in multicriterion control and programming problems. *Nonlinear Anal. Theory Meth. Appl.* 4: 51-65.

Stadler, W. 1984. Multi-criteria optimization in mechanics (a survey). *AMR* 37: 277-286.

Stadler, W. (ed.) 1988. *Multi-criteria optimization in engineering and in the sciences.* New York: Plenum Press.

Svanberg, K. 1987. The method of moving asymptotes – A new method for structural optimization. *Int. Journal Num. Meth. Engineering* 24: 359-373.

Thanedar, P.B. et al. 1986. Performance of some SQP algorithms on structural design problems. *Int. Journal Num. Meth. Engineering* 23: 2187-2203.

Vanderplaats, G.N. 1984. Numerical optimization techniques for engineering design. New York: McGraw-Hill.

Walker, R.J. 1960. An enumerative technique for a class of combinatorical problems. In *Proc. of Symposia in Appl. Math. Amer. Math. Soc. Providence, R.I.* 10: 91-94.

Yu, P.L. 1974. Cone convexity, cone extreme points and nondominated solutions in decision problems with multi-objectives. *J. Optim Theory Appl.* 14: 319-377.

Yu, P.L. 1985. *Multiple-criteria decision making: Concepts, techniques, and extensions.* New York: Plenum Press.

Zadeh, L. 1963. Optimality and non-scalar-valued performance criteria. *IEEE Trans. Auto. Contr.* AC-8.

Zhou, J.L. & Tits, A.L. 1991. User's guide for FSQP version 2.4: A Fortran code for solving optimization problems, possibly minimax, with general inequality constraint and linear equality constraints, generating feasible iterates. Systems Research Center, University of Maryland, Technical Report SRC-TR-90-60rle.

Zhou, J.L. & Tits, A.L. 1992. An SQP algorithm for finely discretized SIP problems and other problems with many constraints. Technical report TR-92-125-ISR, Institute for Systems Research, University of Maryland.
Zhou, J.L. & Tits, A.L. 1993. Nonmonotone line search for minimax problems. *J. Optimiz. Theory and Appl.* 76: 455-476.
Zoutendijk, G. 1960. *Methods of feasible directions.* Amsterdam: Elsevier.

REFERENCES TO CHAPTER 7

Adeli, H. 1986. Artificial intelligence in structural engineering. *Engineering Analysis* 3(3): 154-160.
Adeli, H. (ed.) 1988. *Expert systems in construction and structural engineering.* London-New York: Chapman and Hall.
Andriole, S.J. (ed.) 1985. Applications in artificial intelligence. Princeton: Petrocelli Books.
Balasubramanyan, K. 1990. A knowledge based expert system for optimum design of bridge trusses. *University Microfilms International, Dissertation Information Service.* Ann Arbor, Michigan, No. 8812223.
Buchanan, B.G. & Shortliffe, E.H. 1984. *Rule-based Expert Systems – The MYCIN Experiments of the Stanford Heuristic Programming Project.* Reading, Massachusetts: Addison-Wesley.
BS 2573 1983. *Rules for the design of cranes. Part 1: Specification for classification, stress, calculations and design criteria for structures.* London: British Standard Institution.
BS 5400 1983. *Steel, concrete and composite bridges. Part 3: Code of practice for design of steel bridges.* London: British Standard Institution.
Dym, C.L. & Levitt, R.E. 1991. *Knowledge based systems in engineering.* New York: McGraw-Hill Inc.
Eurocode 3 1992. *Design of steel structures. Part 1.1.* Brussels: CEN European Committee for Standardization.
Farkas, J. 1984. *Optimum design of metal structures.* Chichester: Ellis Horwood Ltd.
Farkas, J. & K. Jármai 1994. Savings in weight by using CHS or SHS instead of angles in compressed struts and trusses. *Proc. 6th Int. Symposium on Tubular Structures, Melbourne.* Rotterdam: Balkema.
Forsyth, R. & Rada, R. 1986. *Machine learning: Applications in Expert Systems and Information Retrieval.* New York: Halsted Press.
Garrett, J.H. 1990. Knowledge-based expert systems. Past, present and future. *IABSE Surveys S-45/90,* Zürich, pp. 21-40.
Gero, J.S. (ed.) 1987. *Expert systems in computer-aided design.* Amsterdam: Elsevier.
Hanna, A.S. 1996. SELECTCRANE: An expert system for optimum of crane selection. *Computing in Civil Engineering:* 958-963.
Harmon, P. & King, D. 1985. *Artificial Intelligence in Business.* New York: John Wiley.
Harmon, P. & Sawyer, B. 1990. *Creating expert systems for business and industry.* New York: John Wiley & Sons Inc.
Hajela, P. & Berke, L. 1991. Neural network based decomposition in optimal structural synthesis. *Computing Systems in Engineering* 2(5/6): 473-481.
Hayes-Roth, F., Waterman, D.A. & Lenat, D. (eds) 1983. *Building Expert Systems.* Reading, Massachusetts: Addison-Wesley.
Hayward, S.A. 1985. Is a decision tree an expert system. In Bremer, M.A. (ed.), *Research and Development in Expert Systems.* Cambridge: Cambridge University Press.
Jármai, K. 1989. Single- and multicriterion optimization as a tool of decision support systems. *Computers in Industry* 11: 249-266. Elsevier Applied Science Publishers.
Jármai, K. 1990. Decision support system on IBM PC for design of economic crane girders. Thin-walled Structures 10: 143-159. Elsevier Applied Science publishers.

Jármai, K., Farkas, J. & Mészáros, L. Optimum design of main girders of overhead travelling cranes using an expert system. The First World Congress of Structural and Multidisciplinary Optimization, May 28-June 2, 1995. Goslar, WCSMO-1, Pergamon, Elsevier Science, Oxford, Edited by N. Olhoff, G.I.N. Rozvany. 1996, pp. 939-944.

Kraft, A. 1984. XCON: An expert configuration system at Digital Equipment Corporation. In Winston, P.H. & Prendergast, K.A. (eds), *The AI Business: The Commercial Uses of Artificial Intelligence*. Cambridge, Massachusetts: MITS Press.

LEVEL 5 OBJECT 1990. *Reference Guide, FOCUS Integrated Data and Knowledge-based Systems*. Information Builders, 1250 Broadway, New York.

Personal Consultant Easy 1987. Getting Started. *Reference Guide Texas Instruments Incorporated*, Austin, Texas.

Restak, R.M. 1984. *The Brain*. New York: Bantam Books.

Rummelhart, D.E. & McClelland, J.L. (eds). 1986. *Parallel distributed processing explorations in the Microstructure of Cognition. Volume 1: Foundations*. Cambridge, Massachusetts: The MIT Press.

Waterman, D.A. 1986. *A Guide to Expert Systems*. Reading, Massachusetts: Addison-Wesley.

REFERENCES TO CHAPTER 8

Bo, G.M., P.M. Capurro & I. Daddi 1974. Sulla razionalizzazione dei profili ad I per flessione e taglio. *Construzione Metalliche* 26:102-112.

British Standard BS 8118: Part 1 1991. *Structural use of aluminium*.

Farkas, J. 1984. *Optimum design of metal structures*. Budapest: Akadémiai Kiadó, Chichester: Ellis Horwood.

Farkas, J. 1991. Fabrication aspects in the optimum design of welded structures. *Structural Optimization* 3: 51-58.

Farkas, J. & K. Jármai 1995. Multiobjective optimal design of welded box beams. *Microcomputers in Civ. Eng.* 10: 249-255.

Farkas, J. & K. Jármai 1996. Optimum design of welded bridges. In *Proc. IIW Int. Conference on Welded Structures in particular welded bridges, Budapest*. Scientific Society of Mechanical Engineers: 299-310.

REFERENCES TO CHAPTER 9

Amir, H.M. & Hasegawa, T. 1994. Shape optimization of skeleton structures using mixed-discrete variables. *Structural Optimization* 8: 125-130.

American Petroleum Institute (API) 1989. Draft Recommended Practice 2A-LRFD, first ed.

Belenya, E.I. 1975. *Prestressed metal structures* (in Russian). Moskva, Stroyizdat.

Chen, W.F. & Sugimoto, H. 1987. Analysis of tubular beam-columns and frames under reversed loading. *Eng. Struct.* 9: 233-242.

Chen, W.F. & Sohal, I.S. 1988. Cylindrical members in offshore structures. *Thin-walled Structures* 6: 153-285.

Chu, Kuang-Han & S.S. Berge 1963. Analysis and design of struts with tension ties. *J. Struct. Div. Proc. ASCE* 89: 127-163.

Duan, L. & Chen, W.F. 1990. Design interaction equations for cylindrical tubular beam-columns. *J. Struct. Eng. ASCE* 116(7): 1794-1812.

Dutta, D. & K-G. Würker 1988. *Handbuch Hohlprofile in Stahlkonstruktionen*. Köln: TÜV Rheinland.

Eurocode 3 1992. *Design of steel structures: Part 1.1*. Brussels: CEN European Committee for Standardization.

Farkas, J. 1990. Minimum cost design of tubular trusses considering buckling and fatigue constraints. In Niemi, E. & P. Mäkeläinen (eds), *Tubular Structures*: 451-459. London-New York: Elsevier.

Farkas, J. 1992. Optimum design of circular hollow section beam-columns. In *Proceedings of the Second International Offshore and Polar Engineering Conference, San Francisco, 1992*. ISOPE, Golden, Colorado, USA: 494-499.

Farkas, J. & Jármai, K. 1994. Savings in weight by using CHS or SHS instead of angles in compressed struts and trusses. In Grundy, P., Holgate, A. & Wong, B. (eds), *Tubular Structures VI. Proceedings of the 6th International Symposium, Melbourne, 1994*: 417-422. Rotterdam-Brookfield: Balkema.

Farkas, J. & Jármai, K. 1995. Stability constraints in the optimum design of tubular trusses. *Stability of steel structures*. Preliminary Report of the International Colloquium, European Session, Budapest, Hungary II: 187-194.

Gerard, G. 1962. *Introduction to structural stability theory*. New York: McGraw-Hill.

Gioncu, V. 1994. General theory of coupled instabilities. *Thin-Walled Structures* 19: 81-127.

Han, D.J. & Chen, W.F. 1983. Buckling and cyclic inelastic analysis of steel tubular beam-columns. *Eng. Struct.* 5: 119-132.

Hasegawa, A., H. Abo et al. 1985. Optimum cross-sectional shapes of steel compression members with local buckling. *Proc. JSCE Structural Engineering/ Earthquake Engineering* 2:121-129.

Jain, A.K., Goel, S.C. & Hanson, R.D. 1980. Hysteretic cycles of axially loaded steel members. *J. Struct. Div. Proc. ASCE* 106: 1777-1795.

Jármai, K. & J. Farkas 1996. Optimum design and imperfection-sensitivity of centrally compressed SHS and CHS aluminium struts. *Tubular Structures VII*: 469-474. Rotterdam-Brookfield: Balkema.

Khot, N.S., Berke, L. 1984. Structural optimization using optimality criteria methods. In Atrek, E., Gallagher, R.H. et al. (eds), *New directions in optimum structural design*: 47-74. Chichester, New York: Wiley & Sons.

Kmet', S. 1989. Relative deformation of a cable as a function of time and stress under nonlinear creep. *Civil Eng.* 10: 534-540.

Lai, Y.F.W. & Nethercot, D.A. 1992. Design of aluminium columns. *Engineering Structures* 14(3): 188-194.

Lee, S. & Goel, S.C. 1987. Seismic behaviour of hollow and concrete filled square tubular bracing members. *Research Report* UMCE 87-11. Department of Civil Engng, University of Michigan, Ann Arbor.

Liu, Zh. & Goel, S.C. 1988. Cyclic load behavior of concrete-filled tubular braces. *J. Struct. Eng. ASCE* 114: 1488-1506.

Maquoi, R. & J. Rondal 1978. Analytical formulation of the new European buckling curves. *Acier-Stahl-Steel* 1: 23-28.

Matsumoto, T., Yamashita, M. et al. 1987 Post-buckling behavior of circular tube brace under cyclic loadings. In *Safety Criteria in Design of Tubular Structures, Proc. Int. Meeting, Tokyo, 1986*: 15-25. Architectural Institute of Japan.

Mauch, H.R. & L.P.Felton 1967. Optimum design of columns supported by tension ties. *J. Struct. Div. Proc. ASCE* 93: 201-220.

Nonaka, T.1977. Approximation of yield condition for the hysteretic behavior of a bar under repeated axial loading. *Int. J. Solids Struct.* 13: 637-643.

Ochi, K., Yamashita, M. et al. 1990. Local buckling and hysteretic behavior of circular tubular members under axial loads. *J. Struct. Constr. Engng AIJ* 417: 53-61 (in Japanese).

Packer, J.A., J. Wardenier et al. 1992. *Design guide for rectangular hollow section joints under predominantly static loading*. Köln: TÜV Rheinland.

Papadrakakis, M. & Loukakis, K. 1987. Elastic-plastic hysteretic behavior of struts with imperfections. *Eng. Struct.* 9: 162-170.

Petersen, Ch. 1980. *Statik und Stabilität der Baukonstruktionen*. Braunschweig-Wiesbaden: Vieweg.

Popov, E.P. & Black, R.G. 1981. Steel struts under severe cyclic loadings. *J. Struct. Div. Proc. ASCE* 107: 1857-1881.

Prathuangsit, D., Goel, S.C. & Hanson, R.D. 1978. Axial hysteresis behavior with end restraints. *J. Struct. Div. Proc. ASCE* 104: 883-896.

Rondal, J. & Maquoi, R. 1981. *On the optimal thinness of centrically compressed columns of SHS.* Lecture Notes, Technical University of Budapest, Faculty of Civil Eng., Dept. of Mechanics.

Rondal, J., K-G. Würker et al. 1992. *Structural stability of hollow sections.* Köln: TÜV Rheinland.

Saka, M.P. 1980. Shape optimization of trusses. *Journal of Structural Division Proc. ASCE* 106(ST5): 1155-1174.

Saka, M.P. 1990. Optimum design of pin-jointed steel structures with practical applications. *Journal of Structural Division Proc. ASCE* 116(10): 2599-2620

Schock, H-J 1976. Untersuchungen zur Tragfähigkeit von seilverspannten Druckstäben. Dissertation. Forschungsberichte aus dem Institut für Tragkonstruktionen, Universität Stuttgart No.4. (Abstract: Stahlbau 1978: 183-187.)

Shanley, F.R. 1960. *Weight-strength analysis of aircraft structures.* New York: Dover Publ.

Shibata, M. 1982. Analysis of elastic-plastic behavior of a steel brace subjected to repeated axial force. *Int. J. Solids Struct.* 18: 217-228.

Sohal, I.S., Duan, L. & Chen, W.F. 1989. Design interaction equations for steel members. *J. Struct. Eng. ASCE* 115(7): 1650-1665.

Supple, W.J. & Collins, I. 1980. Post-critical behaviour of tubular struts. *Eng. Struct.* 2: 225-229.

Thompson, J.M.T. 1972. Optimization as a generator of structural instability. *International Journal of Mechanical Sciences* 14: 627-629.

Thompson, J.M.T. & Hunt, G.W. 1973. *A general theory of elastic stability.* London: Wiley.

Tvergaard, V. 1973. Imperfection-sensitivity of a wide integrally stiffened panel under compression. *International Journal of Solids and Structures* 9: 177-192.

Usami, Ts. 1989. Optimum design of locally buckled steel compression members. *Proc. 4th Int. Colloquium on Structural Stability, Asian Session ICS SAS'89:* 388-397.

Van der Neut, A. 1973. The sensitivity of thin-walled compression members to column axis imperfections. *International Journal of Solids and Structures* 9: 999-1011.

Vanderplaats, G.N. & Moses, F. 1972. Automated design of trusses for optimum geometry. *Journal of Structural Division Proc. ASCE'98* ST3: 671-690.

Wardenier, J., Kurobane, Y. et al. 1991. *Design guide for circular hollow section joints under predominantly static loading.* Köln: TÛV Rheinland.

Zayas, V.A., Mahin, S.A & Popov, E.P. 1982. Ultimate strength of steel offshore structures. In *Behaviour of offshore structures, Proc. 3rd Int. Conference, Boston* 2: 39-58. Washington: Hemisphere Publ. Co.

REFERENCES TO CHAPTER 10

Almar-Ness, A., P.J.Haagensen et al. 1984. Investigation of the Alexander L. Kielland failure – Metallurgical and fracture analysis. *J. Energy Resources Technology Trans. ASME* 106: 24-31.

Farkas, J. 1983. Optimum design of crane runway girders. *Publ. Techn. Univ. Heavy Industry Miskolc Series C.* 37: 233-246.

Farkas, J. 1991. Fabrication aspects in the optimum design of welded structures. *Struct. Optimization* 3: 51-58.

Gregor, V. 1989. The effect of surface preparation by TIG remelting on fatigue life. *Zváracské Správy- Welding News* 39: 60-65.

Ingraffea, A.R., K.I.Mettam et al 1987. An analytical and experimental investigation into fatigue cracking in welded crane runway girders. Structural Failure, Product Liability and Technical Insurance. *Proc. 2nd Internat. Conference University of Vienna 1986,* Rossmanith, H.P. (ed.). Inderscience Enterprises, pp.201-223.

Maeda, Y. & I. Okura 1983. Influence of initial deflection of plate girder webs on fatigue crack initiation. *Eng. Struct.* 5: 58-66.
Van Wingerde, A.M., Packer, J.A. & Wardenier, J. 1995. Criteria for the fatigue assessment of hollow structural connections. *J. Construct. Steel Research* 35: 71-115.

REFERENCES TO CHAPTER 11

Dutta, D. & Würker, K-G. 1988. *Handbuch Hohlprofile in Stahlkonstruktionen.* Köln: TÜV Rheinland GmbH.
Farkas, J. & Jármai, K. 1994. Savings in weight by using CHS or SHS instead of angles in compressed struts and trusses. In Grundy, P., Holgate, A. & Wong, B. (eds), *Tubular Structures VI. Proceedings of the 6th International Symposium, Melbourne, 1994*: 417-422. Rotterdam-Brookfield: Balkema.
Rondal, J., Würker, K-G. et al. 1992. *Structural stability of hollow sections.* Köln: TÜV Rheinland.
Wardenier, J., Kurobane, Y. et al. 1991. *Design guide for circular hollow section joints under predominantly static loading.* Köln: TÜV Rheinland.

REFERENCES TO CHAPTER 12

American Petroleum Institute 1987. API Bulletin on design of flat plate structures. Bul. 2V. 1st ed.
Birchfield, J.R. 1981. Welded machines thrive on tough mining. *Welding Design and Fabrication* 54(7): 47-54.
Chalmers, D.W. 1986. Structural design for minimum cost. In *Advances in Marine Structures '90 Int. Conference, Dunformline*: 650-669. London: Elsevier.
Drews, P. & G. Starke 1990. Robot welding systems. In *Advanced Joining Technologies Proceedings Int. Institute of Welding Congress on Joining Research*: 83-91. London: Chapman & Hall.
Evans, H.R. & N.E. Shanmugam 1984. Simplified analysis for cellular structures. *J. Struct. Eng. ASCE* 110: 531-543.
Farkas, J. 1976. Structural synthesis of welded cell-type plates. *Acta Technica Acad. Sci. Hung.* 83: 117-131.
Farkas, J. 1982. Minimum cost design of welded square cellular plates. *Publ. Techn. Univ. Heavy Ind. Miskolc Ser.C. Machinery* 37: 111-130.
Farkas, J. 1985. Discussion to Evans 1984. *J. Struct. Eng. ASCE* 111: 2269-2271.
Fern, D.T. & R.B.G. Yeo 1990. Designing cost effective weldments. In *Welded Structures '90. Int. Conference, London*: 149-158. The Welding Institute, Abington Publ.
Forde, B., Y.Ch. Leung & S.F. Stiemer 1984. Computer-aided design evaluation of steel structures. In *Proceedings IABSE Congress, Vancouver*: 421-428.
Haroutel, J. 1982. Soudage laser de structures sandwich métalliques du type Norsial. *Soudage et techn. Conn.* Jan.-Fevr. 25-31.
Hicks, J.G. 1979. *Welded joint design.* London: Granada.
Horikawa, K., T. Nakagomi et al. 1992. The present position in the practical application of arc welding robots and proposals on joint design for robot welding of steel structures. *Welding World* 30:256-274.
Isler, H. 1977. Erfahrungen mit selbsttragenden Kassettenplatten aus GF-UP. *Plasticonstruction* 7: 118-122.
Malin, V. 1985-86. Designer's guide to effective welding automation. Part 1-2. *Welding J.* 64(Nov): 17-29, 65(June): 287-310.
Pedersen, P.T. & N-J.R. Nielsen 1986. Structural optimization in ship structures. In *Computer aided optimal design: structural and mechanical systems. Proc. NATO Advanced Study Inst. Troia, Portugal*: 921-941. Berlin: Springer.

Pettersen, E. 1979. *Analysis and design of cellular structures*. University of Trondheim, Norwegian Institute of Technology.

Ramirez, J.C. & A. Touran 1991. An integrated system computer system for estimating welding cost. *Cost Engineering* 33(8): 7-14.

Rao, K.P. 1988. Buckling of composite sandwich rectangular panels (grid core). *J. Reinforced Plastics and Composites* 7: 72-89.

Sahmel, P. 1978. Statische und konstruktive Probleme bei Hilfsvorrichtungen zum Transport schwerer Behaelter. *Fördern und Heben* 28: 844-847.

Schade, H.A. 1941. Design curves for cross-stiffened plating under uniform bending load. *Trans. Soc. Naval Archit. Marine Eng.* 49: 154-182.

Sen, P., W.B. Shi & J.B.Caldwell 1989. Efficient design of panel structures by a general multiple criteria utility. *Engineering Optimization* 14: 287-310.

Shanmugam, N.E. & H.R. Evans 1984. A grillage analysis of the nonlinear and ultimate load behavior of cellular structures under bending loads. *Proc. Inst. Civ. Eng. Part 2* 71:705-719.

Shanmugam, N.E. & T. Balendra 1986. Free vibration of thin-walled multi-cell structures. *Thin-walled Struct.* 4: 467-483.

Timoshenko, S. & S. Woinowsky-Krieger 1959. *Theory of plates and shells*. 2nd ed. New York: McGraw Hill.

Tsai, J.C. & R.S. Orr 1977. Special criteria developed for the design and analysis of floating nuclear plant containment structures. In *Trans.4th Int. Conference Struct. Mech. React. technol. San Francisco* J: J 2/6: 1-10.

Usami, Ts. & Y. Fukumoto 1982. Local and overall buckling of welded box columns. *J. Struct. Div. Proc. ASCE* 108: 525-542.

Williams, D.G. 1969. Analysis of doubly plated grillage under inplane and normal loading. Ph.D.thesis. Imperial College, London.

REFERENCES TO CHAPTER 13

Adeli, H. & Balasubramanyan, K.V. 1988. *Expert Systems for Structural Design*. Englewood Cliffs, N.J.: Prentice Hall

Cohn, M.Z. & Dinovitzer, A.S. 1994. Application of structural optimization. *Journal of Structural Engineering ASCE* 120: 617-650.

Durfee, R.H. 1987. Design of a triangular cross-section bridge truss. *Journal of Structural Engineering ASCE* 113: 2399-2414.

Farkas, J. & Jármai, K. 1995. Fabrication cost calculations and minimum cost design of welded structural parts. *Welding in the World* 35: 400-406.

Farkas, J. & Jármai, K. 1995. Multiobjective optimal design of welded box beams. *Microcomputers in Civil Engineering* 10: 249-255.

Farkas, J. & Jármai, K. 1995. Stability constraints in the optimum design of tubular trusses. In *Stability of Steel Structures, International Colloquium, Preliminary Report Vol.II. Budapest*: 187-194.

Farkas, J. & Jármai, K. 1996. Minimum cost design of Vierendeel SHS trusses. In *Tubular Structures VII*: 463-468. Rotterdam-Brookfield: Balkema.

Ferscha, F. 1987. Querschnittsoptimierung biegesteifer, geschweisster Stahlstabwerke. *Stahlbau* 56: 313-318.

Jármai, K. 1989. Single- and multicriterion optimization as a tool of decision support systems. *Computers in Industry* 11: 249-266.

Jármai, K. & Farkas, J. 1994. Application of expert systems in the optimum design of tubular trusses of belt-conveyor bridges. In Grundy, P. et al (eds), *Tubular Structures VI*: 405-410. Rotterdam-Brookfield: Balkema.

Konishi, Y. & Maeda, Y. 1976. Total cost optimum of I-section girders. Tenth IABSE Congress, Preliminary report, Tokyo, 1976. Zürich, Secretariat of IABSE: 189-194.

Martin, L.H. & Purkiss, J.A. 1992. *Structural design of steelwork to BS 5950*. London-Melbourne: Edward Arnold.

Memari, A.M., West, H.H. & Belegundu, A.D. 1991. Methodology for automation of continuous highway bridge design. *Journal of Structural Engineering ASCE* 117: 2584-2599.

Negrao, J.H.O. & Simoes, L.M.C. 1994. Three-dimensional nonlinear optimization of cable-stayed bridges. In Topping, B.H.V. & Papadrakakis, M. (eds), *Advances in Structural Optimization*: 203-213. Edinburgh: Civil-Comp Press.

Ohkubo, S. & Taniwaki, K. 1991. Shape and sizing optimization of steel cable-stayed bridges. In Hernandez, S. & Brebbia, C.A. (eds), *Optimization of Structural Systems and Industrial Applications*: 529-540. Southampton-Boston: Computational Mechanics Publ.

Packer, J.A., Wardenier, J. et al. 1992. *Design guide for rectangular hollow section joints under predominantly static loading*. Köln: Verlag TÜV Rheinland.

Suruga, T. & Maeda, Y. 1976. Planning of floor system at long span suspension bridges. Tenth IABSE Congress, Preliminary Report, Tokyo, 1976. Zürich, Secretariat of IABSE: 149-154.

Touran, A. & Ladick, D.R. 1989. Application of robotics in bridge deck fabrication. *Journal of Construction Engineering and Management ASCE* 115: 35-52.

REFERENCES TO CHAPTER 14

Bodt, H.J.M. 1990. *The global approach to welding cost*. The Hague: The Netherlands Institute of Welding.

COSTCOMP 1990. *Programm zur Berechnung der Schweisskosten*. Düsseldorf: Deutscher Verlag für Schweisstechnik.

Farkas, J. 1985. Discussion to Elastic behaviour of isolated column-supported ringbeams by Rotter, J.M. *J. Constructional Steel Research* 4(1984): 235-252, *J.C.S.R.* 5: 239-242.

Farkas, J. 1991. Techno-economic considerations in the optimum design of welded structures. *Welding in the World* 29: 295-300.

Farkas, J. 1992. Cost comparisons of plates stiffened on one side and cellular plates. *Welding in the World* 30: 132-137.

Farkas, J. & Jármai, K. 1994a. Minimum cost design of laterally loaded welded rectangular cellular plates. *Structural Optimization* 8: 262-267.

Farkas, J. & Jármai, K. 1994b. The effect of the low cost technologies on the design of welded structures. *Welding in the World* 34: 123-130.

Gaylord, E.H. Jr. & Gaylord, Ch.N. 1984. *Design of steel bins for storage of bulk solids*. Englewood Cliffs, New Jersey: Prentice Hall, Inc.

Martens, P. (ed.) 1988. *Silo-Handbuch*. Berlin: Ernst & Sohn.

Ott, H.H. & Hubka, V. 1985. Vorausberechnung der Herstellkosten von Schweisskonstruktionen. *Proc. Int. Conference on Engineering Design ICED, 1985. Hamburg*: 478-487. Zürich: Ed. Heurista,.

Pahl, G. & Beelich, K.H. 1982. Kostenwachstumsgesetze nach Ähnlichkeitsbeziehungen für Schweissverbindungen. VDI-Bericht Nr. 457: 129-141. Düsseldorf.

Teng, J.G. & Rotter, J.M. 1992. Recent research on the behaviour and design of steel silo hoppers and transition junctions. *Journal of Constructional Steel Research* 23: 313-343.

Trahair, N.S., Abel, A. et al. 1983. *Structural design of steel bins for bulk solids*. Sydney: Australian Institute of Steel Construction.

Name index

339

Subject index

343